Chemical Cosmology

Jan C.A. Boeyens

Chemical Cosmology

Springer

Jan C.A. Boeyens
University of Pretoria
Dept. Chemistry
0002 Pretoria
South Africa
jan.boeyens@up.ac.za

ISBN 978-90-481-3827-2 e-ISBN 978-90-481-3828-9
DOI 10.1007/978-90-481-3828-9
Springer Dordrecht New York Heidelberg London

Library of Congress Control Number: 2010935166

Cover design: WMXDesign GmbH

Printed on acid-free paper

Springer is part of Springer Science+Business Media (www.springer.com)

Preface

I can afford to wait a century for readers ...
– Johannes Kepler

The incentive to write about Chemical Cosmology came with the realization that the natural synthesis of the chemical elements and their isotopes is hopelessly misrepresented in modern cosmology. On closer examination one finds that many other concepts, such as chirality, periodicity, chemical stability and equilibrium, understood in detail by the chemists of the world, are simply ignored by cosmologists, despite their potential relevance to cosmic models.

Apart from Svante Arrhenius and Harold Urey, very few chemists have ventured into the field of cosmology. This reluctance of chemists to contribute to the cosmological debate is no doubt due to the perception that modern cosmologies are soundly based on the concepts of general relativity, a theory not actively pursued, or even understood by most chemists. On making the effort to examine the relativistic details of cosmological theories there awaits the sobering discovery that some of the fundamental, even elementary, dictates of relativity are ignored as blatantly as the fundamentals of chemistry. In the Introduction to their excellent book on the topic, Adler et al. (1965) emphasize that:

> "[a] reduction of gravitational theory to geodesic motion in an appropriate geometry could be carried out only in the four-dimensional continuum of relativity theory."

We shall demonstrate that all of the currently competing models, including the standard big-bang model, fail on this score.

My dispute with big-bang cosmology is a continuation of the arguments, launched by Kepler and Riemann against medieval Aristotelean physics and Ptolemaic cosmology, best described as a hotch-potch of magic and dogma. The dogma consists of the belief in an unreachable perfect heaven and a corrupt world. Heavenly bodies are endowed with the perfect geometry of a circle and quintessential purity, compared to the polygonal nature of mundane four-element matter. No more than a blurred image of divine knowledge, which exists in the celestial sphere, may be invoked by magical practice in the secular world. This is the basis of astrology.

Euclidean geometry is a well-known product of this dogma. Its axiomatic foundation is firm and not subject to further analysis. Alchemy is another example. Kepler had the temerity, not only to reject the dogma of circular orbits, but also to demonstrate a proportional relationship between the Platonic polyhedra and a limiting sphere. The magic is gone. Riemann broke the final schackle by showing that geometry can exist without axioms. The dogma is also gone.

Riemannian geometry produced the general theory of relativity and Kepler's golden proportion can now be shown to feature in the ordering of matter, from the sub-atomic to the cosmic scale. We contend that big-bang cosmology, which is based on the same principles as alchemy and Euclidean geometry should make place for a model based on chemistry and Riemannian geometry.

I developed an interest in the field on noticing that cosmological theories of nucleogenesis are totally out of line with reality. The important clue came from the observed periodicity of the stable nuclides, which is governed by simple concepts of number theory. Any atomic nucleus consists of integral numbers of protons and neutrons such that the ratio Z/N is always a rational fraction, which can be ordered in Farey sequences and mapped by Ford circles. This ordering predicts a unique periodic function, which is readily demonstrated to predict the correct observed cosmic abundances of the elements.

According to standard cosmology only the light elements, H and He, were formed in significant amounts during an explosion at the beginning of time. All heavy nuclides were formed later on by nuclear fusion reactions in stars. It is obvious that such a chain of reactions cannot produce heavy nuclides in the observed periodic abundances.

The validity of the general periodic function extends into the periodicity of the chemical elements under extreme conditions, showing up a cosmic periodic law that includes all elements and anti-elements. The obvious link of this function with the curvature of space-time provided a mechanism for the production of nuclides in an equilibrium process, within a black hole, that matches the observed abundances.

An important property of the extended periodic law is that its geometrical mapping requires an involuted four-dimensional space, recognized as the real projective plane. By assuming the same topology for space-time, several paradoxes of standard cosmology are immediately resolved, including redshift as an intrinsic chemical property of the electronic configuration of atoms in strong gravitational fields. It provides a direct solution of the unsolved problem of anti-matter in the cosmos. It explains the nature of quasars, active galactic nuclei, cosmic radio sources, γ-ray bursts, the X-ray and microwave background radiation and cosmic self-similarity, as evidenced by the proposed structure of the solar system.

Despite our choice of title, this book, which aims to highlight scientific findings that should contribute to an understanding of the world, is strictly speaking not another book on cosmology.

The mistaken belief that Cosmology is a branch of science has resulted in total stagnation of the subject. Cosmology is not part of science, but of metaphysics. Metaphysics is the intellectual pursuit that strives to identify the ultimate realities at the basis of knowing the whole world. Western metaphysics has for mellenia been conditioned by the ancient Babylonian myths, as recorded in the Hebrew Bible, starting with the phrase:

> In the beginning ...

What follows is largely immaterial – there can be no existence before the beginning of time. The logical consequence, after centuries of agonizing, is the absurdity known as standard big-bang cosmology.

Nothing demonstrates the confusion of Western thinking better than the monumental study of James Frazer (1922). All cultures and tribes are shown to recognize the cyclic nature of natural phenomena. Without this mental jump cosmology becomes meaningless. With it, all cosmological riddles and paradoxes are resolved. The necessary assumption is a closed cosmic system, and the implications are immense.

The single most important factor that argues against this assumption is the evidence of our human senses. The casual observer experiences the world as flat, allowing for a few bumps and dips. Idealized into a logical construct this concept led to the development of flat-world geometry in the hands of Euclid and others. The primary concept of a straight line suggests infinite extension in two directions, with the corollary of infinite parallel lines. The whole edifice is simplified by the definition of a metric scale that allows comparison of distances and angles and the classification of triangles. Formalized in terms of a small number of theorems, Euclidean geometry provided the tools for architects and construction engineers, with spectacular success. In essence, the practice is based on what is known universally as the theorem of Pythagoras.

It is ironic that the first instance where Euclidean geometry proved inadequate was in land surveying, the initial purpose of geometry. Later came the need of Renaissance artists to make perspective drawings. During the same era the inability of the natural number system to avoid infinity singularities also became apparent. Whereas geometry managed to solve its problems by the introduction of fictitious ideal points at infinity, number theory remains incomplete.

The root of the problem lies in the reasonable assumption that the axioms of Euclidean geometry are equally valid on both cosmic and microscopic scales. It is now known to break down in both instances. Discovery of the electromagnetic field has shown that the familiar ideas on interaction between mass points need adjustment at both large and small separations, where they lead to singularities. Any cosmology that fails to recognize these complications must fail.

The correct starting point for a new cosmology is to accept that the universe is closed in space and time. The mathematical description of such a closed system automatically introduces the notions of periodicity and scale independent symmetries. The incorporation of scientific data into the model should be consistent with appropriate mathematics.

The mysterious similitude between science and mathematics is more than a coincidence and provides the most powerful means of theoretically exploring unknown science. This is of particular relevance in situations where experimental discovery of empirical facts are problematic – nowhere more so than in cosmology, which should embrace all of science, and hence all of mathematics. The discovery of new mathematics is driven by the requirement to eliminate exceptions. The search for generally valid subtraction led to the discovery of negative numbers and zero. General root extraction revealed the existence of both irrational and complex numbers. Division by zero remains poorly understood, with infinity an exceptional number. Although it remains algebraically elusive, it becomes a regular number in projective geometry. The distribution of prime numbers is an unsolved problem in all of mathematics.

Given the current state of knowledge the best possible cosmology must be consistent with complex algebra, projective geometry, advanced number theory, and all empirical science. As a by-product of only elementary-particle physics, the law of gravity and Euclidean geometry, standard cosmology is totally inadequate by these criteria. The current debates between big-bang cosmologists and dissidents are almost irrelevant as a guide towards a better understanding of the world, unless the fundamentally relevant principles are taken into account.

Many authors have discovered parts of the jig-saw puzzle, without recognizing the grand pattern and their findings, largely forgotten, remain dormant. Judging by the difficulties experienced by some of them to get their work accepted for publication, in the teeth of hostile peer review, countless others have probably, either published important ideas in obscure journals, or failed to publish at all.

Empirical evidence at variance with standard cosmology is, likewise, totally ignored. Even the most fundamental of empirical observations, known as universal CPT (charge conjugation–parity–time inversion) symmetry, which dictates equal amounts of matter and antimatter in the cosmos, is dismissed out-of-hand. Less well known, but of equal importance, cosmic self-similarity, is not considered at all.

From a chemical perspective the most important cosmological evidence includes the relationship between the periodicity of matter, prime numbers, Farey sequences, other aspects of number theory, cosmic abundance of the elements and nucleogenesis. These emerging periodic patterns are diametrically opposed to accepted explanations based on standard cosmology, but well in line with Veblen's projective relativity theory, Gödel's solution of the general relativistic field equations and Segal's chronometric alternative to Hubble's law.

Synthesis of these notions leads to an alternative cosmology that has little in common with the standard model. It may not have the same dogmatic appeal as the big bang, but it unifies a growing body of discordant empirical evidence, based on all the matter and energy in the universe, the Copernican principle and sound mathematics, without singularities. It has the potential directly to address the basic notions of chirality, astronomical anomalies, holistic self-similarity and other cosmological puzzles, such as Zwicky's paradox, in addition to traditional celestial mechanics. In effect, it calls for the re-examination and re-interpretation of the cosmic whole.

With so many, apparently unrelated, aspects of relevance to the argument it is necessary to understand the riddles that remain unsolved in traditional Western cosmologies. Fundamental to the resolution of these cosmological riddles and paradoxes is the structure of space and its interaction with matter. To achieve this the appropriate geometry, physics and chemistry are called for. The cosmological demands of mathematics and physics have been examined repeatedly, while chemistry has been largely ignored. At the time that modern cosmology was born an eminent science writer, (Sullivan, 1933), concluded that:

> "...the atomic theory is the one fundamental contribution that chemistry has to make towards an understanding of the structure of the universe. And it is only in its simple form [...] that the atomic theory is really helpful to the chemist. For this reason we are not justified in expecting great theoretical advances from chemistry of the future. The direction of any possible theoretical advance in chemistry lies so far as we can see, through physics."

The main purpose of the present work is to also emphasize the cosmological implications of chemistry.

The idea of the vacuum as an interface first came to me in Daresbury while staying over in a hotel, richly decorated with images and themes from Lewis Carroll's *Alice in Wonderland* and *Through the Looking-Glass*. Eventual publication (Boeyens, 1992) happened after several rejections and what kept the idea alive was the response from a number of prominent scientists. The following are verbatim extracts from private correspondence, as dated:

Olof Sundèn, 5 November 1992: You say that 'vacuum' is an interface between different states of matter (matter and antimatter), that the universe itself is the double cover of the manifold and that its involution serves to position conjugated forms of matter and antimatter on opposite sides of the interface .... I talk about space and antispace .... Out there on the other side 'time' changes character and appears as a conjugated antispace and even as a condensed conjugate antiparticle.

Hans-Jürgen Treder, 20 February 1993: .... you are right if you say that quantum field theory and relativistic cosmology have problems with respect to their reciprocal interpretation.

W. Kilmister, 24 February 1993: I thought that the first two sections (apart from the Intro) were extremely well done.

István Hargittai, 2 March 1993: It sounds very interesting what you wrote about...

Halton Arp, 3 March 1993: I had the rare experience, as an observer and empiricist, to be able to anticipate most of your conclusions before I came to them in your text!

Aldus Salam, 9 March 1993: I learnt a lot from it.

Peter Bussey, 7 April 1993: I did not understand a single sentence properly.

Ken Barker, 12 April 1993: .... the general thrust of the paper is most impressive.

J. Hošek, 15 June 1993: Concerning the "vacuum" which, apparently, is something very deep in your considerations. In ordinary sense, the vacuum is simply the ground state of a quantum system, which is both intuitively, and operationally well understood.

Peter Carruthers, 20 August 1993: I have been thinking for a long time that we need to rethink the vacuum.

Bruno Crosiguani, 2 September 1993: I have really enjoyed reading it since it is beautifully written and clear. I have found your model quite appealing and I believe it could have some connection with the work I did on the fine-structure constant....

Milo Wolff, 30 September 1993: You may find yourself in the science history books.

The subsequent discovery (Boeyens, 2003) of the grand periodicity of atomic matter put these speculations into sufficient perspective to allow definite conclusions about the projective topology of space-time and the universe. In the final analysis, all conclusions reached in this work can be reduced to the gauge principle, as summarized in Appendix B. Some readers may like to set the scene by reading this appendix before the main text.

I must finally acknowledge the encouragement by Sonia Ojo of Springer, the friendly information retrieval system of Sunette Steynberg and her staff, the mine of information published by William Corliss (1994), many stimulating discussions with Demetrius Levendis, Casper Schutte and Johan van Staden. I could never do this without the constant support of my wife, Martha.

Pretoria
May 2010.

References

Adler, R., M. Bazin & M. Schiffer, (1965): *Introduction to General Relativity*, McGraw-Hill, NY.

Boeyens, J.C.A. (1992): *The Geometry of Quantum Events*, Specul. Sci. and Techn., 1992 (15) 192–210.

Boeyens, J.C.A. (2003): *Periodicity of the stable isotopes*, J. Radioanal. Nucl. Chem., 2003 (257) 33–43.

Corliss, W.R. (1994): *Science Frontiers: Some anomalies and curiosities of Nature*, Sourcebook Project, Glen Arm, USA.

Frazer, J.G. (1922): *The Golden Bough*, Macmillan, London.

Sullivan, J.W.N. (1933): *The Limitations of Science*, The Viking Press, reprinted (1953) by Mentor Books, NY.

Contents

Chapter 1

Introductory Synopsis

Any cosmological theory must inevitably rest on many diverse concepts and observations. The reader, not familiar with the central thesis, could find it difficult to appreciate the relevance of several apparently unrelated aspects without some guidance. In response we start with a brief outline of standard cosmology, highlighting some paradoxical features, followed by an outline of the ideas that led towards an alternative model. Not to loose the thread of the argument, the final conclusion, that the observable universe is topologically closed, is stated up front. Each of the relevant aspects that support this central postulate will be repeated in more technical detail in the chapters to follow.

The main building blocks of the proposed new model are the relationship between geometry, numbers and space; the theory of relativity; and the periodicity of atomic matter. Taken together, these considerations indicate a cosmic symmetry that defines a harmonious holistic system that embraces all objects from the subatomic to extragalactic scales. The common geometrical factor is the ubiquitous golden parameter, $\tau = 0.61803...$

For the benefit of those readers who are familiar with the paradigms of modern science, but find the arguments and methodologies of cosmology totally bewildering, it is necessary to provide a historical survey of how two pursuits, so disparate, could develop from their common root of astronomy. Naked-eye astronomy never was powerful enough to establish a definitive model of the cosmos and to eliminate fanciful notions based on personal conviction or dogmatic prescription.

No progress was possible until the defects in Aristotelean physics could be demonstrated experimentally and the discovery of the telescope showed up astronomical features at variance with the Ptolemaic system. This progress was too slow to cause an immediate paradigm shift and to break the unity of scientific astrophysics and astronomy on one hand, with astrology and

J.C.A. Boeyens, *Chemical Cosmology*, DOI 10.1007/978-90-481-3828-9_1,

cosmology, with their baggage of dogma and mysticism.

The history of this awkward co-existence is traced into the modern era which holds the prospect of science-based cosmology – the main object of this work.

1.1 Cosmological Riddles

The word "somehow", or any equivalent, such as "by guess or by God", should be banished from scientific discourse, to be replaced by the phrase: "I don't know". Most of the so-called cosmological paradoxes exist because of half-baked explanations that gained respectability by frequent repitition. One of the most annoying examples is the way in which the chirality problem is confidently swept under the carpet, so well that most commentatators are unaware of it.

1.1.1 Chirality

The world, meaning the entire cosmos or universe, consists of matter and energy and the study of their distribution through space and time is what we consider to be the purpose of cosmology. Until one has a fair understanding of this it is premature to speculate about the origin or end of the world or time.

Whereas energy appears to be distributed almost continuously, compared to matter, the distribution of matter presents the more tractable problem. When delving into the nature of matter, in its most elementary state, it is encountered in two antagonistic forms with equal probability. These two forms, called *matter* and *anti-matter* are best considered as the complementary forms of elementary stuff in two symmetry-related modes of existence. Symmetry-related units of matter and antimatter have the same mass but opposite electric charges and behave in opposite sense with respect to time. This relationship obeys the dictates of what is probably the most fundamental symmetry in Nature, known as CPT (charge conjugation-parity-time inversion) symmetry. Wherever it has been investigated, all natural phenomena have been found to obey CPT symmetry without exception.

The symmetry forms known as matter and anti-matter are said to be of opposite chirality, or handedness. The most common form of antimatter occurs on earth, during the decay of some radioactive atomic nuclei, as positively charged electrons, called *positrons*. An encounter between positron and electron leads to mutual annihilation

$$p^+ + e^- \to \gamma \,, \tag{1.1}$$

converting their mass into energy in the form of gamma rays. Virtually all mass on earth occurs in the form of matter, which means that any approaching anti-matter, for instance in cosmic radiation, is extremely short-lived.

We seem to have arrived at a full-blown paradox: Symmetry predicts equal quantities of matter and antimatter, but one form seems to predominate. Where then is the missing antimatter? The answer given by cosmologists is that "somehow there was a slight imbalance in the beginning", with the result that the minor component was obliterated to produce the photons that now fill all space. There is no evidence to support this circular argument.

It is much more likely that the chirality problem hangs together with the strange property of space, recognized as the well-known right-hand rule of electromagnetic interaction. This rule reveals that space itself is chiral. If CPT holds it implies an equal segment of space, of opposite chirality and populated by antimatter. Cosmological models that fail to account for this chirality are incomplete.

In anticipation we point out that the four-dimensional Minkowski space of relativity is naturally segmented into time-like and space-like regions, separated by a conical light surface. It could imply two equal massive regions of opposite chirality, separated by an achiral electromagnetic interface. Maybe there is a logical resolution of the chirality paradox after all.

1.1.2 Self-similarity

The concept of self similarity is akin to Coleridge's principle of beauty as "this unity in multaety".

Certain structures, when examined on different scales from small to large, always appear exactly the same. Such structures are said to be self-similar or to be endowed with the symmetry of self-similarity. Self-similar structures are found to be invariably associated with the geometrical relationship known as the *golden ratio*, or what Johannes Kepler (1571 – 1630) referred to as the "divine proportion", adding:

> "Geometry has two great treasures: one is the Theorem of Pythagoras; the other, the division of a line into extreme and mean ratio.
>
> I believe that this geometrical proportion served as idea to the Creator when He introduced the creation of likeness out of likeness, which also continues indefinitely".

Kepler's obsession with the golden ratio kept him occupied in an effort to demonstrate the self-similar structure of the solar system and the music of the spheres. His biggest problem probably was insufficient data.

A later generation of German astronomers, including Titius of Wittenberg (1729 – 1796) and Johann Bode (1747 – 1826) discovered a numerical regularity that interrelates the mean orbital distances of the planets from the sun. Their formula

$$d = a + bc^n$$

with $a = 0.4$, $b = 0.3$, $c = 2$ and $n = -\infty, 0, 1, 2, \ldots$, accounts for the orbits of all planets from Mercury ($n = -\infty$) to Uranus ($n = 6$), with a gap at $n = 3$. The formula gained credibility with the discovery of the asteroid belt of minor planets, precisely in the Bode–Titius gap. It is fashionable to dismiss the recognition of harmonious celestial ratios as coincidental numerological delusions, not to distract the attention of serious cosmologists. However, the increasing evidence of cosmic self-similarity cannot be ignored indefinitely. As an important step in this direction the Bode–Titius law is revisited as a special case of general commensurability in the solar system, intimately related to the golden ratio and the topology of space-time.

1.1.3 Dark Matter

All sensible cosmologies strive to understand and interpret the observable material universe. Observation documents the distribution of matter, classified according to perceived masses, distances and relative motion. Interpretation reduces observation to theoretical notions on the interaction between condensed and separated massive bodies. The end product is a catalogue of celestial bodies and the necessary interactions that dictate their relative equilibrium distribution in space.

It is difficult to explain how a cosmology, which claims that only one percent of cosmic matter is observable by any means, can be considered a cosmology at all. But that is precisely the current status of the standard big-bang theory. What it means is that the theoretical model has no observational basis. If cosmology pretends to explain this *dark matter* it should do better than characterize it as WIMPs – Weakly Interacting Massive Particles.

The notion of dark matter first emerged as a proposed resolution of Zwicky's paradox. Today, together with its cousin, dark energy, it is the panacea that cures all big-bang ailments (Turner, 2000). It even works where inflation fails and because it defies detection can safely be assigned any desired attribute, without fear of contradiction.

1.1.4 Singularities

The solution to many mathematical equations, such as

$$x(r) = \frac{c}{r - r_0} \tag{1.2}$$

goes infinite for specific values of some variable, for instance at $r = r_0$, said to be a singular point of the equation. Where a function, which is associated with some observable quantity, behaves in a wild manner close to a singular point, the physical interpretation of the situation becomes problematical. Well-known examples include the electrostatic and gravitational potentials, which are represented by functions of the form (1.2) with $r_0 = 0$. Common-sense suggests that, either something extraordinary happens at such singular points, or else the validity of the model function breaks down close to the singularity.

In field theory singularities are interpreted as *sources* of the field. It seems reasonable to identify an electric charge as the source of an electrostatic field; the source of a gravitational field becomes more difficult to visualize, but a primeval singularity in space-time, as the source of the ultimate cosmic field, challenges credulity. However, singularities in space-time have been mooted as mass points or as bridges between multi-connected universes. The centre of a black hole coincides with another singularity that awaits reasonable interpretation.

The more likely conclusion is that singularities in both mathematics and physics are avoided in Nature and signal the failure of a model. The most conspicuous failures are the inverse-square laws of gravitational and electromagnetic interaction that predict infinite attraction at infinitesimal separation, $\lim_{r\to 0} K/r^2 = \infty$. Imagine all matter in the universe to be concentrated at such a singular point – it should be infinitely stable. A cosmology, like a field theory, based on singularities, should be treated with suspicion. In the present context our immediate aim is cosmology without singularities.

1.2 Authority in Science

The progress of science is often hampered by the reputation of great scientists. During his lifetime the Greek philosopher Aristotle (384–322 BCE) acquired a reputation of infallibility that persisted for millenia. His four-element theory virtually precluded the development of Chemistry and to question his mechanics was tantamount to heresy. In the same breath one can mention Claudius Ptolemæus who flourished in Alexandria during the second century CE. He produced the definitive description of the cosmos, with the Earth at its centre, and his authority rivalled that of Aristotle – all part of the "...authority to which human nature instinctively clings", (Curtis,1934).

The heroic revolution that rescued modern science from the shackles of authority spanned centuries, over the eras of renaissance, reformation, industrial revolution, enlightenment, and continues to this day. In the hands of too many pseudo-scientists the Hebrew Bible as scientific authority remains at loggerheads with modern concepts such as biological evolution, and still dictates the philosophy behind standard cosmology and the theory of intelligent design.

Scientists of the Enlightenment during the 17th and 18th centuries insisted that knowledge is generated through reason and transcends authority. The resultant growth of science during the next two centuries is history. The new science reached its pinnacle in two theories of Physics: the theory of relativity and of the quantum. The price that was paid for the spectacular progress was the sanctification of a new breed of infallible authorities.

Popular media spread the myth that the theory of relativity was understood by no more than a handful of prodigies, of whom Einstein was the foremost. Today he is the new Aristotle. People in all walks of life get away with unsubstantiated assertions attributed to him, and scientists are no exceptions to this. When Einstein first introduced the cosmological constant it was claimed to modify the gravitational field equations in perfect analogy with an extension to the classical Poisson equation. It was pointed out (Harvey & Schucking, 2000) a quarter of a century later that Einstein's analogy contained an elementary error, but that a long sequence of eminent authors and a "generation of physicists have parroted this nonsense".

We are told (Michelmore, 1963) that on his first encounter with the big-bang theory Einstein applauded Lemaître, after his lecture in California, with the words:

> "This is the most beautiful and satisfactory explanation of creation to which I have ever listened".

According to another historian (Luminet, 2007) :

> "When Lemaître met Einstein for the first time at the 1927 Solvay Conference, the famous physicist told him: 'Your calculations are correct, but your physical insight is abominable!' "

In an unpublished letter to the New York Times[1] Irving Segal, who worked

[1] Computer file retrieved by Aubert Daigneault.

as Oswald Veblen's assistant next door to Einstein for several years after the war, wrote on 3 March 1998:

> What is on record is that Einstein was strongly opposed to the Expanding Universe models of the Russian meteorologist Alexander Friedman and of the Belgian Abbé George Lemaître in the 1920's

Other biographical evidence suggests that all of these accounts could be factual. Despite the quality of his work, it therefore reflects very negatively on Einstein as an authority and his personal opinions are immaterial in the face of scientific logic.

Einstein is by no means the only latter-day scientific authority whose stature stifles legitimate enquiry. It was stated by Murray Gell-Mann (1979) that

> "Niels Bohr brainwashed a whole generation of theorists into thinking that the job was done 50 years ago".

There are many more, but the point has been made: Although it's fair to recognize scientists for their seminal discoveries and ideas, these should constantly be re-examined and re-interpreted where necessary.

No discipline has suffered more under the tyranny of authority than modern cosmology. Accounts without number testify to discrimination against scientists who hold alternative views or propose revolutionary new ideas. Even Gauss did not make his discovery of the principles of non-Euclidean geometry known for he feared "the outcry of the Bœotians"; there were only a few people who could understand the essence of his discovery. In the same spirit many scientific views are withheld or suppressed in the mainstream media. In reaction the present study is not beholden to any cosmological model. There are simply too many unknowns, uncertainties and misconceptions that militate against any definitive theory. That includes the present author who stands to be corrected on many issues which may nevertheless stimulate healthy scientific debate.

1.3 Geometry and Number

The word "geometry" features in most languages as a concept to relate the sizes of different objects. Of the smaller number of individuals with more advanced mathematical skills the overwhelming majority would consider geometry as equivalent to the ancient discipline, described more correctly as Euclidean geometry. That is the version of geometry codified by Euclid in classical times and for centuries considered to be the only and absolutely irrefutable way of measuring the whole world.

Some measure of unease must have crept in with the realization that the surface of the planet was not flat but spherical. However, treated as an awkward special case it did not require revision of the basic axioms and theorems of geometry. Particularly awkward was the fact that two great circles in a spherical surface, the equivalent of straight lines in a flat surface, intersect each other not in one, but in two points. It was Felix Klein (1849–1925) who realized that each spherical point determines a unique antipodal point and that nothing would be lost by identifying each pair of antipodal points, and calling it a *point*. This way points and lines of spherical geometry can be described by the same conventions of flat plane geometry. This modified version of spherical geometry, now known as *elliptic* geometry, is only one of several non-Euclidean geometries that emerged during the nineteenth century[2]. In the same spirit the exceptional situation of non-intersecting parallel lines can be avoided by the addition of an extra point at infinity for each direction, as first proposed by Johannes Kepler. This procedure defines the projective plane in which every two lines intersect once, at either an ordinary point or at infinity.

One of the most significant results of those non-Euclidean geometries, which consider space to be continuous and isotropic, is that it must then also be homogeneous and of constant curvature.

Of crucial importance is that the introduction of ideal points at infinity, in order to avoid the exceptional status of parallel lines, implies that space must close on itself in a way that makes the infinity concept redundant. Without infinity the number system is also closed by identification of the points ∞ and $-\infty$. A closed numerical interval is periodic and a closed space must contain periodic matter. A valid cosmological model must reflect these values.

Projective geometry is a concept even less familiar than non-Euclidean geometry. It is generally recognized only in perspective engineering drawings and works of art. Being central to our main thesis an entire chapter is devoted to this topic.

[2]A brief historical record is given by Coxeter (1998).

1.4 Physical Theory

Like Euclidean geometry the perception that we live in a three-dimensional world went unchallenged for millenia. The first indication that the perception could be wrong followed the formulation of the special theory of relativity.

1.4.1 Special Relativity

The science writers of the previous century painted the theory of special relativity as a sensational departure from rational thinking. They tried to imply that no more than twelve scientists, all of them somewhat mentally challenged, could hope to understand the theory. The attitude of serious scientists softened over the years, but many intellectuals of different pursuasion still find the idea highly suspicious. It is not uncommon for erudite individuals in the softer sciences, engineering and medicine to perform their own thought experiments to challenge Einstein's conclusions. These efforts are often ridiculed by physicists in the know, rather than responding in a responsible manner by informing non-experts about the merits of the theory.

It is probably true that commonsense thought experiments about trains going through a station with flashing lights and reliable clocks can never reveal the intricacies of special relativity. On the other hand a mathematical treatment, even at the pedestrian level presented in these pages, is of no use to the uninitiated. It is ironic that without grasping the fundamental ideas of relativity theory the large-scale structure of the universe must remain hidden and cosmologies handed down by the experts can never be critically evaluated by the masses nor the popular commentators. It so happens that the simplifying assumptions, made as a matter of course by most cosmologists, deviate so much from basic theory that most of the scientific content is lost. In the end, cosmologies appeal to the masses not because of their scientific content, but because of ideology. Things have changed very little since pre-Copernican times. We are, for all practical purposes, back to earth-centred cosmology, based on Euclidean geometry.

In order to convince the sceptics it is necessary to formulate the fundamental principle of relativity in words that make it comprehensible to any intelligent person, without appealing to mathematics.

The Basic Idea

The greatest discovery of 19th century science was the electromagnetic field. Only a negligible number of people in the modern world are not touched by some aspect of this discovery. Most people are familiar with some electric or magnetic phenomenon and a substantial majority appreciate that these phenomena are intimately linked together. A smaller, but meaningful, number of people actually understand that natural wave motion in the form of visible light, microwaves, X-rays and radio waves are manifestations of a combined electromagnetic field.

More than words are needed to relate this understanding to the theory of relativity. A magnetic field permeates different material media in a characteristic, measurable way. Even the vacuum has a characteristic permeability, which is represented by the symbol μ_0. An electric field penetrates dielectric media in the same way and also has a characteristic permittivity for each medium, including the vacuum, where it is represented symbolically as ϵ_0. Maxwell made the discovery that a wave disturbance of electromagnetic origin is described by a differential equation with the product $\mu_0\epsilon_0$ as a constant coefficient that links its rate of propagation to variation in field strength in the vacuum, as given by a wave equation. Dimensional analysis shows that the coefficient $\mu_0\epsilon_0$ is inversely proportional to the square of a velocity. Symbolically, $\mu_0\epsilon_0 = 1/c^2$.

Maxwell's theory was confirmed experimentally by measuring the velocity of light, which turned out to be the same as the calculated value of the constant $c = 1/\sqrt{\mu_0\epsilon_0}$. It is crucial to understand that c is a natural constant, determined by the electromagnetic properties of free space – μ_0 and ϵ_0. There is no argument about this – like μ_0 and ϵ_0, c is an invariant property of space. Most textbooks on special relativity introduce the constancy of c as a brilliant, but falsifiable, conjecture of Einstein. It is nothing of the kind. The only unexpected counterintuitive consequence was that the observed velocity of a light ray should be independent of the relative velocity of the light source, with respect to an observer. It simply means that the propagation of electromagnetic phenomena does not obey the laws of relative motion, valid for the familiar mechanical systems of the time, known as Galilean relativity.

The Lorentz transformation is the reformulation of the Galilean laws of relative motion to be consistent with both mechanics and electromagnetic wave motion.

The massive volume of hot air, by way of a century's worth of comment on aether drift and time dilation, is totally irrelevant to the central issue. Even typographical errors in Einstein's papers have been quoted as evidence against the validity of of the theory of relativity.

The Implications

The implications of special relativity are enormous; least of which is the predicted equivalence of matter and energy according to what is perhaps the best known equation of theoretical physics:

$$E = mc^2.$$

A more revolutionary implication is that the Lorentz transformation is mathematically equivalent to a rotation in four-dimensional complex space, called *Minkowski* space. In addition to the orthogonal directions x, y, z of familiar three-dimensional space, Minkowski space has a fourth orthogonal direction associated with time flow. Whereas the length of a vector is invariant on rotation in three-dimensional space, it is no longer the case in Minkowski space. The situation is analogous to rotations in two and three dimensions respectively. The length of the rotating two-dimensional vector is observed to be invariant only in the plane of rotation. On rotation out of the preferred plane the vector, seen in projection, appears to contract. The apparent contraction is compensated by an increase of the third out-of-plane component. Our inability to visualize 4-space has the same effect on our perception of Lorentz transformation. The length of a moving object appears to contract in the direction of motion, while the time flow in the fourth dimension appears to dilate. This phenomenon depends on the same mathematics as Einstein's famous equation, which is nowhere in doubt.

A further implication is that, instead of a space interval ds, calculated from

$$ds^2 = dx^2 + dy^2 + dz^2 ,$$

and an absolute time interval dt, the relativistic description of two events links the two variables by defining a single space-time distance

$$d\sigma^2 = c^2dt^2 - dx^2 - dy^2 - dz^2 .$$

Two infinitesimally close points can be connected by a light signal only if $d\sigma^2 = 0$, which defines a *light cone.* Events within the light cone, where $d\sigma^2 > 0$, are known as *time-like* and represent uniform motion at velocities $v < c$. *Space-like* events with $d\sigma^2 < 0$ and $v > c$ are considered unphysical.

1.4.2 General Relativity

Special relativity is a purely kinematical theory without directly involving mass or force. It only deals with uniform motion without the complication of acceleration. The dynamic theory that takes these factors into account is known as *General Relativity.* Whereas special relativity reveals the four-dimensional nature of space-time general relativity demands non-Euclidean geometry. It follows from one of Einstein's famous thought experiments:

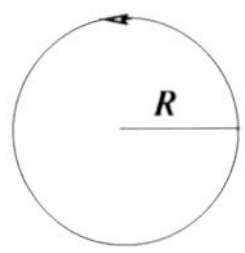

Consider a frame of reference in uniform rotation. It is noted that sustained circular motion requires constant acceleration towards the centre. Lorentz transformation dictates the contraction of a measuring rod in the direction of motion in this rotating frame. In a second, accelerated frame of reference, with the same origin, there is no contraction and the ratio of the circumference to the diameter of a reference circle remains $S/2R = \pi$. For the same circle, observed in the stationary frame, $S'/2R' > \pi$, since the radial measurement is not affected by the motion. To account for this effect it is necessary to realize that Euclidean geometry does not apply in the stationary frame of reference.

The alternative non-Euclidean geometry bears the same relation to Euclidean geometry as the geometry of curved surfaces bears to the geometry of the flat plane. The distance between two infinitesimally close points in the plane is given by the equation

$$ds^2 = dx^2 + dy^2 .$$

In a curved surface such a distance is defined in Gaussian coordinates as

$$ds^2 = g_{11}dx^2 + 2g_{12}dxdy + g_{22}dy^2 .$$

If the g-coefficients are given as functions of x and y the surface is fully determined geometrically.

Extended to the four dimensions of the relativity theory of curved space-time the element of arc ds is given by

$$ds^2 = \sum_{\mu,\nu=1}^{4} g(x)_{\mu\nu} dx_\mu dx_\nu .$$

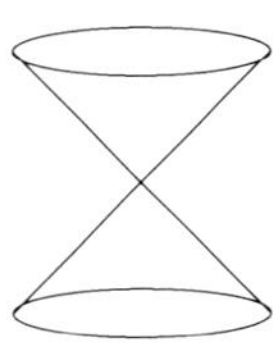

The next vital assumption, based on experimental observation, is the equivalence of inertial and gravitational mass. This means that a gravitational field must distort the Euclidean geometry of space and that the matter content of space must be in balance with the non-Euclidean geometry of space, known as its *curvature*. The field equations of general relativity summarize these assumptions mathematically.

More than a theory of relative accelerated motion, general relativity is the theory of the gravitational field. Compared to special relativity it is incomplete in the sense of ignoring electromagnetic effects. The reason for this is not too difficult to find. In Minkowski space the world lines of mechanical objects lie within a so-called time cone, whereas electromagnetic phenomena happens in the surface of the time cone. The field equations of general relativity only incorporate a mechanical energy-momentum tensor, in balance with a curvature tensor. In fact, Minkowski space is pseudo-Euclidean, four-dimensional and affine. It is, by definition, devoid of matter and, strictly speaking, special relativity is therefore only suitable to simulate the kinematics of massless entities in a local environment. In contrast, general relativity dictates that ponderable matter only exists in curved space which, in its simplest form, is closed.

The only way in which to transform Minkowski space into a closed manifold is by adding a point at infinity to each coordinate axis, to produce a multiply-connected non-orientable hypersurface, known as a projective plane. General relativity should therefore ideally be formulated as a field theory in projective, and not in affine, space.

Relativistic Cosmology

Cosmological models obtained through the simplifying assumption of an affine-connected manifold must, by definition, be no more than a crude approximation. The one feature in common to all possible models is the balance between a curvature tensor and a stress tensor, as specified by Einstein's field equations, in their most general form,

$$G_{\mu\nu} + \Lambda g_{\mu\nu} = (\text{const})T_{\mu\nu}, \tag{1.3}$$

where Λ is an arbitrary constant and g is the metric tensor. In free space with zero energy density $T_{\mu\nu} = 0$, requires $\Lambda = 0$ in order for the equations to reduce to $G_{\mu\nu} = 0$, as demanded by general relativity.

The cosmological constant Λ has been a bone of contention for many years. Historians find no evidence that Einstein ever referred to it as his "greatest blunder", although he conceded that it has no place in expanding-universe cosmologies. In Einstein's original proposal of a closed static universe the cosmological constant played an important role.

In order to define the energy-momentum tensor, matter is assumed to be uniformly distributed, as in an ideal gas, with stellar velocities orders of magnitude less than the velocity of light. However, a stable Boltzmann distribution of this type cannot persist in an open system from which entire stars may escape. Einstein therefore proposed a spatially closed continuum of constant curvature. The cosmological constant served to define both the mean density of the equilibrium distribution and a radius of the closed spherical space, calculated from the boundary surface of a four-dimensional hypersphere.

The assumptions on which the model rests are too crude to be realistic. In particular, the assumption of a universal time coordinate directly contradicts the basis of general relativity. To avoid the problem de Sitter repeated the calculation based on relativistically curved space-time, with the surprising result of an empty universe with variable radius. The traditional interpretation of this result as an expanding universe is not unique. It could just as well imply space-time with continuous curvature, characteristic of projective space. In the event, both solutions were soon superseded by an expanding-universe cosmology based on a Doppler interpretation of galactic redshifts.

Spectroscopic observation indicates a relationship between redshift and intergalactic distance, which, interpreted as a Doppler effect, implies a rate of recession between galaxies that depends on their mutual separation. To account for this relationship, known as Hubble's law, a metric tensor, in which the time coordinate is separated from a monotonically expanding three-dimensional Euclidean space, is assumed.

Although the model accounts for Hubble's law the necessary assumptions are not supported by general relativity, as emphasized in an unbiased authoritative assessment (Adler, et al., 1965):

> "... the use of a distinguished time-coordinate marks the abandonment of a completely covariant treatment of the cosmological problem. This is the price one has to pay to simplify the cosmological models and to describe reality in convenient mathematical terms".

> " ... the exploding models found in the case $\Lambda = 0$ are not compatible with our present knowledge of the world. The discontinuous birth which they imply is quite unrealistic and due to the neglect of the role of the pressure".

Nevertheless, these exploding models, with many fanciful embellishments, have by now developed into a standard cosmology, better known as the hot big bang. It enquires into the early history of a constantly expanding cosmos and reaches the conclusion that it must converge back to an instant of infinite density created by a singularity at the beginning of time. There is no scientific support for this scenario. To continue with the assessment:

> "... general relativity can be used to analyze the main features of the universe in the large, but (that) the scarce and unprecise

observational facts available do not allow us to test any models in detail ..."

1.4.3 Quantum Theory

At the same time that European physics extended Newton's mechanics by taking electromagnetic effects into account, another modification was necessitated by the strange behaviour of sub-atomic systems. Although the resulting new relativistic and quantum theories developed from the work of essentially the same group of scientists, common ground between the two systems remained excruciatingly elusive. By 1935 Albert Einstein, one of the leading minds in both pursuits, reached the conclusion that the two theories were largely incompatible. Whereas relativity, based on constant c, was a local theory, quantum mechanics allowed non-local (instantaneous) interaction between distant events. Einstein and his collaborators concluded that the problem lay with quantum theory being incomplete, but subsequent work has now confirmed the quantum theory experimentally.

It is almost incomprehensible how most physicists have been indoctrinated to ignore this dilemma and how cosmology remains indifferent to it. As both theories are demonstrably valid, the difficulty clearly lies elsewhere, meaning that quantum and relativistic signals are propagated differently through space. The theory of relativity defines the ultimate velocity of *mechanical* objects to be less than c, whereas quantum effects are initiated by a scalar wave function whose phase velocity may exceed the velocity of light. It is difficult to reconcile such behaviour with the three-dimensional affine space of everyday experience and whatever configuration of space-time is needed to accommodate the quantum potential field, it must have profound cosmological implications. Not only do we require space to be four dimensional and non-Euclidean, as indicated by the theories of relativity, but the conveniently assumed affine metric should be abolished in order to explain quantum effects.

Not surprisingly, the physical interpretation of quantum mechanics has remained problematical to this day. The theory came into being as a means to address a number of problems related to the nature of radiant energy and sub-atomic matter, and their interaction. It unfolded against the background of the perennial debate on the particle or wave nature of radiation. The general perception is that Newton championed a corpuscular interpretation of light against the wave model of Huygens. There is no documentary proof of any such debate and although Newton assumed that light must be material in some sense, he was equally aware of its wave-like mode of propagation (Hall, 1963). The real confusion started with Einstein's interpretation of the photoelectric effect as an interaction between particulate electrons and *photons*, which implies that light consists of a stream of particles. By then it was already firmly established that the energy of a light beam was directly related to its frequency, which is a property of waves. The compromise resolution of the problem constitutes the quantum notion of wave/particle duality, which is more of a standoff than an explanation.

This fateful compromise, now known as the Copenhagen interpretation of quantum theory, contributed to the mindless acceptance of the flawed big-bang theory as respectable science. Rather than searching for the origin of matter in something more general it perpetuated the notion of elementary particles, found wanting even in classical times. It sanctioned the marriage between a cosmology without scientific support and a theory without experimental backing, underpinned by a mathematical singularity.

Reassessment of the situation suggests that the interpretation of, not only elementary matter, but also electromagnetic interaction as standing waves, explains the photoelectric effect in a more natural way and removes any discrepancy between the quantum and relativity theories. Negation of the wave nature of elementary matter has hit the science of chemistry particularly hard. Futile efforts to reduce chemical phenomena to the probability distribution of structureless objects has led to complete theoretical stagnation, despite the rich culture of nineteenth-century theories of chemical affinity and structure.

1.5 Chemical Cosmology

The most compelling evidence of an intimate relationship between the number system and the cosmic distribution of matter is provided by the experimentally established laws of chemical periodicity.

Chemistry is the science that studies the nature and interconversions of material systems. Observed chemical patterns should therefore be reducible to the structure and properties of space-time and the assumptions of cosmology. However, unlike the theory of relativity, the fundamental periodic laws of chemistry find no detectable basis in standard cosmology.

As discrete numbers of nucleons are involved in the constitution of nuclides the periodicity of atomic matter is readily simulated in terms of the elementary number theory of rational fractions, Farey sequences and Ford circles.

The fractional ratio of protons:neutrons provides a reliable measure of nuclide stability and converges from unity to the golden ratio with increasing atomic mass. Rather than forming a smooth triangle, the limiting stability boundaries have a saw-tooth appearance with inflection points that define an easily recognizable periodic pattern.

Plots of Z/N *vs* Z, N and A converge to Z', $N' = Z'/\tau$ and $A' = Z'/\tau^2$ respectively. Not only are the separate periodic functions in terms of Z, N and A intimitely related, but, in each case, also reduce to closed functions, consistent with equal fields of stability for matter and anti-matter.

Sampled at any ratio of $Z/N < 1$ the predicted periodicity in terms of N or A remains invariant. In contrast, the periodic function, defined by atomic number Z, assumes distinctive canonical forms at $Z/N = 0, 0.58, \tau$ and 1, obviously related to the variability of extranuclear electronic configurations with environmental pressure.

Conditions that favour the ratio of unity, characteristic of the most stable nuclide ^{4}He, also known as an α-particle, appear as the most favourable for nucleogenesis, in a chain reaction of α-addition, within black holes. The expected abundance of nuclides as produced in such an equilibrium process, while at variance with standard cosmological predictions, matches the observed cosmic abundances remarkably well.

The predicted periodicity at $Z/N = \tau$ corresponds with the empirically derived periodic table of the elements, naturally associated with the uniform curvature of low-density space-time. By inference $Z/N = 0.58$, the value at which the periodicity based on the quantum-mechanical energy spectrum of hydrogen is projected out, corresponds to flat empty space.

The symmetry that combines the different periodic arrangements of the elements is summarized best by mapping to a projective plane, a two-dimensional section of which is a Möbius band. This construct is an attractive model for a closed universe in which the conjugate chiral forms of matter are separated in a natural way.

The symmetry of the overall periodic system further highlights the importance of the irregular response of atomic energy levels to environmental pressure, as known from computational simulation. The general raising of electronic levels under high pressure predicts spectroscopic redshifts in the light, received from highly curved regions of space-time, such as galactic centres or quasars. This effect, which is totally ignored in standard cosmology would result in a vastly reduced size of the universe, if taken into account.

The close link of elemental periodicity with the golden ratio, which can scarcely be accidental, suggests an intimate relationship between atoms, botanical structures and spiral galaxies. The intermediate position of the solar system in this self-similar chain prompted a re-investigation of the commonality between atoms and celestial structures, previously invoked by Nagaoka and Bohr to formulate their atomic models.

Application of the same techniques, involving rational fractions, as used in the analysis of elemental periodicity, shows that the (in)famous Bode–Titius law is a special case of a general pattern that involves all planets, satellites, asteroids and rings in the solar system in a common commensurable distribution. At the galactic level this self-similarity predicts a periodic gravitational field that manifests as quantized redshifts in galactic light, without any reasonable basis in standard cosmology. In a universe without singularities one is forced to consider that the self-similar pattern, extrapolated both ways, demands a closed projective topology of space-time.

1.6 Projective Relativity

Projective Minkowski space is generated by adding a point at infinity to each possible set of parallel lines. Local observation cannot distinguish between affine and projective Minkowski spaces. In particular, the difference between projective and affine time parameters is partly responsible for the observation of spectroscopic red shifts in galactic light, giving rise to the illusion of an expanding universe. In fact, Segal's chronometric analysis of galactic redshifts even shows that Hubble's law is a poor approximation of the phenomenon.

Theories like those of Lemaître or Friedmann, which predict an expanding universe, are all based on forcing an affine metric, such as the Robertson-Walker metric, on the projective geometry of space-time. This has the effect of splitting local Minkowski space into separate space and time coordinates, without the natural complex relationship that ties space and time together.

One solution of the field equations of general relativity, not of this type, was obtained by Gödel, who used a metric that mixes space and time coordinates and therefore does not allow the definition of absolute time. Not surprisingly, there is no longer a dynamic redshift for distant objects, but instead, it predicts a rotation of matter relative to a compass of inertia. The Gödel geometry, although not formally projective, has many features in common with, and is readily expanded into, four-dimensional projective geometry.

The appearance in the projective geometry of n-dimensional space, of points, lines and planes at infinity, is recognized in describing each point by $n + 1$ homogeneous coordinates. When formulated projectively the points, vectors and tensors of general relativity therefore have the appearance of being defined in 5-dimensional affine space. In fact, the five-dimensional unified gravitational and electromagnetic fields of Kaluza and Klein are readily demonstrated to be contained in four-dimensional projective relativity, except for the unphysical compacted fifth dimension. In addition, the electromagnetic field appears as a gauge field and hence quantum wave mechanics emerges as a natural feature of general projective relativity.

A remarkable property of Minkowski space is the nature of its isotropic lines, which are perpendicular to themselves, and define a set of points, all at zero distance apart. These lines also represent the geodesics of light rays and give a new interpretation to the nature of photons, as standing waves. In this sense the emitter and absorber of a light ray are in virtual contact. The contact is established as the absorber responds to a spherical signal eminating from the emitter by returning an advanced wave that reaches the emitter at the time of emission. (The return signal is propagated as a negative function of time.) The result is a standing wave, also known as a virtual photon between absorber and emitter. Transmission, in the form of an actual photon happens when the emitter exists at a higher energy level.

Translation along projective geodesics produces an involution that inverts CPT symmetry which, when acting on matter, induces an inversion of chirality and conversion into anti-matter. The two-dimensional analogue is translation along a Möbius strip, or more realistically, along its double cover, which avoids direct contact between antipodal points. As a cosmological model it is therefore the double cover of a projective manifold, rather than the actual mathematical space that must be assumed. No other cosmological model can account for the chirality of space or explain the apparent cosmic imbalance between matter and antimatter.

The interface between antipodal regions of the cosmic double cover is interpreted to represent the physical vacuum that carries the electromagnetic field. An interfacial potential prevents contact between antipodes, except when the general curvature of space is catastrophically disturbed by astronomically large masses, known as black holes. Rupture of the interface then corresponds to what is traditionally considered to be a singularity at the centre of a black hole. But no longer is there an infinite accumulation of matter, as an outlet is provided through the interface. An environment like this is precisely what is needed for the synthesis of new atomic matter according to an equilibrium mechanism, consistent with the observed periodicity of nuclear abundances in the cosmos.

The subject of projective relativity theory was studied intensively during the 1920's by several mathematicians, but ignored by theoretical physicists. The significant discovery that five-dimensional unification of the gravitational and electromagnetic fields is more precisely formulated in four-dimensional projective space does not feature in modern field theory following the demonstration by Bergmann (1976) and others that the projective field equations can be reduced to those of Kaluza. Thus they discarded the baby with the bathwater. In a parallel situation matrix and wave mechanics were demonstrated to be mathematically equivalent, but without recognizing the informative physical basis of wave mechanics, an unworkable interpretation was concocted. The appeal of projective relativity goes beyond formal unification. Its cosmological implications derive from the underlying geometry which does not appear in Kaluza theory.

The nature of electromagnetism and, in fact, all of modern field theory is rationalized in terms of a gauge principle. The idea arose from an early attempt by Weyl to unify the gravitational and electromagnetic fields, based on the assumption that the length of a vector may change under displacement, due to a change of scale or gauge. In refined forms of the theory, rather than a change in physical dimensions, a complex phase change, which has no observable effects, and may cause the appearance of a compensating field, is assumed. All field theories rely on such an assumption without any rationale of the gauge principle, although it was demonstrated by Veblen that it occurs as a natural feature of projective geometry.

Veblen's classic monograph of 1933, written in German, is virtually unknown to modern physicists. This is a pity as it demonstrates that all gauge theories are a natural feature of projective spaces. As a result it produces all versions of relativistic quantum theory without further assumption. Transformation to the affine space in which Schrödinger's and Dirac's equations are traditionally formulated is shown to involve a parameter now recognized as the ubiquitous golden ratio of number theory, as a conversion factor. This observation, for the first time, serves to identify the mysterious cosmic self-similarity among growth structures as a property of space-time.

Noting that a universe with projective topology offers the only explanation of the chirality of space and the symmetry between matter and antimatter it becomes almost self-evident that a cosmological model should reflect all of these various properties. Friedmann expanding universe cosmology based on the Robertson-Walker metric does not meet this requirement. The only solutions of the cosmological problem that approximate projective geometry are de Sitter's original solution and Gödel's solution for a rotating, non-expanding universe. Combined with Segal's chronometric analysis of galactic spectroscopic redshifts, a fully self-consistent cosmology based on projective geometry emerges. Segal's analysis depends on exactly the same relationship between local measurement of time in affine space, compared to the time scale of an underlying projective space with five homogeneous point coordinates.

While studying projective relativity it became necessary for me to translate Veblen's monograph into English. As a credible version of projective cosmology must primarily be based on this work, a translation of the (80 page) monograph is included as an Appendix to the present work. For ease of reference the pagination of the original German version is strictly adhered to in the translation.

The synthesis of general relativity and quantum theory is embodied in the gauge principle that emerges as a natural feature of projective relativity and explains the unification of the electromagnetic and gravitational fields. A brief introduction to the concept of gauge invariance is provided in a second Appendix.

1.7 References

Adler, R., M. Bazin & M. Schiffer, (1965): *Introduction to General Relativity*, McGraw-Hill, NY.

Bergmann, P.G. (1976): *Introduction to the theory of relativity*, Dover edition, NY, p.268.

Coxeter, H.S.M. (1998): *Non-Eucidean Geometry*, 6th edition, Math. Ass., USA.

Curtis, L. (1934): *Civitas Dei*, Macmillan, London.

Gell-Mann, M. (1979): in D. Huff & O. Prewett, *The Nature of the Physical Universe*: 1976 Nobel Conference, 1979, New York, p.29.

Hall, A.R. (1963): *From Galileo to Newton*, Collins, London.

Harvey, A. & E. Schucking (2000): *Einstein's mistake and the cosmological constant*, Am. J. Phys., 2000 (68) 723 – 727.

Luminet, J-P. (2007): *The Rise of Big Bang Models, from Myth to Theory and Observations*, Astrophys., 2007, arXiv:0704.3579vi.

Michelmore, P. (1963): *EINSTEIN – Profile of the Man*, Frederick Muller, London.

Turner, M.S. (2000): *More than meets the eye*, The Sciences, 2000 (40) 32–37.

Chapter 2

History

2.1 Introduction

Cosmology is the speculative component of astrophysics, which is the science that interprets astronomical data by using the laws of physics and chemistry. Historically the three concepts, cosmology, astrophysics and astronomy, which included astrology, were integrated into a single pursuit, driven by religious dogma. As scientific data gradually became incorporated, religious considerations diminished in importance, but have still not been eliminated completely from the mix.

Before the invention of telescopes astronomy depended on naked-eye observations, using instruments like quadrants, gnomons, sextants and astrolabes. The Great Pyramid of Giza, in its original truncated form, is claimed (Tompkins, 1978) to have served as a naked-eye observatory. The sophistication of pre-dynastic Egyptian mathematics and astronomy, before its mysterious decline, was only matched again in the Western world during the 17th century CE[1]. Although it led to the development of incredibly accurate calenders, the evidence remained insufficient to challenge philosophical models of the cosmos.

Classical Greek science developed from the transfer of basic Egyptian know-how by visiting scholars, such as Pythagoras (*ca.*550 BCE), who worked as a priest in Egypt for more than twenty years. Of decisive importance was the recognition of a spherical Earth, which inaugurated a dramatic change in cosmological thinking. Immediately the observed cyclic motion of the heavenly bodies started to make logical sense, eventually culminating in the

[1]Common Era.

J.C.A. Boeyens, *Chemical Cosmology*, DOI 10.1007/978-90-481-3828-9_2,

Ptolemaic cosmology, which remained unchallenged in the western world for a thousand years.

Apart from Pythagoras, new ideas emerged in the Ionian commercial centres, inspired by Thales of Miletus and his intellectual descendants. The correct prediction of a solar eclipse enabled them to claim that the movement of heavenly bodies was dictated by natural laws, rather than the whims of gods.

The innovations of Thales and his followers are still rated as one of the greatest achievements of human intellect, but without experimental support could never develop into scientific cosmology. Thales himself had a problem to identify a means of support for his disc-like earth. His successor, Anaximander, saw the need to abandon the notion of absolute up and down, as being directed relative to the centre of the spherical earth, that hence needs no support. The difference between heaven and earth was still ascribed to the separation between the cold and wet that condensed into a solid earth with oceans, against the partially evaporated hot and dry, becoming air and mist surrounded by a fiery cosmic sphere. The remaining breathing holes through the rotating mist are visible to mankind as the sun, moon and stars.

A major complication with this cosmic model was the Pythagorean demand of perfect circular orbits for all celestial bodies. The more accurate astronomical observations became, the more complicated the required corrections to restore the circular pattern. The authority of Pythagoras and Thales must have been enormous, especially as Plato also endorsed the geometrical perfection of the cosmic composition. The planets, including the sun and the moon, like the fixed stars, circled the earth on concentric spheres with numerical regularity, each contributing its harmonious pitch to the music of the spheres.

The enduring proposition of Pythagoras was that fundamental reality is not to be found in the elementary composition of matter, but in an underlying structure, that imparts a distinctive character to all things, and which can only be understood in terms of numbers. This conclusion generalized the discovery of an exact correlation between the lengths of musical strings and the notes they produce. This proposition was re-interpreted by Plato who substituted Form, for Number, as intelligible structure, instead of material elements. In these terms creation consists in bringing together the eternal forms with imperfect matter to produce the temporal world. By this procedure a demiurge fashioned the physical world on the model of forms, using space as his material[2]. This is not creation *ex nihilo*, out of nothing.

The evidence that musical scales and planetary motion both obeyed the same numerical rules was simply too attractive to discount as coincidence and nobody dared to contradict the interpretation of the masters.

2.1.1 A Heliocentric Model

Trying to bring the observed motion of the planets better into line with their philosophy, Philolaus, later in the 5th century BCE, modified the geocentric view and posited a model in which the earth, a counter-earth, the seven planets and the fixed stars all circle about an (unseen) central fire. The earth's revolution around the fire was believed to occur on its own axis during the period of a day and a night. However, the planetary orbits still deviated from ideal circles and the model was forgotten until it was mentioned again by Plato. When a lesser-known philosopher, Aristarchus of Samos developed the concept of a heliocentric universe (270 BCE), his ideas also fell victim to the authority of the great men, not to be revived again for almost two thousand years.

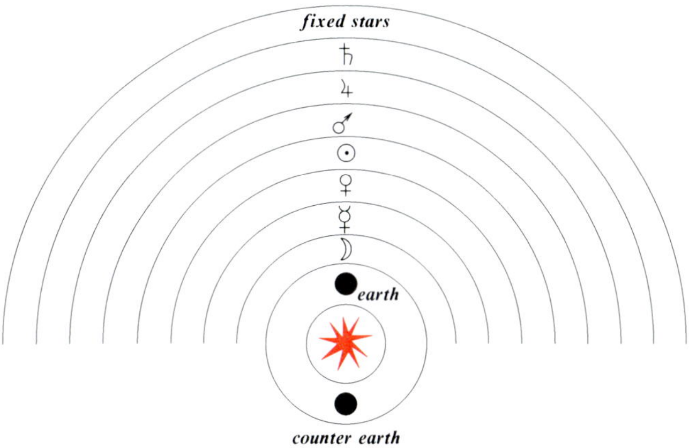

The world was to fall into the same trap again many times. Reluctance to question the pronouncements of Aristotle, the Hebrew Bible, Georg Stahl – the champion of the phlogiston theory – and many other authorities whose ideas became outdated, continues to plague the development of science.

[2]Plato, *Timæus*, 4th century BCE dialogue.

Equally distinguished savants like Galileo, Charles Darwin and Lavoisier suffered abuse for their temerity to contradict the demi-gods of science.

To conclude the story of Ionian and Pythagorean cosmology, their ideas were slowly degraded by those who lacked the ingenuity, but not the desire, to refine the original model. All manner of irrelevant data, supposedly in support of the perfect geocentric universe, were added to strengthen the argument, making it progressively more difficult to revise the basic model, in the teeth of public opinion. In effect, a variety of sequences were postulated to be sympathetically linked to the seven planets, as shown in the Table, reconstructed from a variety of sources (MacNeice,1964; Graves, 1966; Priesner & Figala, 1998). In the sense that cosmology should be consistent with all available scientific knowledge these correlations may be seen as an effort to demonstrate exactly that. Superficially it shows support of the model from number theory, geometry, astronomy, alchemy, natural history, poetry and theology. In reality the scientific content of these correlations is nil, being no more than a summary of mystic traditions, secret formulae and recipes for horoscopy. The serious astronomy of classical Greece became hopelessly

Number	Planet	Symbol	Metal	Gemstone	Colour	Letter*	Tree #	Bird #	Day	Archangel
1	Moon	☽	Ag	Ruby	Orange	S	Willow	Hawk	Mon	Gabriel
2	Mercury	☿	Hg	Beryl	Green	N	Ash	Snipe	Wed	Raphael
3	Venus	♀	Cu	Lapis Lazuli	Blue	H	Hawthorn	Night-crow	Fri	Sariel
4	Sol	☉	Au	Sard	Red	B	Birch	Pheasant	Sun	Uriel
5	Mars	♂	Fe	Chrysotile	Yellow	L	Rowan	Duck	Tue	Michael
6	Jupiter	♃	Sn	Chalcedony	White	D	Oak	Wren	Thu	Raguel
7	Saturn	♄	Pb	Garnet	Violet	F	Alder	Gull	Sat	Remiel

* *Original seven consonants, invented by the three fates*

\# *Ogham alphabet*

entangled with astrology and dogma without producing further insight into the nature of the cosmos, despite the rhetoric of the famous classical philosophers, Socrates, Plato and Aristotle. Plato mooted the possibility of using various combinations of circular motion to predict planetary positions, without proposing the model as a physical description of the universe. However, in time this model led to the notion of epicycles, on the basis of which Ptolemy formulated the first standard model of cosmology that remained unchallenged in the western world from the second to the 17th century.

2.2 The Almagest

The cosmological model, which is generally known as the Ptolemaic system was formulated in the second century by Claudius Ptolemæus in Alexandria. It was published in his 13-volume treatise, which became known to the Arabs as the *Almagest*, the name, meaning *The Greatest*, still used today.

To prove that the earth was fixed at the centre of the universe Ptolemy used the observation that all falling objects drop to the centre of the earth. He accepted the traditional order of celestial objects, rotating on crystal spheres around the earth, but realized that the implied circular motion had to be replaced by auxiliary circular motion along epicycles on circular deferents. Deferents are large circles centred on the earth and epicycles are small circles whose centres moved around the circumferences of the deferents. The sun, moon and planets move around the circumferences of their own epicycles. This model was still insufficient to account in full for the motion of the

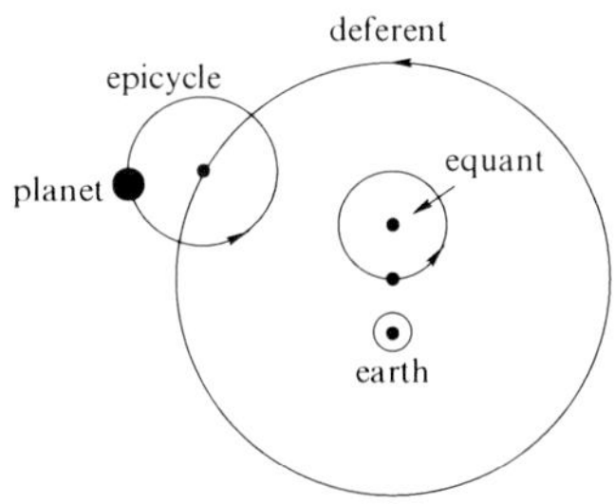

planets. The final refinement introduced by Ptolemy was to move the centre of the deferent away from the earth and to define an imaginary point, called the *equant*, on the diameter that contains the centres of both earth and deferent, at the same distance as the earth, but on the opposite side of the centre. To describe the orbit of a planet the centre of the deferent is now assumed to move along the circle centred on the equant and going through the midpoint between earth and equant. The oscillating deferent carries the planetary epicycle through the same rocking motion, with the planet following the circular motion along the circumference of the epicycle.

This ingenious construction served several purposes. Perhaps the most important was to restrict the motion of each heavenly body to an orbit, defined by three perfect circles, while accounting for the irregular observed motion of the wandering planets through the skies. Of almost equal importance was the mathematically precise description of the periodic motion of celestial objects, that the model provided.

Beyond the outermost planet, the crystal sphere that carries the fixed stars, was located, showing the placement of the twelve constellations of the zodiac. For the first time it became possible to draw up accurate maps of the sky to coincide with important moments or events, past, present or future, in the life of individuals or nations. This is the basis of astrology, in which Ptolemy was as expert as in astronomy. For the purpose of our discussion this issue is of no importance, but for millions, over millenia, it provided a method to avoid disaster, make informed decisions and foretell the future.

Outside the sphere of the fixed stars Ptolemy proposed other spheres, ending with the *primum mobile* to represent the prime mover of Greek philosophy.

2.3 Medieval Cosmology

For a millenium, after the conversion of Constantine's mother, thinking in the western world was dominated by the spread of Christianity, orchestrated by the philosopher-saints, Augustinus and Thomas Aquinas. Their main concern was to align the Ptolemaic system, Aristotelian philosophy and the practice of astrology with their holy scriptures.

The major stumbling block that inhibited the development of common-sense cosmology in the middle ages was the dogmatic adherence to Aristotelian physics, and in particular, the theory of motion, based on the idea that terrestrial and celestial bodies moved in fundamentally different ways. This difference relates to the natural distribution of the four elements of which the world was made up, and shown in the diagram below.

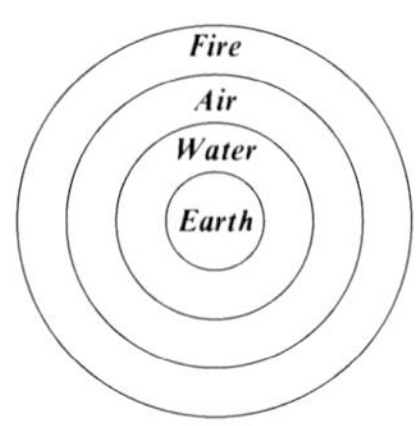

Each element has a tendency to move towards its own sphere. If, by some unnatural action, an earthly object is moved beyond the sphere of earth, it strives to return where it belongs. The bigger the object the faster it falls back. On approaching its own sphere more closely the more friendly environment stimulates the natural motion as the object accelerates towards its target. By this theory an object which is launched in a horizontal direction should drop down to its natural place as soon as the thrust is discontinued. The observed inertia whereby the object maintains its horizontal motion was explained by assuming that the air continues to push the object in its initial direction.

The same argument implied that celestial objects should accelerate towards higher spheres. Beyond the ring of fire the realm of Ptolemaic planetary spheres commenced. The celestial bodies carried by these spheres consisted of more subtle matter than terrestrial objects. The god-like spirits which controlled the seven planetary spheres were now identified with the seven archangels. The fundamental characteristic of these crystal (transparent) spheres was that they moved in the perfect mode of permanent, uniform, circular motion.

The fixed stars were supposed to move with the eighth sphere, and beyond that was heaven, the ninth immovable sphere. Aristotle identified this sphere as the prime mover, responsible for initiating the motion of all the other spheres, down to all things that move on earth. Thomas Aquinas identified the prime mover as God, who was responsible for the creation of an orderly, harmonious universe. The natural tendency of celestial objects to approach the highest sphere provided the mechanism which enabled the ascension of sanctified beings such as Enoch, Elijah, Jesus, Maria and Mohammed.

The astronomy behind the Ptolemaic system was of secondary importance, if not totally irrelevant, to the philosophers of the middle ages. An idealized version of the system, stripped of irrelevant epicycles and deferents, was elevated into revealed dogma and vigorously defended against heresy. An important distinction was made between the perfection in heaven, with all celestial bodies in perfect circular motion, and the irregular motion of earthly objects. Escape from this dogmatic strait-jacket of Thomism was a long and arduous process. The authoritarian philosophy of biblicized Aristotelianism obviated the need of scientific enquiry and the activities of free-thinkers and heretics were jealously controlled by the catholic church.

What is commonly referred to as the Copernican revolution, started with the birth of Copernicus and lasted until the death of Newton – a period of 250 years. This period of renaissance corresponds to the transition from medieval cosmology to the first scientific description of the solar system. It is instructive to compare the lifespans of the major players:

Nicolaus Copernicus	1473 – 1543
Tycho Brahe	1546 – 1601
Galileo Galilei	1564 – 1642
Johannes Kepler	1571 – 1630
Isaac Newton	1642 – 1727

It was Copernicus who revived the heliocentric model proposed almost two thousand years earlier by Aristarchus. As before, the proposal failed to generate meaningful interest outside of a small circle of initiates. So little in fact that Copernicus postponed publication of his ideas literally to his dying day and, not to give offence, he dedicated his work to the pope. Perhaps he should not have bothered because it still raised no comment until 1616, when it was condemned by the church, save as a calculating device (Hall, 1963). Although Copernicus rejected the Ptolemaic system, he accepted the Aristotelian theory of motion, and only managed to account for the observed planetary positions by a combination of deferents and epicycles centred on the stationary sun. It was the moving earth that annoyed his peers.

Tycho Brahe, the great astronomer of the next generation clearly appreciated the simplicity of the heliocentric model but, even more conservative than Copernicus, refused to consider a moving Earth. As a compromise he made the five planets spin around the sun, while the sun and moon revolved about the earth. This device predicted the same relative motion of the heavenly bodies as the Copernican model.

Tycho's devotion to astronomy was stimulated first by the prediction of a solar eclipse and secured by his personal observation of the famous new star (nova) of 1572 in Cassiopeia. He managed to demonstrate that the shining object, brighter than Venus and visible in broad daylight, was one of the fixed stars in the process of disintegration. This observation was a serious blow to the Aristotelian concept of the immutability of the heavens.

He made it his life's work to correct the astronomical tables by more accurate observations. He devoted enormous effort to the enigmatic motion of the planet Mars and towards the end of his life invited his younger collaborator, Johannes Kepler, to use these data to find the correct description of the Martian orbit.

2.3.1 Johannes Kepler

Kepler studied the problem for twenty years before coming up with a solution, only to find that nobody, not even Galileo, took him seriously. However, Kepler himself was convinced that it was impossible to fit the orbit to Tycho's data by any possible combination of circles. He found the answer by assuming an elliptic orbit.

The Kepler solution for Mars was shown to work for all planets, including Earth, assuming the Copernican heliocentric model. Kepler formulated his results in the form of three laws (Stephenson, 1960):

1. The law of elliptic orbits: The planets move in ellipses about the sun

which is located at one focus of the ellipse.

2. The law of areas: The line joining any planet to the sun sweeps out equal areas in equal intervals of time.

3. The harmonic law: The square of the period of any planet about the sun is proportional to the cube of its mean distance from the sun.

In symbols the harmonic law is formulated as $T^2/r^3 = k$.

Nothing demonstrates the power of authority and dogma better than the failure of Kepler's description of the solar system to convince the renaissance world.

Whereas the dogma of uniform circular motion demanded that the moving planet traverse equal arcs in equal intervals of time, Kepler discovered the new uniformity of equal areas in equal times to be valid for any ellipse, including the circle. By adding an additional focus to planetary orbits the need of epicycles was eliminated and the wandering planets were shown, for the first time, to follow harmonious paths around the sun. This discovery should have destroyed the Ptolemaic system for good. It failed. Not even Kepler's discovery of a new supernova, that remained visible for 17 months, was sufficient to shake the world's faith in a permanent sphere of fixed stars.

Kepler's adherence to the Pythagorean ideal of an harmonious planetary system and astrology has been commented on negatively by several modern critics as misguided mysticism. My reading of Kepler's ideas is different.

He was familiar with the five regular Platonic solids. Like Pythagoras he accepted a fundamental relationship between physical reality and number, manifesting itself in constructs such as the tetraktys and the golden ratio, shown below. He surmised that the distances of the heavenly bodies from

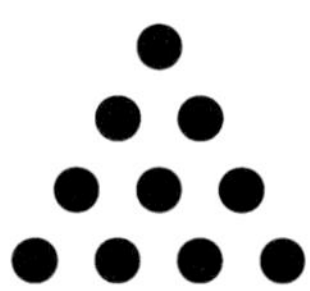

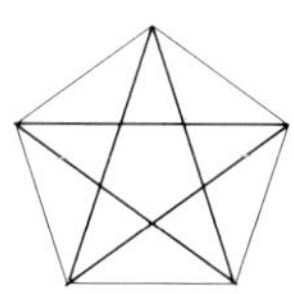

each other somehow correspond to musical intervals as in the Pythagorean harmony of the spheres. Although a final solution to the problem eluded him, he stated that "this geometrical proportion (golden ratio) served the Creator as an idea when He introduced the continuous generation of similar objects from similar objects" (Coxeter, 1989). He rejected the idea of an infinite universe, showed how to relate affine geometry to the more general projective geometry and explained the symmetry of snow flakes by the close-packing of equal spheres, self-similar to the regular bodies inscribed within planetary orbits.

2.3.2 Galileo

It seems odd that Kepler, who openly supported the Copernian system in writing, never had to face an inquisition, while Galileo was condemned for what appears to be legitimate astronomical observations. A likely reason is that the Aristotelian perfection of the heavens was dogmatically more important than the computational basis of astronomical and astrological predictions, whether Ptolemaic or Copernican. Galileo offended the dogma by reporting his suspect telescopic observations of irregular features on the perfect surface of the moon and spots on the face of the sun. He compounded this sin by argueing that there was no difference in type between terrestrial and celestial motion. Instead, he distinguished between natural motion, which is uniform and circular, and forced motion, which is accelerated and rectilinear. Accordingly it is logical that the earth should move in a circle, like the other planets, and for the solar system to be centred on the sun. Bodies that fall back to the surface of the earth are not seeking the centre of the universe, but are moving by their natural tendency to become re-united with their own.

Galileo was right about many things, like contradicting Aristotle on the motion of different size bodies in free fall. They all fall equally fast. He was totally wrong in others. In a dispute with Kepler he argued that tidal motion was due to the rotation of the earth, rather than "the moon's dominion over the waters"[3].

Galileo formulated some mechanical principles that enabled Newton to establish the physical basis of Kepler's laws. By noticing that a ball which rolls on a plane that slopes downward is accelerated and one on a plane that slopes upward is retarted, he argued that motion along horizontal planes should be uniform and perpetual. This view represents an important advance

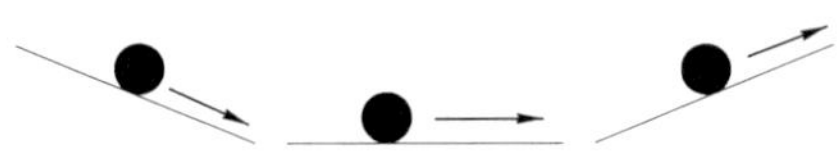

[3]Translations of Galileo's work in English include: Drake (1953); Crew & de Salvio (1914)

on Aristotle, but falls short of Newton's law of inertia. When Galileo refers to a horizontal plane the motion is still implied to be circular, around the globe. The only objects in rectilinear motion are those found outside their proper places, arranged badly, and therefore in need of being restored to their natural state by the shortest path. Galileo's proposition that the distance (s) traversed during uniform accelerated motion is proportional to the square of the elapsed time (t^2) has a simple geometrical proof.

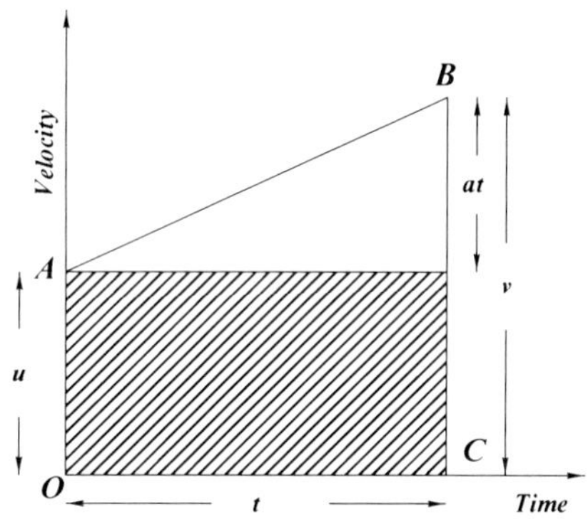

Galileo's Acceleration Formula The acceleration (a) represents the velocity increase in unit time. The total increase over a period t is therefore given by at. Hence the final velocity is $v = u + at$, where u is the initial velocity. A graph of the acceleration is given by the straight line AB, on axes of velocity *vs* time. The distance traversed at uniform velocity u over a period t is given, by definition, as $s = ut$, which corresponds to the area of the shaded rectangle. The total area OABC represents the distance traversed in the accelerated motion, *i.e.*

$$s = ut + \frac{1}{2}at^2$$

By equating uniform and circular motion it is natural for the earth to move in a circle, not an ellipse, but without a law of gravitation Galileo could never present a coherent discussion of centrifugal effects. Objects kept on flying away from the rotating earth as from a potter's wheel.

It seems that Galileo distanced himself from Kepler to demonstrate his respect for traditional dogma, but, by denying the difference between heaven and earth, this concession was not enough to protect himself against the Inquisition.

2.4 The Mechanical Universe

Once Galileo had demonstrated the inconsistencies in Aristotle's theory of motion and Kepler had formulated the laws of planetary motion, the only remaining obstacle to the development of a mathematical model of the solar system was the law of inertia, still based on uniform circular motion. This final problem was removed by the French mathematician and philosopher, Descartes.

2.4.1 Descartes

René Descartes (1596 – 1650), who also published under the latinized name Renatus Cartesius, operated on the assumption that understanding of the physical world was contingent on the formulation of a sound metaphysical model or logical framework, drawing its first principles from reason. As the most fundamental principle he accepted a conservation law that demands the quantity of motion in the universe to remain constant. He equated matter with extension, which must therefore fill all space.

Motion in such a plenum would be impossible if all matter were rigid. In addition to opaque gross bodies he therefore postulated a transparent aether with the ability to transmit motion between solid objects, and material light, eminating from the sun and stars, plastic enough to fill all gaps between particles of matter and aether. As the universe is completely filled, motion consists of a rearrangement that involves constant impacts of particle on particle. To include aether and light with ordinary matter, these particles were assumed infinitely divisible.

Under these conditions any movement tends to create a vortex. The solar system is such an aetherial vortex with the sun at its centre. The planets are swirled around the sun by the aether. A planet, in turn, is at the centre of a subsidiary vortex, which may carry satellites, like the moon, around. On a larger scale the universe is made up of solar vortices centred on stars, like the sun, and fitting together like a foam, to fill all space.

The aether of a rotating vortex tends to recede from the central star under centrifugal pressure which becomes visible as light, emitted by the star and reflected by the planets. Because of its confinement in the foam-like structure the aether exerts an inward pressure on the planets. The reason why planets are not pushed into the central star is because they are in a state of uniform linear motion. Under aetherial pressure the direction of this motion changes continually and in the state of balance the planet stays on a closed characteristic orbit around the sun. Although Newton scoffed at the Cartesian model he gave the law of inertia in this exact same form,

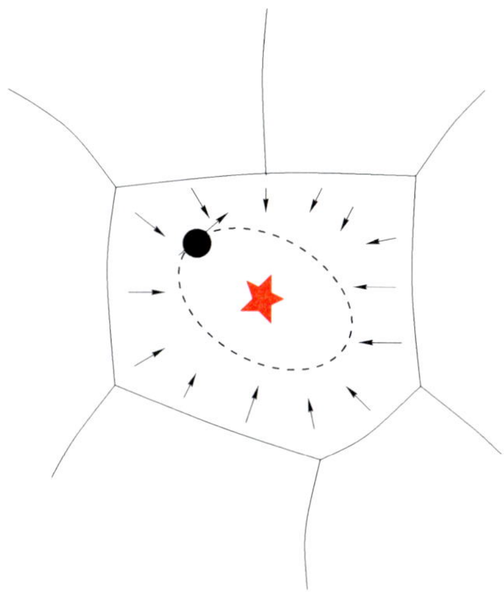

replacing the inward pressure by the law of gravitation. Newton's third law is a paraphrase of the Cartesian conservation of motion.

On a local scale gravity was explained by fast-moving aether particles pushing down on more sluggish earthly matter, the same mechanism that shapes spherical water droplets.

Descartes described the sunspots discovered by Galileo as an accumulation of coarse material, which could eventually prevent emission of light and collapse of the vortex into a planet, to be captured into a neighbouring stellar vortex.

Cartesian cosmology, the first of several mechanical models of the universe, was taken seriously, especially in continental Europe, for a whole century. It was a theory of everything and a good example of rampant speculation, unchecked by experimental observation. Across the English channel Francis Bacon (1561 – 1626) inspired a generation of scientists to rely on empirical evidence, immune against experimentally untestable models, such as the Copernican system. The future of science clearly lay somewhere between these two extremes. The simplicity and beauty of a scientific hypothesis, if not in contradiction with observation, should be of decisive importance when comparing rival theories.

2.4.2 Newton

By the middle of the seventeenth century the final stage of renaissance cosmology by the synthesis of empirical data and reasoned conjecture was in sight. Tycho Brahe had generated the reliable data and Descartes had provided ideas to spare. Kepler was responsible for systemising Tycho's data into a useful form and Galileo started the revolt against authority that liberated Descartes' fertile mind. The final thrust required the refinement of Cartesian ideas to extract those features in common with Kepler's model.

This final thrust, which centred around Isaac Newton, turned into an unseemly squabble with his contemporaries Robert Hooke (1635 – 1703), Edmond Halley (1656 – 1742), Wilhelm Leibniz (1646 – 1716), Christiaan Huygens (1629 – 95), Robert Boyle (1627 – 91), Christopher Wren (1632 – 1723) and whomsoever else contributed to the effort. Newton despised them all.

The important principles that enabled Newton's final formulation: the law of inertia, gravity without aether, the inverse-square law and reciprocal reaction, were in the minds of all scientists of that era. The central contentious issue was unmediated action at a distance. It was Newton who, like Kepler before him, took the plunge of proceeding with the mathematical description without physical understanding. Newton, unlike Kepler, shared his insight with others like Hooke, Halley and Wren, whom he obliquely acknowledged by stating, "If I have seen further it is by standing on ye shoulders of Giants", (Hall, 1963). It is because of this collaboration that the ideas did not remain dormant like those of Kepler. Newton never acknowledged Leibniz, who developed the calculus which transformed Newtonian mechanics into the grand edifice that emerged in continental Europe.

It was widely appreciated by Newton and his contemporaries that the compressive effect of the Cartesian aether was equivalent to some hypothetical force, often likened to gravity, that pulls a planet towards the central star and bends its rectilinear motion around into a closed curve. The magnitude of such a force had to be sufficient to balance the centrifugal, potter's wheel, effect and so prevent the rotating planet from escaping. It became vitally important to calculate the centripetal acceleration that was needed to stabilize such a closed orbit.

Newton's Inverse-square Law Newton's calculation (Stirling, 1935) of the correct formula enabled the synthesis with Kepler's model.

If a body at P moves in a circle with uniform velocity v it travels from P to B in the time $T = (arcPB)/v$. Had there been no force on it acting towards O it would have reached the point A at the same time T. To stay on the circular path it may therefore be considered having travelled to A and, because of the central force, it fell the distance AB towards O. By the

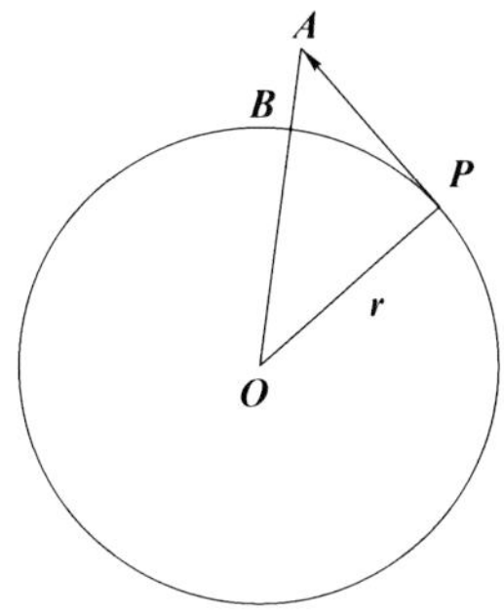

theorem of Pythagoras:

$$(AP)^2 + r^2 = (OA)^2 = (r + AB)^2 = r^2 + 2r.AB + (AB)^2$$

For a small interval in time the square $(AB)^2$ becomes so small that, in the limit, it can be considered equal to zero, while AP and the $arcPB$ become equal in length. The result is that $AB = (BP)^2/2r$. From Galileo's formula the uniform acceleration moves the object through a distance $s = (1/2)at^2$ and $t = BP/v$, such that

$$\frac{(BP)^2}{2r} = \frac{a}{2}\left(\frac{BP}{v}\right)^2$$

or $a = v^2/r$.

For a satellite to complete one revolution along the circumference of the circle, $2\pi r$, in time T, the velocity is given by $v = 2\pi r/T$ and the acceleration follows as $a = v^2/r = 4\pi^2 r/T^2$. From Kepler's third law the relationship between orbital period and distance for a planet on a closed orbit is given by the formula $T^2/r^3 = k$, a constant. On combining the two equations, by eliminating T^2, it follows that $a = 4\pi^2/(kr^2)$. Stated in words, the acceleration is inversely proportional to the square of the distance from the centre.

Clearly, if this calculated acceleration is ascribed to gravity, it becomes possible, by inverting the calculation, to derive Kepler's laws of planetary motion from an inverse-square law of gravitation. That was Newton's assumption of celestial mechanics: the same law of gravity, recognized by Galileo for falling objects on earth, also accounts for the harmonious motion of heavenly bodies. In its most general form the model applies to any pair of massive objects, moving relative to each other as if a force, proportional to the product of their masses and inversely proportional to their separation, pulls them together. Mathematically,

$$F = \frac{Gm_1m_2}{r^2}$$

On earth, the gravitational force, $F = mg$, where m is the mass of the gravitating object and g is the acceleration due to gravity, in line with Galileo's observations.

Newton's Laws

In order to give a mathematically precise account of these observations Newton gave canonical definitions of the concepts mass (m), momentum ($p = mv$), inertia, force ($F = ma$) and centripetal force ($F = mv^2/r$), together with the three famous laws of mechanics:

1. Law of Inertia: Every body continues in its state of rest or of uniform motion in a right line, unless it is compelled to change that state by forces impressed upon it.

2. Law of Force: The change of motion is proportional to the motive force impressed; and is made in the direction of the right line in which that force is impressed.

3. Law of Reaction: To every action there is always opposed an equal reaction; or, the mutual actions of two bodies upon each other are always equal and opposite.

In the case of elliptical orbits, Newton's mechanics relates the two parameters that specify an ellipse to the dynamic parameters that define orbital motion. The size of an ellipse depends on its semimajor axis, its shape depends on the eccentricity, and these respective parameters are proportional to the energy and angular momentum of the rotational motion.

It has been stated (Hall, 1963) about Newton's published work:

> No other work in the history of science equals the *Principia* either in originality and power of thought, or in the majesty of its achievement.... Order could be brought to celestial physics only once, and it was Newton who brought order.

However, like all scientific achievements, it raised as many questions as answers. While the concept of force was introduced to explain planetary motion, the question of interaction at a distance remained unanswered. The mathematical law of gravity that pertains to falling apples as well as solar satellites, had no clear physical basis. Was this divine mathematics or an innate property of matter to act at a distance? As Newton's reservations against the latter interpretation were gradually forgotten, so was the mystery around the law of gravity, as it became generally accepted and commonplace. Yesterday's heresy is today's dogma.

2.5 Cosmic Evolution

The development in celestial mechanics after Newton was largely in the hands of the French mathematician Pierre-Simon Laplace (1749–1827). The stability of the solar system was the major unsolved problem. Neither Kepler's laws nor Newton's mechanics could be applied successfully to more than a single orbit at a time. The universal law of gravitation must clearly apply to any pair of celestial bodies and with several planets and moons circling the sun it is inevitable that mutual perturbations of the predicted perfect elliptical orbits should occur. Newton himself could never precisely model not even the lunar motion and concluded that divine intervention was periodically necessary to maintain the equilibrium of the solar system.

Without this luxury Laplace set out to assess the natural stability of the system. When asked by Napoleon to clarify the role of God in this, he replied: "I have no need of that hypothesis". The first important result, which he demonstrated mathematically, was that the irregularities in the eccentricities and inclinations of planetary orbits oscillate about fixed values, without amplification, and hence never deviate too far from the ideal orbits. He could therefore theorize that the solar system remains indefinitely stable. Like some self-correcting clockwork, driven by the universal force of gravitation, the solar system was concluded to be inherently stable and predictable. Laplace saw no reason why the whole universe should not be dynamically stable in the same sense. He claimed that:

> "An intelligence knowing, at any given instant in time, all forces acting in nature, as well as the momentary positions of all things of which the universe consists, would be able to comprehend the motions of the largest bodies in the world and those of the smallest atoms in one formula, provided it were sufficiently powerful to subject all data to analysis: to it, nothing would be uncertain, both future and past would be present before its eyes".

Not everybody agreed with this deterministic philosophy, but the model proved hard to refute. It prevailed long enough to stimulate the world view of a well-organized cosmos, designed with a mathematical precision that guarantees its faultless operation. At about this time (1772) the independent observation by three German astronomers (Johann Bode in Berlin, Johann Titius in Wittenberg and Christian Wolfe in Halle) of a peculiar numerical regularity in the mean distances of the known planets from the sun, became known as the Bode –Titius law.

Starting with the sequence of numbers: 0, 3, 6, 12, 24, 48, 96, ..., in which each new term beyond 3 is obtained by doubling the previous term, adding 4 to each term, and divide by 10, the sequence of planetary distances is reproduced in astronomical units. 1AU equals the distance of planet Earth from the sun. Hence:

Planet	Observed distance	Distance according to B–T law
Mercury	0.39	(0+4)/10=0.4
Venus	0.72	(3+4)/10=0.7
Earth	1.00	(6+4)/10=1.0
Mars	1.52	(12+4)/10=1.6
?		(24+4)/10=2.8
Jupiter	5.20	(48+4)/10=5.2
Saturn	9.54	(96+4)/10=10.0

The subsequent discovery (1801) of the asteroid belt, consisting of thousands of minor planets on an orbit that corresponds with the gap in the Bode –Titius table elevated the law from a curiosity into a serious scientific observation. The physical basis of this law remained a mystery for more than two centuries.

Theologians of the day soon elaborated on the Laplace model by arguing that any system which functions like a machine must be the product of intelligent design, with a specific purpose in mind. The prime mover, or first cause, was back in another guise. To emphasize the fact that cosmology is the product of all human knowledge it is pointed out that astronomy and mathematics by themselves cannot successfully counter this teleological argument. It was left to a naturalist, Charles Darwin (1809–82) to show that, what may be perceived as purposeful design in Nature, arose through chance mutations that persist in organisms that happen to be more compatible with their environment. The theory even allows for the origin of life in lifeless matter and obviates the need of a designer of the world.

The first known instance where evolutionary development featured in cosmology came with the model, proposed by the philosopher Immanuel Kant in 1755, who postulated the origin of the solar system in a rotating nebula, rather than in creation. The idea was further developed on scientific grounds by Laplace in 1796. It gained general acceptance at about the time that the theory of biological evolution started to develop.

Laplace postulated a vast disc-shaped mass of cold gas and dust in slow rotation. As this nebula contracted under the mutual gravitation of its parts, its rate of rotation increasd to conserve angular momentum.

Motion on a circular orbit depends on three parameters: the mass of the object (m), its linear velocity (v) and its distance from the attracting centre (r). In such a system Newton's laws require that the momentm, $p = mv$,

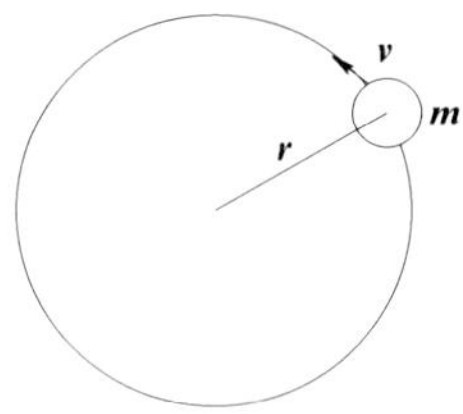

and the angular momentum, $L = mvr$, remain constant. Should the radial distance therefore decrease, the angular momentum must increase, speeding up the rotation and *vice versa*. The effect is illustrated by a ballerina folding her arms in a pirouette.

To continue with Laplace's model: As the rate of rotation increased to the point where the centrifugal force at the periphery exceeded gravitation, a ring of material separated from the main mass, eventually contracting towards a point. As the process continued, more planets condensed in the outer regions, while the inner region contracted to form the sun. The satellites were formed by condensation from the contracting planets. Excess material between the planets turned into comets and meteors.

The unequal distribution of angular momentum between the planets and the sun remains an unsolved problem, but not considered serious enough to finally abandon the nebular hypothesis. Rival theories such as the disintegration of one of a binary pair of stars to form the planets also leave many questions unanswered and a quantitatively correct model has not been formulated. However, after the time of Newton observational astronomy moved out of the solar system to examine the stars in the Milky Way, also known as the Galaxy, based on the Greek word for milk.

From his observations William Herschel (1738-1822) concluded that stars in the Galaxy were distributed in a disc-like structure, centred on the Sun. It remained the accepted picture until early in the twentieth century, when, based on the study of globular star clusters, Harlow Shapley proposed (1912) a disc-like Milky Way surrounded by a spherical arrangement of clusters and

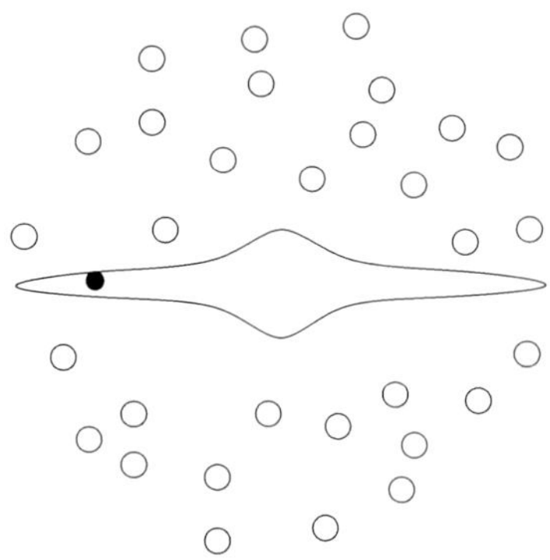

with the Sun in the galactic arm as shown by the black circle. A large number of hazy nebulae were assumed to lie within the Galaxy. The idea, proposed by a few contemporary astronomers, that the nebulae were galaxies of stars like the Milky Way was either ridiculed or ignored. Even Shapley shared the majority view that all nebulae are part of the Milky Way.

The true nature of Shapley's disc became clear with the discovery of Lindblad and Oort in 1926 that stars in the Milky Way are moving in a pattern, consistent with rotation of the Galaxy, like the solar system, around a central attracting mass. It was confirmed much later (1952) that the Milky Way indeed has a rotating spiral structure.

These new observations introduced a dramatic change in general cosmology. Not that it settled the dispute between evolution and design. Creation scientists responded by accepting stellar evolution as part of creation, starting with a single miraculous event, and guided forward by design, as described by the novelist Edgar Alan Poe (Beaver, 1976) in his essay, Eureka. Biological evolution, which gave impetus to evolutionary cosmology, is now hotly disputed again, following the dramatic developments in 20th century astronomy, which spawned a cosmology based on orthodox creation theory. No sooner had astronomy therefore managed to get rid of the mystic elements of astrology, which assumes a causal relationship between terrestrial and celestial events, when it became infected by a more virulent disease – creationist cosmology. Cosmology therefore came full circle since the time of Galileo by finding a new basis, which is compatible with religious dogma, and intolerant of heretics. As is often the case, this massive paradigm shift is credited to a single individual, in this case the astronomer, Edwin Hubble (1889–1953).

2.6 An Expanding Universe

The astronomical evidence that prompted abandonment of the concept of an equilibrium universe, in favour of an expanding one, is largely due to the work of Edwin Hubble, but without the explosive development of physics in the first quarter of the 20th century, his work would not have been possible. As with Newton and Darwin, new ideas were in the air, and Hubble was the one to complete the synthesis of ideas that created a new world view. Some of the other important players include James Clerk Maxwell (1831–79), Johann Balmer (1825–1898), Max Planck (1858–1947), Albert Einstein (1879–1955) and Niels Bohr (1885–1962). Without the insight of these scientists Hubble's observations make no sense.

The most important development was the ability to make reliable estimates of astronomical distances. To establish the distance to a star, astronomers rely on the phenomenon of stellar parallax. The first successful measurement of this type was performed at the Cape Observatory in the early 19th century.

Parallax is the apparent shift in the position of an object against a distant background because of a change in the observer's point of view.

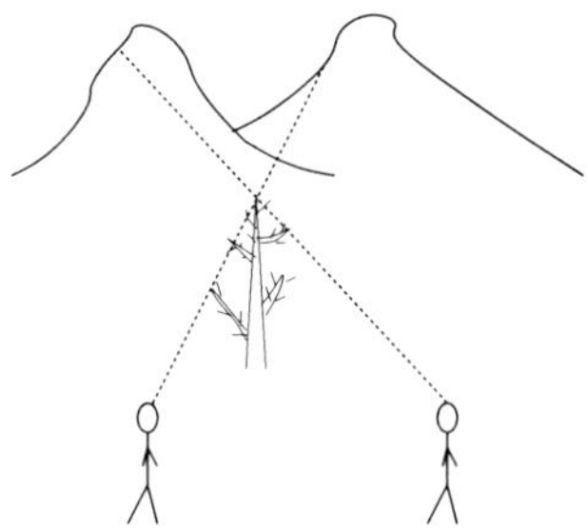

Stars exhibit the same phenomenon. As the earth orbits the sun nearby stars appear to move back and forth against the background of more distant stars. The parallax of the star is the half angle p which defines the shift of the star's apparent position viewed from opposite sides of the earth's orbit around the sun. To derive the distance d from the angular measurement it is necessary to know the distance between planet Earth and the sun, which defines the astronomical unit (AU) of distance. If the angle p is measured in seconds of arc, the distance d, defined as

$$d = \frac{1}{p}$$

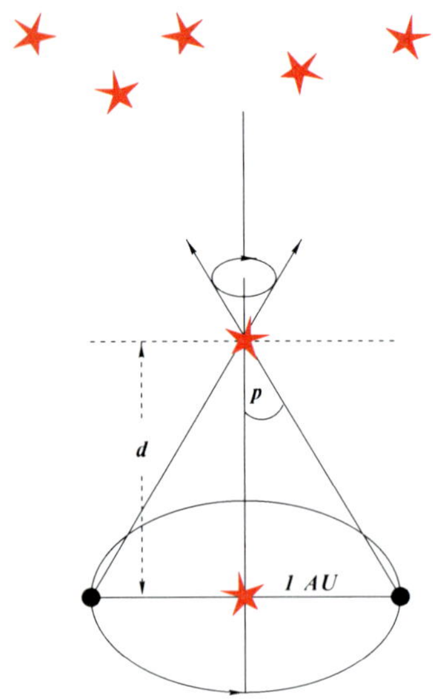

is obtained in *parsecs*. One parsec = 3.26 light years and 1 ly$\simeq 10^{13}$km.

The distance between Earth and Sun is obtained by measuring the parallax of another planet and use of Kepler's third law for the orbital periods of the planets. The closer an object is to the observer, the larger its parallax. On occasion the minor planet Eros approaches the earth more closely than any of the major planets and measurement of its parallax displacement during diurnal rotation of the earth at such time provides one of the best estimates of the AU.

Stellar parallax is measured most reliably by photographic methods. The most extensive set of early measurements were conducted from the Johannesburg station. The position of a star of interest is measured relative to a number of more distant reference stars over a period of months. The method has a serious limitation – for stars beyond a certain maximum distance the paralactic angle becomes too small to measure.

An alternative method based on following certain stars, known as Cepheid variables, allows measurement of much larger distances. The brightness of a Cepheid star fluctuates with a characteristic period of several days.

On assuming an inverse-square relationship between luminosity and distance and starting from a nearby Cepheid, a crude estimate of the distance to more remote Cepheids is derived by measuring their periods of fluctuation. Finally, the distance of any object, such as a star cluster or a nebula, which contains a Cepheid can be obtained by this method.

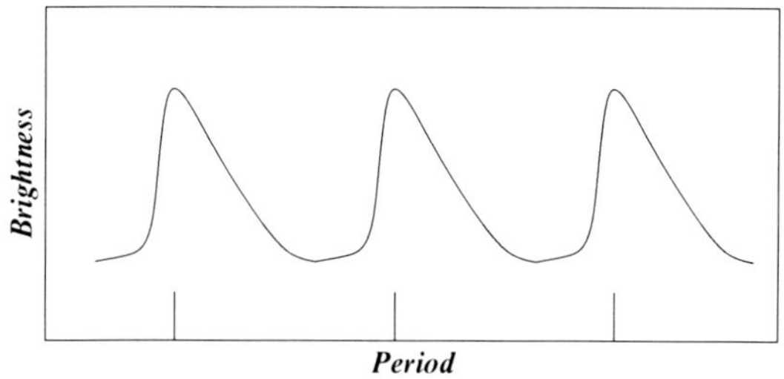

These new methods of measuring astronomical distances inaugurated modern astronomy and astrophysics and the interpretation of such measurements in terms of an expanding universe led on to the modern standard cosmology.

2.7 References

Beaver, H. (ed.) 1976):*The Science Fiction of Edgar Alan Poe*, Penguin Books, NY.

Coxeter, H.S.M. (1989):*Introduction to Geometry*, 2nd ed., Wiley, USA.

Crew, H. & A. de Salvio (1914):*Dialogues concerning Two New Sciences*, (1638), translated by H. Crew and A. de Salvio, New York.

Drake, S. (1953):*Dialogues concerning the Two Chief Systems of the World*, (1632), translated by S. Drake, Berkeley.

Graves, R (1966): *The White Goddess*, American ed., Farrar, New York.

Hall, A.R. (1963): *From Galileo to Newton*, Collins, London.

MacNeice, L. (1964): *Astrology*, 1964, Aldus, London.

Priesner, C. & K. Figala (1998): *Alchemie*, Beck, München.

Stephenson, R.J. (1960): *Mechanics and the Properties of Matter*, 2nd ed., Wiley, New York.

Stirling, S.G. (1935):*Mechanical Properties of Matter*, Macmillan, London.

Tompkins, P. (1978): *Secrets of the Great Pyramid*, Penguin Books, Harmondsworth, England.

Chapter 3

World Geometry

Modern geometry developed from the work of Euclid which gave rise to two self-contained geometries, known as *absolute* geometry and *affine* geometry. Euclidean geometry depends on five postulates:

1. A straight line may be drawn from any point to any other point.

2. A finite straight line may be extended continuously in a straight line.

3. A circle may be described with any centre and any radius.

4. All right angles are equal to each other.

5. If a straight line meets two other straight lines so as to make the two interior angles on one side of it together less than two right angles, the other straight lines, if extended indefinitely, will meet on that side on which angles are less than two right angles.

Absolute geometry accepts the first four postulates but does not assume an unique parallel. Affine geometry excludes postulates three and four, but the unique parallel plays a leading role. The concepts in common are:

> The *points* A, B, ... and the relation of *intermediacy* $[ABC]$, which is interpreted to show that B is between A and C.
>
> If A and B are two distinct points, the segment AB is the set of points P for which $[APB]$.
>
> The *interval* $\overline{AB}$ is the segment AB plus the end points A and B.
>
> The *ray* A/B is the set of points P for which $[PAB]$.

J.C.A. Boeyens, *Chemical Cosmology*, DOI 10.1007/978-90-481-3828-9_3,

The *line* AB is the interval $\overline{AB}$ plus the two rays A/B and B/A.

Parallelism is introduced by stating that for any point A and any line r, not through A, in the plane Ar, there is at least one line through A, in the plane Ar, which does not meet r.

A triangle $\triangle ABC$ is the union of three lines: $\triangle ABC = AB \cup AC \cup BC$.

A set of lines is coincident (concurrent) if all the lines intersect at the same point.

Triangles $\triangle ABC$ and $\triangle A'B'C'$ are in *perspective* if the lines AA', BB', CC' that join coprresponding vertices on the two triangles are coincident.

3.1 Affine Geometry

Each geometry has specific rules whereby to identify equivalent objects (Coxeter, 1989). In Euclidean geometry two equivalent, or congruent, triangles must have the same area, sides of the same length and corresponding angles the same. To determine that two figures belong to different congruence classes, it is sufficient to find one geometric property which they do not have in common. For instance the two triangles of Figure 3.1(a) have the same area because the lengths of their bases are the same and they have the same perpendicular height. However, the fact that they have different angles shows that they are not Euclidean equivalents. In non-Euclidean geometries without an angular measure these two triangles are congruent.

Not all geometries are more flexible than Euclidean geometry. For instance, it may be required that the sides of equivalent triangles make the same angles with a given line as in 3.1(b). The two triangles are said to have the same orientation. Should one of them be rotated in the plane they are no longer considered congruent, but remain equivalent in the Euclidean sense. One example is a pair of triangles in a vector field. Equivalent vectors must have the same orientation.

In another case the requirement that all lengths should be the same may be dropped. The two triangles of 3.1(c) may now be considered congruent in a geometry that permits transformations which involve proportional magnification (or contraction) in addition to rigid transformations (translation, rotation and reflection), defining a *similarity* geometry.

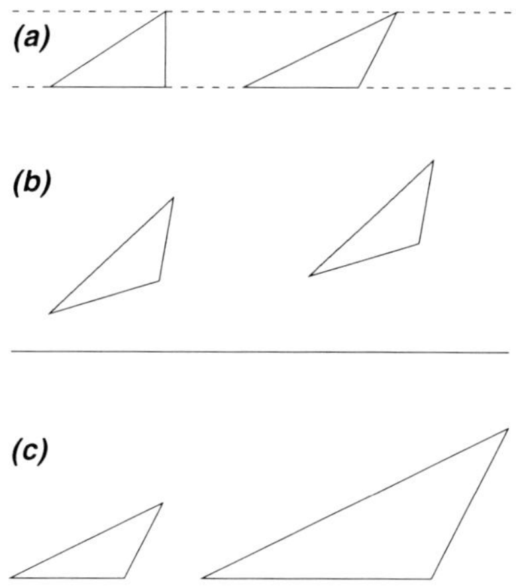

Figure 3.1: *(a) A pair of affine equivalent triangles; (b) congruent triangles in the same orientation; (c) similar triangles*

The transformations (Figure 3.2) allowed within a given geometry are summarized algebraically by a *mapping*. A two-dimensional translation is given by

$$T : (x, y) \mapsto (x + a, y + b)$$

[Point (x, y) *maps* to point $(x + a, y + b)$]

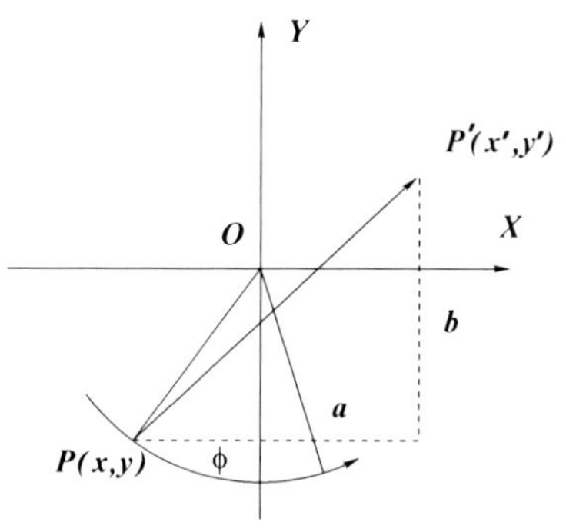

Figure 3.2: *Translation and rotation transformations*

Rotation in the X, Y plane through an angle ϕ about the origin O is given by:

$$R : (x, y) \mapsto (x \cos \phi - y \sin \phi, x \sin \phi + y \cos \phi) \tag{3.1}$$

To show that length is preserved under T, consider two points $P_1(x_1, y_1)$ and $P_1(x_2, y_2)$. The length of the line P_1P_2 is defined as:

$$d = \sqrt{(x_2 - x_1)^2 + (y_2 - y_1)^2}$$

Under T, the line P_1P_2 is translated to $P_1'P_2'$ with coordinates $(x_1 + a, y_1 + b)$ and $(x_2 + a, y_2 + b)$. The length of $P_1'P_2'$ becomes

$$\sqrt{[(x_2 + a) - (x_1 + a)]^2 + [(y_2 + b) - (y_1 + b)]^2} = d$$

Since T represents any translation in the plane, length is preserved under all such translations. The same calculation using equation (3.1) shows that rotation also leaves P_1P_2 invariant.

A space $\boldsymbol{R^n}$ consisting of all points $(x_1, x_2, \ldots, x_n)$ where the distance between $x = (x_1, x_2, \ldots, x_n)$ and $y = (y_1, y_2, \ldots, y_n)$ is defined by

$$d(x, y) = \left[\sum_{i=1}^{n} |x_i - y_i|^2\right]^{1/2}$$

is termed an *n-dimensional Euclidean space.*

The transformation that transforms any point (x, y) in the plane to another (x', y') in the same plane is given by:

$$\begin{aligned} x' &= a_{11}x + a_{12}y \\ y' &= a_{21}x + a_{22}y \end{aligned} \tag{3.2}$$

Reciprocally, to every point x', y' will correspond an unique point x, y provided the determinant of the transformation

$$A = \begin{vmatrix} a_{11} & a_{12} \\ a_{21} & a_{22} \end{vmatrix} \neq 0$$

This condition allows solution of equations (3.2) for the ratio $x : y$ in terms of $x' : y'$, as follows:

$$\begin{aligned} x &= A_{11}x' + A_{12}y' \\ y &= A_{21}x' + A_{22}y' \end{aligned}$$

The coefficients A_{ij} are the cofactors of the elements a_{ij} respectively in the determinant A. Equation (3.1) shows that a rotation is defined by the matrix

$$R = \begin{pmatrix} \cos\phi & \sin\phi \\ -\sin\phi & \cos\phi \end{pmatrix} \tag{3.3}$$

Transformation in the plane, given by the mapping

$$(x, y) \mapsto (ax + by + c, dx + ey + f) \tag{3.4}$$

where a, b, c, d, e, f are real numbers and $ae \neq bd$ preserves neither length, nor angle, nor shape. Such a transformation is known as an *affinity* and the resulting geometry is known as *affine* geometry.

Any triangle in affine geometry can be defined with coordinates $A(0,0)$, $B(1,0)$, $C(x,y)$. The affine transformation generates $A'B'C'$ such that $A' =$

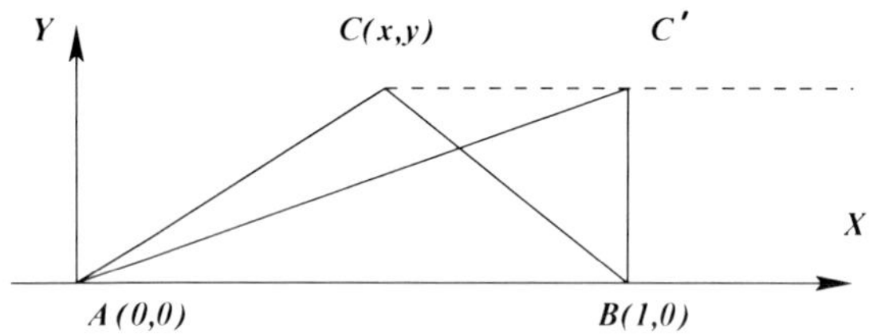

(c, f), placed at (0,0). $B' = (a, d)$. Setting $a = 1$, $d = 0$, $B' = B$, $C' = (x + by, ey)$. Since $ae \neq bd$, $e \neq 0$. For $e = 1$, $C' = (x + by, y)$, *i.e.* on the line CC', parallel to AB. This transformation is known as a *shear*. For $e \neq 1$, CC' is not parallel to AB and the transformation describes a *strain*. In the case of a shear (an example of an *equiaffinity*) the areas of the two triangles are the same, but clearly not for a strain. *Magnifications* and *contractions*, as in similarity geometry, are therefore permitted.

The transformation of a square under shear and strain is shown in Figure 3.3.

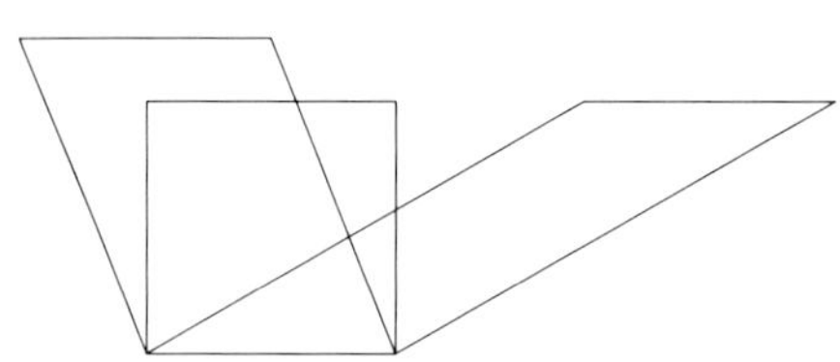

Figure 3.3: *Deformed cube under shear and strain*

The two resulting figures are both equivalent to the original square and to each other. Thus no distinction is made between squares and parallelograms. Further, no distinction is made between circles and ellipses. According to Figure 3.3 lines that were parallel in the original remain parallel although angles between lines are not invariant. It can be shown that this is generally true under affine transformations in the plane.

Let PQ and RS be parallel, with P, Q, R, S at coordinates (x_1, y_1), (x_2, y_2), (x_3, y_3), (x_4, y_4). The two parallel lines have equal slopes:

$$\frac{y_2 - y_1}{x_2 - x_1} = \frac{y_4 - y_3}{x_4 - x_3}$$

Under the transformation, P, Q, R, S map to $P'(ax_1 + by_1 + c, dx_1 + ey_1 + f)$, $Q'(ax_2 + by_2 + c, dx_2 + ey_2 + f)$, $R'(ax_3 + by_3 + c, dx_3 + ey_3 + f)$, $S'(ax_4 + by_4 + c, dx_4 + ey_4 + f)$ respectively. The slope of $P'Q'$ is given by:

$$\frac{d(x_2 - x_1) + e(y_2 - y_1)}{a(x_2 - x_1) + b(y_2 - y_1)} = \frac{d + e\left(\frac{y_2 - y_1}{x_2 - x_1}\right)}{a + b\left(\frac{y_2 - y_1}{x_2 - x_1}\right)}$$

$$= \frac{d + e\left(\frac{y_4 - y_3}{x_4 - x_3}\right)}{a + b\left(\frac{y_4 - y_3}{x_4 - x_3}\right)} = \frac{d(x_4 - x_3) + e(y_4 - y_3)}{a(x_4 - x_3) + b(y_4 - y_3)}$$

which is the slope of $R'S'$.

A square or a parallelogram cannot be transformed, for example, into a trapezium as this would contravene the invariance of parallelism. All triangles are, however, equivalent; no parallel lines are involved, and successive transformations of shear and strain, in addition to the rigid transformations, will transform any given triangle into any other triangle.

Another important invariant under affine transformation is the ratio in which points divide straight line segments. Also, finite configurations remain finite.

3.1.1 Affine Lattices

Crystallography relates the geometrical properties of crystals to lattices with translational and rotational components. In affine geometry where angle is not preserved the same lattices are generated without loss of generality, by translational symmetry only. The simplest two-dimensional example is the repetition of two independent translations X and Y of any given point, as shown in Figure 3.4 .

The grid of lines through points with equal powers of X and of Y is said to form a *tessalation* of congruent parallelograms, called *unit cells*. The unit cell can be defined in many ways, all of them with the same area. A *reduced* cell is obtained by taking Y as the shortest translation between points and

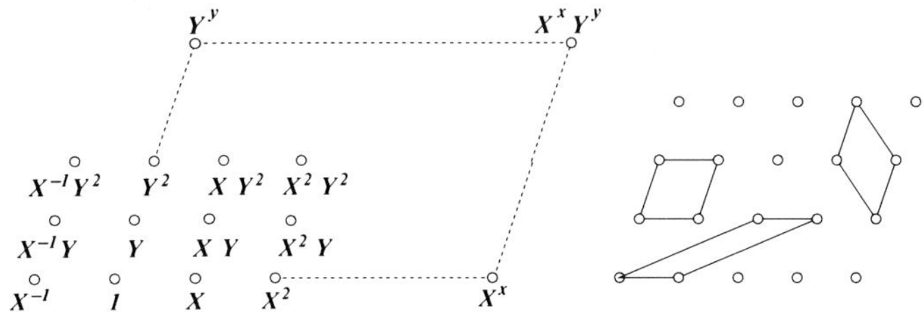

Figure 3.4: *A two-dimensional lattice and unit cells*

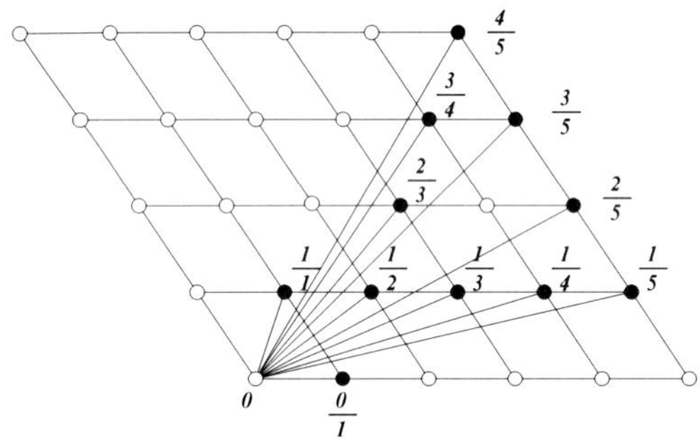

Figure 3.5: *The Farey sequence F_5 as defined by the visible points of a lattice, for convenience shown here with an obtuse angle*

X equal to the next shortest translation in another direction, such that the angle between X and Y is acute. Any lattice point can be chosen as the origin O. The first lattice point, A, along the ray OA is said to be *visible*. The affine coordinates of each lattice point are two integers, counting from 0. As shown in Figure 3.5 any point (l, m) will be visible if l and m are coprime, that is, they have no common divisor greater than 1. The parallelogram formed by the visible points (1,0), (1,1) and (0,1) with the origin defines the unit cell.

The series of fractions m/l form a Farey sequence of order n if m and l are coprime and $0 \leq m \leq l \leq n$. For instance:

$$F_5: \quad \frac{0}{1}, \frac{1}{5}, \frac{1}{4}, \frac{1}{3}, \frac{2}{5}, \frac{1}{2}, \frac{3}{5}, \frac{2}{3}, \frac{3}{4}, \frac{4}{5}, \frac{1}{1}$$

Such a series has the important property of being *unimodular*, which means that consecutive terms have

$$\begin{vmatrix} l & l_1 \\ m & m_1 \end{vmatrix} = 1$$

Proof of this is that such points on a lattice are the only visible ones.

3.2 Projective Geometry

Whereas most theorems of geometry are concerned with the concepts distance, angle and congruence, a smaller number of others are only concerned with the incidence of points and straight lines. This distinction differentiates between the common *metrical* properties of geometry and those, which are independent of measurement, and which reflect the characteristics of what became known as *projective geometry*.

A means of incorporating metric geometry into the more general projective geometry was developed by Cayley, preparing the way for the demonstration that projective geometry includes the affine, Euclidean and non-Euclidean geometries. This aspect is addressed by Veblen in the Appendix.

The other seminal work of Veblen and Young (1910) served as a guide in the following short summary of what is considered essential aspects of projective geometry applied to cosmology.

Projective geometry has developed independent of affine and absolute geometries and in a sense is a combination of the two by avoiding all but the first two of Euclid's axioms. The relation of intermediacy therefore also falls away and segments are not defined.

A fundamental operation in projective geometry is the mapping of points on a line (l) onto another (l') from a central point P, as shown in Figure 3.6.

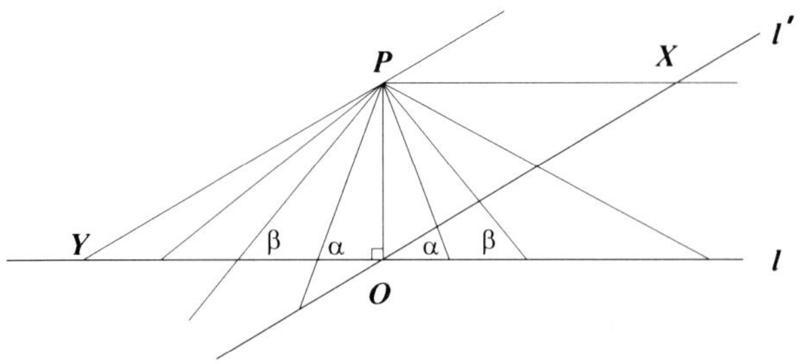

Figure 3.6: *Perspective mapping of points on a line to another line*

Each point on l projects to a unique point on l' and each possible line through P connects two, uniquely defined (homologous) points on l and l', with one exception. The perpendicular on OP through P fails to intersect l and hence the point X has no homologue on l. By the same reasoning the point Y on l has no homologue on l'. This creates an awkward dilemma because there is nothing special about the points X and Y. Points on either side of X are, for instance, successfully mapped to l. The problem lies with the lines XP and YP.

Starting from PO, which is perpendicular to l, the angles of intersection, α, β, $\gamma \ldots$, decrease monotonically until XP is reached with a zero angle of intersection. By continued rotation of the ray X/P the angle increases again, through Y, to a maximum at OP. This situation is reminiscent of the problem with the number system before the discovery of zero, which was solved by the addition of a fictitious integer, 0. The geometrical problem is likewise solved by the addition of ideal points at infinity on the lines l and l'. This discovery is as momentous for geometry as the discovery of zero was for algebra. It means that two lines, traditionally considered as parallel, can now be defined to intersect at an ideal point. The concepts of parallel and endless lines therefore fall away. There is no distinction between the ideal point and the regular points on a line. Any regular point may therefore also be interpreted as a point at infinity.

The mapping shown in Figure 3.6 also illustrates the procedure commonly used in making perspective drawings. In this case the central point is the eye of the artist. The line l represents the scene of interest and l' the canvas. Each radial line registers as a point at P and the plane of projection appears as a line, such as l. All parallel lines share the same ideal point and all ideal points at infinity appear as the ideal line at infinty, or *horizon*, which terminates the plane of projection. In three dimensions all parallel lines contain the same line at infinity and finally, all points and lines at infinity form a single plane at infinity.

The foregoing is interpreted to mean that the projective model of space is closed by a single surface that corresponds to the ideal plane at infinity. In Euclidean geometry this plane appears curved. If we therefore assume that the structure of the cosmos is subject to mathematical analysis and that the mathematics applies without exception, it is a logical necessity that the geometry of space-time be projective.

Points on the projective line are arranged in such a way that their order is cyclic[1]. The Euclidean relations of parallelism, betweenness, order and congruence therefore have no meaning in projective geometry. It can be stated that, without exception, any two lines in the projective plane intersect in a point, any two points in projective space define a line and any two planes

[1]It will in fact be shown that all conics, including the straight line, are equivalent in projective geometry.

intersect in a line.

Radial lines in projection look like points and radial planes look like lines when viewed edge-on, which means that radial dimensions are lost[2]. This correspondence defines the principle of *duality* which asserts that any definition or theorem in projective geometry remains valid on interchanging the words *point* and *line*, as well as the operations:

lie on – pass through

collinear – concurrent (or coincide)

As an illustration of the duality principle, consider the **Axiom**:

There exist four points of which no three are collinear,

which implies the **Dual**:

There exists a *complete quadrilateral* (Figure 3.7) which is a set of four lines (sides) intersecting in pairs in six distinct points (vertices).

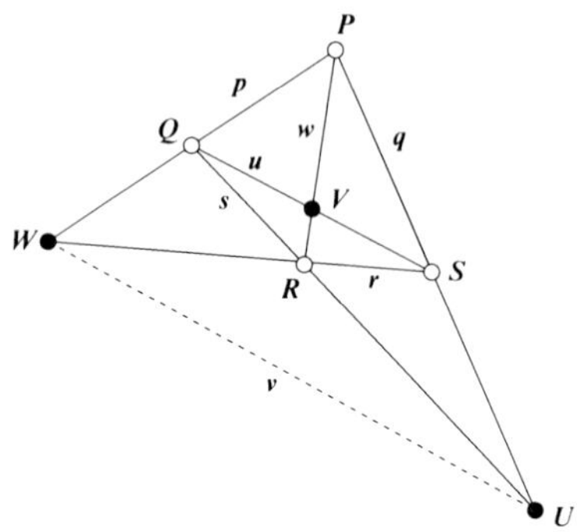

Figure 3.7: *Quadrangle PQRS and its dual quadrilateral pqrs*

Two vertices are said to be *opposite* if they are not joined by a side. The three joins of pairs of opposite vertices are called *diagonals*. If $PQRS$ is a quadrangle with sides:

$$p = PQ\,,\; q = PS\,,\; r = RS\,,\; s = QR\,,\; w = PR\,,\; u = QS$$

[2]It will be shown that physical models formulated in five-dimensional Euclidean (or affine) space are reproduced in four-dimensional projective space without the awkward concept of compacted dimensions.

then $pqrs$ is a quadrilateral with vertices:

$$P = p \cap q \,,\; Q = p \cap s \,,\; R = r \cap s \,,\; S = q \cap r \,,\; W = p \cap r \,,\; U = q \cap s$$

Fano's **axiom** states that the three diagonal points of a complete quadrangle are never collinear. Its dual asserts that the three diagonals of a complete quadrilateral are never concurrent.

The diagonals u, v, w of $pqrs$ can only be concurrent if $V = u \cap w$ lies on the line v, which is forbidden by Fano.

3.2.1 Projective Space

The previous conclusions are put into perspective by reference to Figure 3.6. All points on l are elements of the Euclidean plane $\boldsymbol{R^n}$. The radial lines in $\boldsymbol{R^n}$ that connect the Euclidean points to the point of projection (P) are points in the projective plane $\boldsymbol{P^n}$, which in addition contains the points at infinity, *i.e.*

$$\boldsymbol{P^n} = \boldsymbol{R^n} \cup (\text{points at infinity})$$

To summarize:

A projective point $(\boldsymbol{P^0})$ is a radial line.

A projective line $(\boldsymbol{P^1})$ is the set of radial lines in a projective plane.

A projective plane $(\boldsymbol{P^2})$ is the set of radial lines in a radial three-dimensional space.

Hence:

An n-dimensional projective space $\boldsymbol{P^n}$ is the set of radial lines in $\boldsymbol{R^{n+1}}$.

or:

An n-dimensional projective space is obtained by starting with $\boldsymbol{R^n}$ and completing it by addition of its points at infinity.

The following propositions are stated without proof:

Every pair of points in $\boldsymbol{P^n}$ lies on exactly one line.

Every pair of lines in $\boldsymbol{P^2}$ intersects in exactly one point.

Every pair of planes in $\boldsymbol{P^3}$ intersects in exactly one line.

Any three noncollinear points in $\boldsymbol{P^3}$ lie in exactly one plane.

3.2.2 Barycentric Coordinates

Within traditional usage there is no way in which to define the coordinates of a fictitious ideal point, in the same way as for all other points on a line.

The procedure which is commonly used to establish the centre of gravity between two weighted points, by the lever rule of mechanics, suggests a new way of defining coordinates that would apply to all points, without exception (Coxeter, 1989).

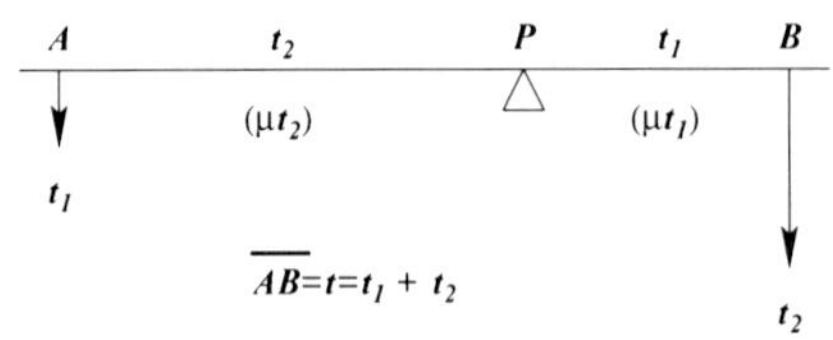

Two weights, t_1 and t_2, suspended at the points A and B are balanced on a support placed at P, the centre of gravity, on condition that

$$\frac{AP}{PB} = \frac{t_2}{t_1}.$$

This balance is not disturbed when both weights are increased by the same factor, $\mu \neq 0$, such that

$$\frac{AP}{PB} = \frac{t_2}{t_1} = \frac{\mu t_2}{\mu t_1}.$$

Irrespective of the measured positions of points A and B on the line, the correct condition of balance will always be described by assigning coordinates (t_1, t_2) to point P, where $\overline{AB} = t_1 + t_2$. Alternatively $\overline{AB} = \mu t_1 + \mu t_2$, with coordinates $(\mu t_1, \mu t_2)$ for point P. The coordinates of point P is independent of measure and conveniently formulated as:

$$x_1 = \frac{t_1}{t_1 + t_2} \qquad x_2 = \frac{t_2}{t_1 + t_2}$$

and normalized to the *homogeneous* coordinates $x_1 = t_1, x_2 = t_2$, provided $t_1 + t_2 \neq 0$. The ratio $x = x_1/x_2$ determines the position of the point on the line.

The homogeneous coordinates x^i of points in projective space therefore have the property of remaining valid when multiplied by an arbitrary factor, k, *i.e.*

$$x^i = kx^i$$

This remarkable property bears great similarity to the original conjecture of Weyl that on displacement a vector may be affected by a change of *gauge*. As with Weyl's proposal, a variable projective gauge acquires physical meaning when interpreted as a phase factor. A phase change associated with the displacement of a vector is known to underpin the appearance of an electromagnetic field as a property of space-time. Such a phase factor is defined for projective vectors by choosing $k = e^{\alpha(x)}$, as first proposed by Schrödinger.

The point P is at A if $t_2 = 0$ and at B if $t_1 = 0$. These special points therefore have coordinates $A(1,0)$, $B(0,1)$.

3.2.3 Linear Transformation

If the point of support is to change from $P(x_1, x_2)$ to $P'(x_1', x_2')$ without disturbing the balance the masses t_1 and t_2 need to be changed accordingly. By an appropriate choice of the coefficients in

$$x_1' = ax_1 + bx_2$$

$$x_2' = cx_1 + dx_2$$

the system can always be balanced at P'. These equations define a *linear transformation* from P to P'.

In terms of the non-homogeneous coordinate $x = x_1/x_2$, the transformation is formulated as

$$x' = \frac{x_1'}{x_2'} = \frac{axx_2 + bx_2}{cxx_2 + dx_2} = \frac{ax + b}{cx + d}\,. \tag{3.5}$$

The inverse transformation

$$x = \frac{dx' - b}{-cx' + a}$$

only exists if $ad \neq bc$.

If four points A, B, C, D on a line have coordinates of $x = \alpha, \beta, \gamma, \delta$ respectively, the ratio

$$\frac{(\alpha - \gamma)(\beta - \delta)}{(\alpha - \delta)(\beta - \gamma)}$$

known as the *cross product* of the four points, is found to be unaltered by the transformation (3.5), and therefore independent of the coordinate system. It will be shown to remain constant in projective mapping.

By the same procedure barycentric coordinates can be set up in the plane of a reference triangle $A_1A_2A_3$. If $t_1 + t_2 + t_3 \neq 0$ masses t_1, t_2, t_3 at the three vertices determine a point P (the centroid) with coordinates (t_1, t_2, t_3).

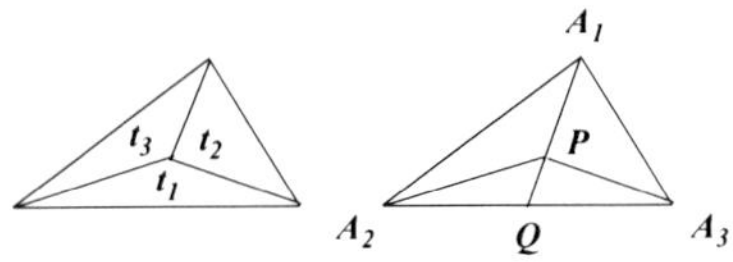

In particular: (1,0,0) is A_1, (0,1,0) is A_2, (1,0,0) is A_3 and $0, t_2, t_3$ is the point on A_2A_3 whose one-dimensional coordinates with respect to A_2 and A_3 are (t_2, t_3), *i.e.* Q. A mass t_1 at A_1 balances a mass of $t_2 + t_3$ at Q to make P the centroid.

The coordinates of P on A_1Q follows from t_1, the mass at A_1 that balances the triangle on P. *Areal* coordinates t_i correspond to the areas t_i, *e.g.*:

$$\frac{t_3}{t_2} = \frac{A_2Q}{QA_3} = \frac{area(A_1A_2Q)}{A_1QA_3} = \frac{PA_2Q}{PQA_3}$$

$$= \frac{A_1A_2Q - PA_2Q}{A_1QA_3 - PQA_3} = \frac{PA_1A_2}{PA_3A_1}$$

with similar definitions of t_1/t_3, t_2/t_1. Each areal coordinate is the area of the triangle PA_2A_3, PA_3A_1 or PA_1A_2 as a fraction of the whole area $A_1A_2A_3 = t_1 + t_2 + t_3 = t \neq 0$. Setting $t = 1$ normalizes the coordinates to $x_i = t_i$, $i = 1, 3$.

3.2.4 Projective Coordinates

The restriction $t_1 + t_2 + t_3 \neq 0$ disappears by adding the line at infinity, $t_1 + t_2 + t_3 = 0$, and all its points at infinity in various directions. Interpret T_1, T_2, T_3 as the distances from A_1, A_2, A_3 to the line $T_1t_1 + T_2t_2 + T_3t_3 = 0$. A parallel line is obtained by adding the same number to each T. These lines intersect where

$$T_1t_1 + T_2t_2 + T_3t_3 = t_1(a + T_1) + t_1 + t_2(a + T_2) + t_3(a + T_3)$$

i.e. $a(t_1 + t_2 + t_3) = 0$, which shows that the point of intersection lies on the line at infinity.

In projective geometry the line at infinity no longer has a special role and barycentric coordinates may be replaced by general projective coordinates (x_1, x_2, x_3), given by

$$t_1 = \mu_1x_1, \qquad t_2 = \mu_2x_2, \qquad t_3 = \mu_3x_3$$

where μ_1, μ_2, μ_3 are constants, $\mu_1\mu_2\mu_3 \neq 0$. Thus (x_1, x_2, x_3) is the centroid of masses $\mu_\alpha x_\alpha$ at $A_\alpha(\alpha = 1, 2, 3)$, and the line at infinity has the equation $\mu_1 x_1 + \mu_2 x_2 + \mu_3 x_3 = 0$.

Barycentric coordinates can be referred to any given triangle with vertices (1,0,0), (0,1,0), (0,0,1) and unit point (1,1,1), the centroid. In contrast, projective coordinates can be applied to any quadrangle: Take three of the four vertices to determine a system of barycentric coordinates and suppose that the fourth vertex is (μ_1, μ_2, μ_3). Converted to projective coordinates the fourth vertex becomes (1,1,1). Whereas all triangles are alike in affine geometry, all quadrangles are seen to be alike in projective geometry.

3.2.5 Theorems of Projective Geometry

A theorem, now considered seminal for projective geometry, was discovered by Pappus of Alexandria in about the year 320:

Theorem of Pappus

> In $\boldsymbol{P}^2$, given points A, B, C on a line l and A', B', C' on a line l', set $P = AB' \cap A'B$, $Q = AC' \cap A'C$, $R = BC' \cap B'C$. Then P, Q and R are collinear.

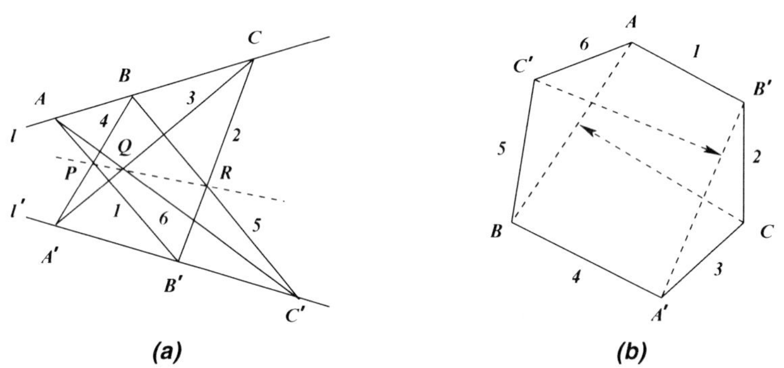

Figure 3.8: *Pappus configuration (a), drawn as hexagon in (b)*

The Pappus configuration can also be defined in terms of a plane hexagon inscribed in two intersecting lines, obtained from the arrangement in Figure 3.8(b). The Pappus configuration is reached by moving points C and C' to the lines AB and $A'B'$ respectively, noting that there is no ordering of points in projective space.

Calling sides such as 1-4, 2-5, 3-6 opposite, the theorem may be stated in the form:

> If a simple hexagon be inscribed in two intersecting lines, the three pairs of opposite sides will intersect in collinear points.

Dual Theorem of Pappus

> If A_1A_4, A_2A_5, A_3A_6 are concurrent and A_2A_3, A_1A_6, A_4A_5 are concurrent, then A_1A_2, A_3A_4, A_5A_6 are concurrent.

The way in which the lines define a hexagon is shown in Figure 3.9.

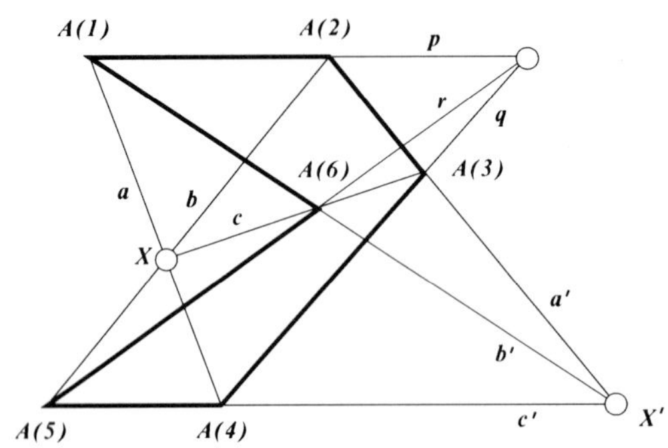

Figure 3.9: *The Pappas dual arrangement*

Alternatively:

> In $\boldsymbol{P}^2$ given lines a, b, c passing through a point X, and a', b', c' passing through a point X', set:
>
> $$p = (a \cap b') \cup (b \cap a') , \quad q = (a \cap c') \cup (c \cap a') , \quad r = (b \cap c') \cup (c \cap b')$$
>
> Then p, q and r are coincident.

Theorem of Desargues

> If the triangles $\triangle ABC$ and $\triangle A'B'C'$ (Figure 3.10) are in perspective then the points $AB \cap A'B'$, $AC \cap A'C'$, $BC \cap B'C'$ where their corresponding sides intersect, are collinear.

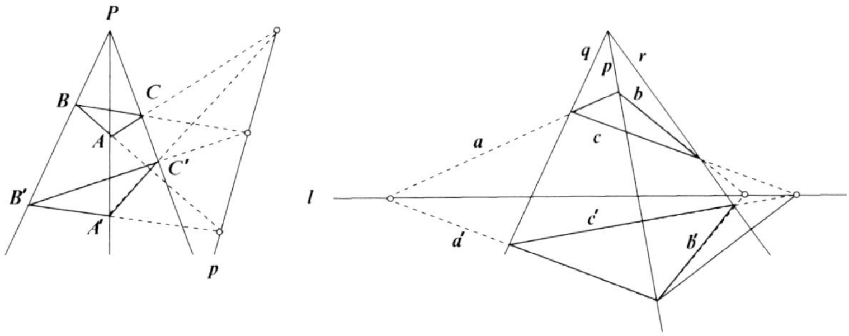

Figure 3.10: *Theorem of Desargues and its dual*

Dual of Desargues

Let $\triangle ABC$ and $\triangle A'B'C'$ be triangles in $\boldsymbol{P}^2$ whose sides are the lines a, b, c and a', b', c' respectively. If the points $a \cap a'$, $b \cap b'$ and $c \cap c'$ all lie on the same line l, then the lines $p = (a \cap b) \cup (a' \cap b')$, $q = (a \cap c) \cup (a' \cap c')$ and $r = (b \cap c) \cup (b' \cap c')$ all pass through the same point.

Pascal's Theorem

If a hexagon is inscribed in a smooth conic, the intersections of opposite sides of the hexagon are collinear.

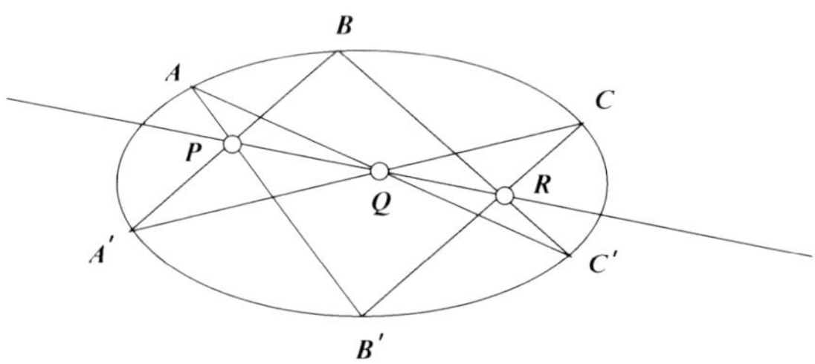

or

If the points A, B, C, A', B', C' lie on a smooth conic then $P = AB' \cap A'B$, $Q = AC' \cap A'C$, $R = BC' \cap B'C$ are collinear.

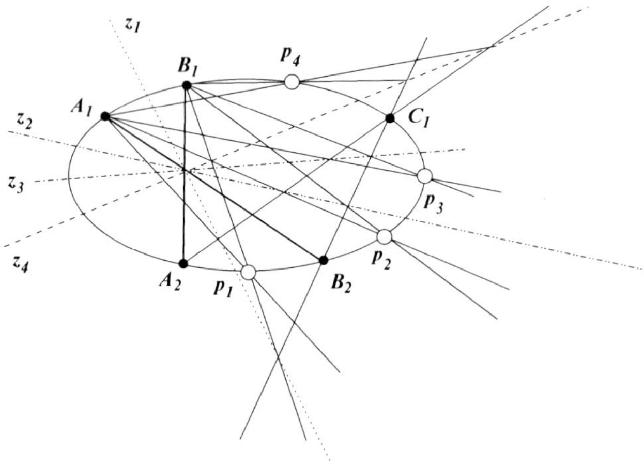

This is clearly the same as the theorem of Pappus if, indeed, all conics are equivalent in projective geometry.

The inverse of Pascal's theorem allows the construction of a conic through five arbitrarily defined points A_1, B_1, C_1, A_2, B_2 as the locus of the point $C_2 = p_i = A_1 \cup (z_i \cap C_1A_2) \cap B_1 \cup (z_i \cap C_1B_2)$ where z_i is a variable line through the point $A_1B_2 \cap A_2B_1$.

3.2.6 Projectivity

The set of all points on a line is said to define a *range*. Dually, a *pencil* is the set of all lines through a point. Where the lines of a pencil join the points of a range to another point, the range corresponds to a *section* of the pencil. The correspondence between two ranges that are sections of one pencil by two distinct lines is called a perspective, written as $X \overset{=}{\wedge} X'$ or $X \overset{O}{\overset{=}{\wedge}} X'$, meaning that, if X and X' are corresponding points of the two ranges, their join XX' continually passes through a fixed point O, called the *centre* of perspectivity.

The product of any number of perspectivities is called a *projectivity*. Two ranges (or pencils) related by a projectivity are said to be *projectively related*: $X \overline{\wedge} X'$.

As shown in Figure 3.11(a):

$$ABCD \overset{O}{\overset{=}{\wedge}} A_0B_0C_0D_0 \overset{P}{\overset{=}{\wedge}} A'B'C'D' \quad , \quad ABCD \overline{\wedge} A'B'C'D'$$

Alternatively, given three points A, B, C on a line, and three distinct points $A'B'C'$ on another line, they can be related by a pair of perspectives as

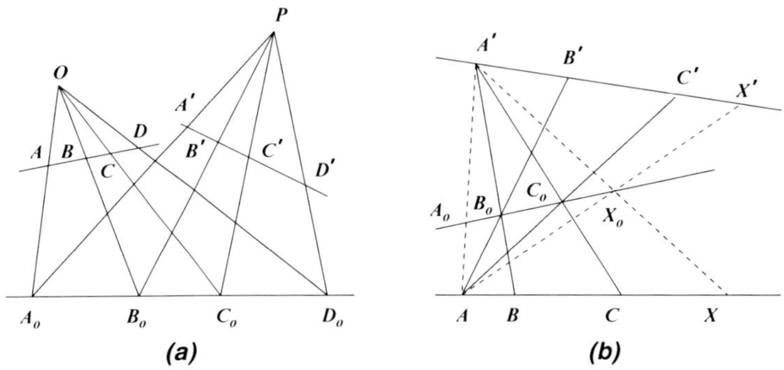

Figure 3.11: *Demonstration of (a) a projectivity $ABCD \barwedge A'B'C'D'$ and (b) a perspective pair*

shown in Figure 3.11(b). The *axis* of the projectivity joins the points $B_0 = AB' \cap A'B$ and $C_0 = AC' \cap A'C$, such that

$$ABC \stackrel{A'}{\overline{\overline{\wedge}}} A_0B_0C_0 \stackrel{A}{\overline{\overline{\wedge}}} A'B'C'$$

For each point X on AB there is a corresponding point X' on AB', obtained by joining A to the point $X_0 = A'X \cap B_0C_0$, to give

$$ABCX \stackrel{A'}{\overline{\overline{\wedge}}} A_0B_0C_0X_0 \stackrel{A}{\overline{\overline{\wedge}}} A'B'C'X'$$

From the theorem of Pappus it follows that the hexagon $ABCA'B'C'$ contains the point $BC' \cap B'C$, and all other cross joins of pairs of corresponding points. Any other pair of perspectives, *e.g.* from B and B', could therefore be used to derive X' from X. In other words:

> Any projectivity relating ranges on two distinct lines is expressible as the product of two perspectivities whose centres are corresponding points (in reversed order) of the two related ranges.

Hence:

> It is possible, by a sequence of not more than three perspectivities, to relate any three distinct collinear points to any other three collinear points.

Finally:

The Fundamental Theorem of Projective Geometry

> A projectivity is determined when three points of one range and the corresponding three points of the other are given.

It tells us that if A, B, C are any three points of one line and A', B', C' are any three points of either the same or another line, a projective mapping may be constructed (as in 3.11(b)) in which the images of A, B, C are A', B', C' respectively. This construction can be made in a great variety of ways; but the fundamental theorem states that the actual mapping will always be the same. Irrespective of how the construction is made, the image of a fourth point D will be a unique point D'. Expressed algebraically this condition follows from the *cross ratio*:

$$[A, B, C, D] = \frac{AC.BD}{AD.BC} = \frac{A'C'.B'D'}{A'D'.B'C'}\,.$$

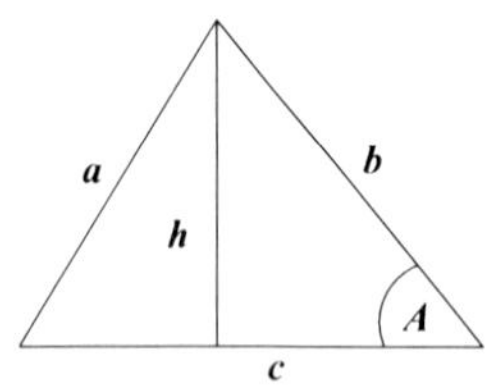

The area of a triangle is calculated as either $\triangle = base \times h/2$ or $\triangle = (c/2).b \sin A$. The triangles AOB, BOC, *etc.* in Figure 3.11(a) have the same vertical height and hence:

$$[A, B, C, D] = \frac{\triangle(AOC) \times \triangle(BOD)}{\triangle(AOD) \times \triangle(BOC)}$$

$$= \frac{OA \cdot OC \sin(A\widehat{O}C) \times OB \cdot OD \sin(B\widehat{O}D)}{OA \cdot OD \sin(A\widehat{O}D) \times OB \cdot OC \sin(B\widehat{O}C)}$$

$$= \frac{\sin(A\widehat{O}C) \cdot \sin(B\widehat{O}D)}{\sin(A\widehat{O}D) \cdot \sin(B\widehat{O}C)}$$

which remains the same for any four points A', B', C', D' into which A, B, C, D may be projected from O, and so

$$[A, B, C, D] = [A', B', C', D']\,.$$

Formal proof of the fundamental theorem, and all other theorems of projective geometry, are given at various levels of rigour in many reference works. Graphical demonstration is condidered adequate for our purposes.

We consider a general perspectivity of three radial lines coincident at P.

Three collinear points at any desired fixed distance apart (*e.g.* 1:1) can always be placed in proper perspective at points A, B, C on l. Parallel displacement of l to l' preserves the ratio $AB/BC = 1$. Another set of points

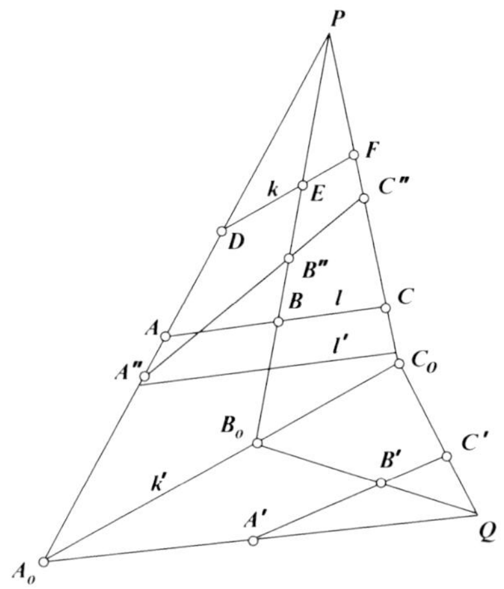

D, E, F arranged, for instance, in golden proportion (1.61803...) generates the same perspective on parallel displacement to k'. A second random perspective on A_o, B_o, C_o from point Q can always be constructed. Three points A', B', C' (or A'', B'', C'') separated in any ratio (2:1 say) can always be placed in perspective on A_o, B_o, C_o. The projectivity $ABC \barwedge A'B'C'$ for any two sets of three points always exists.

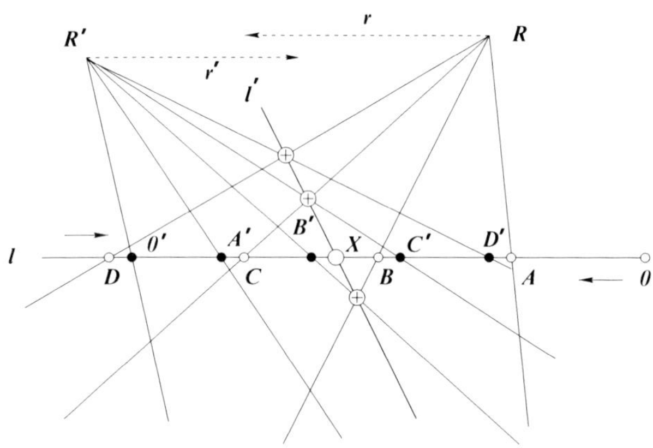

Figure 3.12: *The pencil through R maps the progress of the first runner on l and the pencil through R' that of the second runner. The corresponding points on l' lie on l at X and the point at infinity*

Any triangle in perspective appears as three collinear points. The fundamental theorem therefore confirms that as all triangles are equivalent in affine space all quadrangles are equivalent in projective space.

Double Points

If two ranges are on the same line a point that corresponds projectively to itself is called a *double* (or *invariant* or *fixed* point). To picture a double point we consider two runners going in opposite directions on a linear track. Perspective mapping of their progress along the track l is shown in Figure 3.12. The image points X and X', where the runners cross is a double point. A second double point occurs at infinity as the lines r, r' and l meet.

Two ranges, with reversed sense, on the same line are called *opposite* and have two double points. Direct ranges can have any number of double points from zero to infinity.

3.2.7 Quadrangular Sets

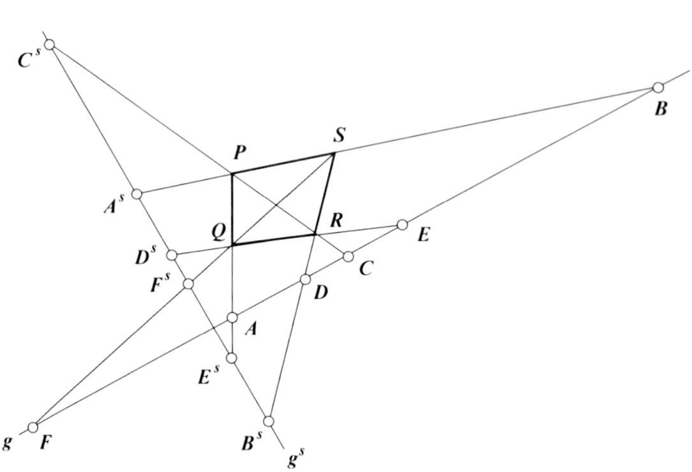

Figure 3.13: *The complete quadrangle and the quadrangular set*

The complete quadrilateral (Figure 3.13) consists of the sides of a quadrangle $PQRS$ together with the diagonals PR and QS. The six sides of a complete quadrangle intersect any line, g, not through a vertex, in a *quadrangular set* of six points, $Q(ABC, DEF)$. It can be shown that each point of Q is uniquely determined by the remaining points.

In order to establish consistent notation the sides of the quadrangle $PQRS$, through a selected vertex (P say), intersect g in points A and B. The diagonal through P intersects g at C. Projecting from the vertex S towards g^s, by the same procedure, Q^s is defined as shown.

To show that each point of Q (F say) is uniquely determined by the remaining points, $A \rightarrow E$, a second quadrangle $P'Q'R'S'$ is constructed with 5 sides passing through $A \rightarrow E$ as shown.

Without considering a sixth point the pairs of triangles $PRS/P'R'S'$ and $PQR/P'Q'R'$, as well as the quadrangles $PQRS/P'Q'R'S'$ in Figure 3.14 are perspective in the line g. By the dual of Desargues it follows that they are also perspective in the point O where all pairs of sides intersect.

As the triangles QRS and $Q'R'S'$ are perspective from the point O Desargues tells us that they are also perspective from the line DE (*i.e.* g) and hence the sides QS and $Q'S'$ must meet in g, *i.e.* at the common point F.

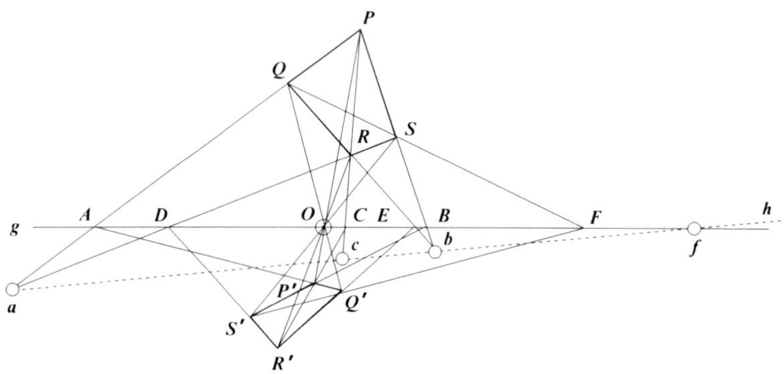

Figure 3.14: *Properties of quadrangular and harmonic sets*

On the special line h the points A, D and B, E coincide at the two diagonal points of $PQRS$, while PC and QF intersect at the third diagonal point. The quadrangular set $Q \rightarrow Q(ABC, ABF) \equiv H(AB, CF)$, in this case, is known as an *harmonic set*, with cross ratio:

$$[a, b, c, f] = \frac{ac.bf}{(-bc).af}$$

The point f is the harmonic conjugate of c with respect to a and b.

Relabel the points of interest for a complete quadrilateral, as shown in Figure 3.15. If AF is considered to be the line at infinity, the pairs $PS||QR$ and $SR||PQ$ become parallel, defining a parallelogram $PQRS$. The harmonic

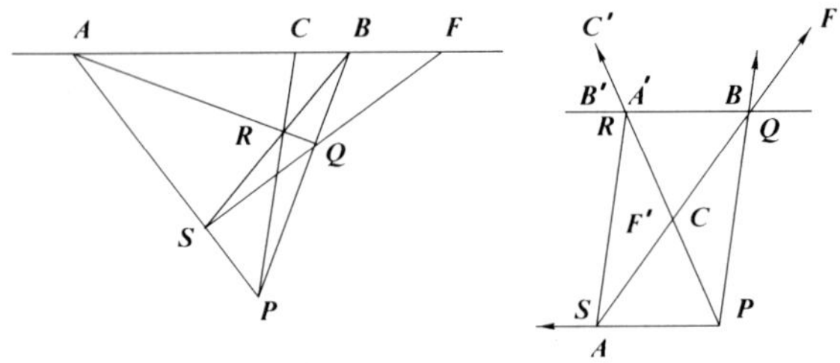

Figure 3.15: *The harmonic ratio* $[A, B, C, F] = -1$

points are defined on the lines: $PS(A)$, $PQ(B)$, $PR(C)$, $SQ(F)$, *i.e.* A at S, C at $PR \cap SQ$ and B at Q, all on the line AF.

Alternatively, the same points lie on $QR(A)$, $SR(B)$; *i.e.* A' at R, B' at r, F' at $RP \cap QS$, on the line $A'C'$. It follows that $AC = -BC$, $AF = BF$ and hence, $[A, B, C, F] = -1$.

This property is exploited in finding the fourth harmonic to a set of three points: Given A, B, and C on a line, draw two lines, l_1 and l_2 through A and another l_3 through C. Let $P = l_1 \cap l_3$, $R = l_2 \cap l_3$, $S = PB \cap l_2$, $Q = RB \cap l_1$, and $F = QS \cap AB$ is the fourth harmonic of a complete quadrilateral $PQRS$. Starting with the same three collinear points A, B, C, the same point F is obtained, independent of the choice of l_1, l_2, l_3.

In the dual construction the sides PR, PS and RS of the triangle PRS are extended to intersect a line in C, B and A respectively. Let $Q = BR \cap AP$. $F = QS \cap AB$ is the harmonic conjugate of C. The harmonic points divide the segment AB internally and externally in the same ratio:

$$\frac{AC}{BC} = -\frac{AF}{BF} .$$

3.2.8 Involution

An involution is a projectivity of period 2, that is, a projectivity which interchanges pairs of points. The quadrangle of Figure 3.13 is modified by labelling $U = AR \cap QS$, $V = AS \cap PR$, $W = PR \cap QS$.

The perspectivities

$$ADCF \overset{S}{\overline{\wedge}} VRCW \overset{A}{\overline{\wedge}} SUFW \overset{R}{\overline{\wedge}} DAFC$$

define the projectivity $ADCF \overline{\wedge} DAFC$. By the fundamental theorem there is only one projectivity $ADC \overline{\wedge} DAF$. Hence, the projectivity that interchanges

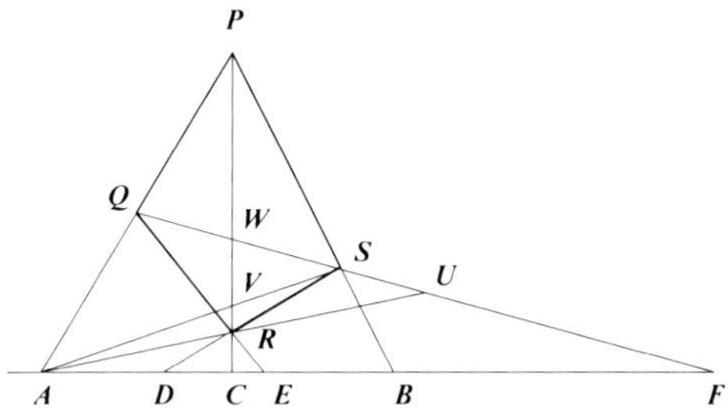

Figure 3.16: *Involution in* $Q(ABC, DEF)$

A and D transforms C into F. It follows that any projectivity which interchanges two points is an *involution*. Now, applied to the point B, the same three perspectivities interchange it with E:

$$B \overset{S}{\doublebarwedge} P \overset{A}{\doublebarwedge} Q \overset{R}{\doublebarwedge} E\,.$$

Since $Q(ABC, DEF)$ as in Figure 3.16 this result leads to a new formulation of the theorem of the quadrangular set:

> The three opposite pairs of sides of a quadrangle meet any line (not through a vertex) in three pairs of an involution, $(AD)(BE)(CF)$.

3.2.9 Conics

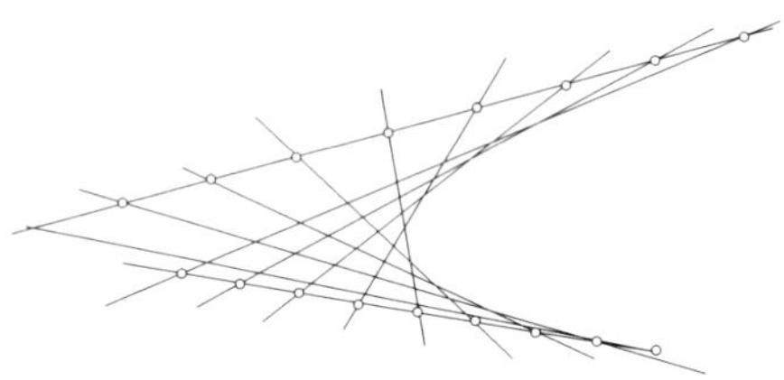

Figure 3.17: *A line conic*

The simplest definition of a (point) conic is the set of all points of intersection of homologous lines of two projective, nonperspective flat pencils which are on the same plane but not on the same point.

Homologous points on l and l' are defined by the perspectivity $l \overset{P}{\doublebarwedge} l'$ $(123\cdots \overset{P}{\doublebarwedge} 1'2'3'\ldots)$. The homologous projective pencils $l \overset{A}{\doublebarwedge}$ and $l \overset{B}{\doublebarwedge}$ intersect,

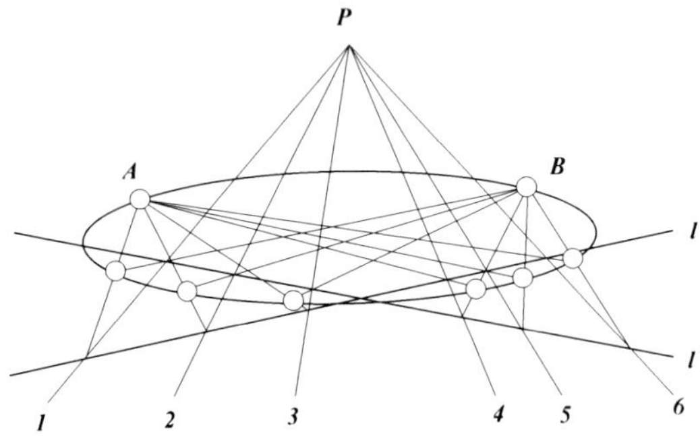

Figure 3.18: *Construction of a general point conic*

as shown in Figure 3.18, to define a conic, symbolically represented by $l \overset{A}{\overline{\wedge}} \cap \overset{B}{\overline{\wedge}} l'$.

The plane dual of a point conic is called a *line conic* – Figure 3.17. We state without proof that the tangents to a point conic is a line conic (Veblen & Young, 1910).

The set of all lines joining pairs of homologous points of two projective pencils on skew lines, as in Figure 3.19, is a *regulus*. The section of a regulus by a plane is a point conic.

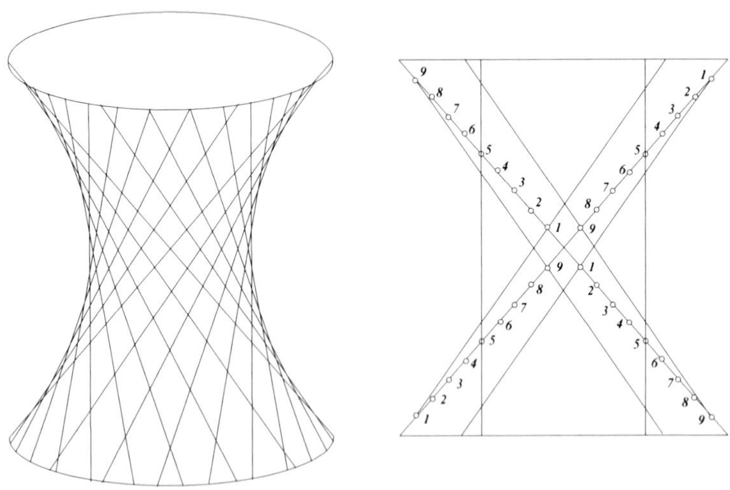

Figure 3.19: *Familiar form of a regulus and its mode of construction*

3.2.10 Collineation and Correlation

Any transformation of two (or three)–dimensional forms, such as the triangles of Figure 3.10, that projectively transforms collinear points of one form into collinear points of the other, is called a *collineation*. Point P, in Figure 3.10, is called the *centre* and line p the *axis* of the collineation, known as a *homology*. A collineation in which the centre coincides with the axis is called an *elation*.

A related type of transformation, which transforms points into lines and lines into points is called a *correlation*. It transforms collinear points into pencils, quadrangles into guadrilaterals, and so on.

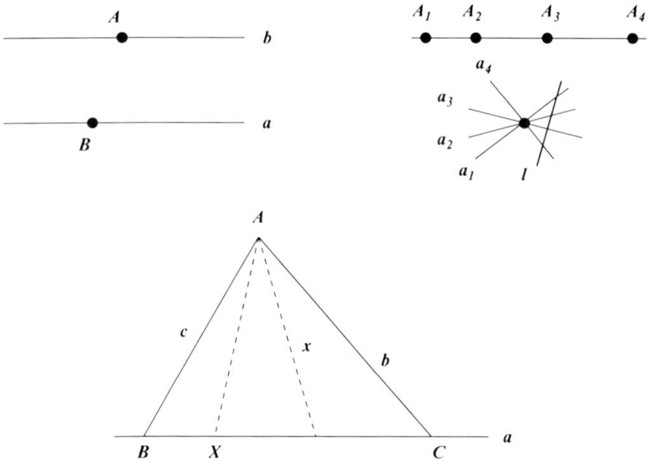

Figure 3.20: *Polarity and Involution*

A set of collinear points A_i on b correlates with a pencil of concurrent lines a_i on B, such that the cross product of the pencil, measured on an arbitrary line l,

$$[a_1, a_2, a_3, a_4] = [A_1, A_2, A_3, A_4]$$

A *polarity* is a projective correlation of period 2. A polarity, Figure 3.20, dualizes incidences: if A, correlated with a, lies on b, correlated with B, then a passes through B. A and B are *conjugate* points, while a and b are conjugate lines. Line a is called the *polar* of A and A is the pole of a. Where A and a coincide the line is *self-conjugate*. It can be shown that a polarity induces an involution on any line that is not self-conjugate.

The projectivity $X \bar{\wedge} x \cap a$ transforms any non-self-conjugate point B into another point $a \cap b \equiv C$, whose polar is AB. The same projectivity transforms C into B. It therefore interchanges B and C and hence is an involution.

3.2.11 The Algebra of Points

Projective mapping is simplified when expressed algebraically in terms of coordinates, as was done for affine transformation before.

Addition

Figure 3.21 shows how the operation of *addition* is defined in projective mapping by reference to three fundamental points P_0, P_1, P_∞, which determine a *scale* on l. Let l_∞ and l'_∞ be two lines through P_∞. A line l_0 through P_0 meets these lines in A and A'. Let P_xA and P_yA' meet l'_∞ and l_∞ in the points X and Y respectively. The point P_{xy} in which XY meets l, is called

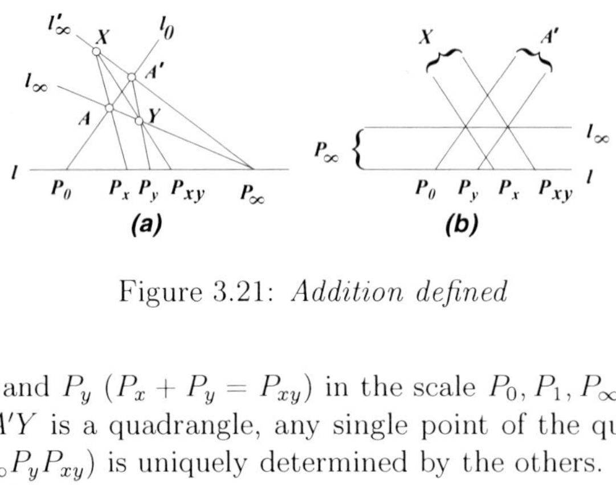

Figure 3.21: *Addition defined*

the *sum* of P_x and P_y ($P_x + P_y = P_{xy}$) in the scale P_0, P_1, P_∞.

Since $AXA'Y$ is a quadrangle, any single point of the quadrangular set $Q(P_\infty P_x P_0, P_\infty P_y P_{xy})$ is uniquely determined by the others.

The projectivity

$$[P_0, P_x] \overset{A}{\bar{\bar{\wedge}}} [A', X] \overset{Y}{\bar{\bar{\wedge}}} [P_y, P_{xy}]$$

shows that when P_0 transforms to P_y, P_x transforms to P_{xy}. Relabel P_y as P_a (Figure 3.22) to show that each point P_x of l corresponds to $P'_x = P_x + P_a$, where P_a is any fixed point ($\neq P_\infty$) of $P_\infty P_0 P_x P_y \bar{\wedge} P_\infty P_0 P_{x+a} P_{y+a}$.

Referred to z each point $P_z \overset{B}{\bar{\bar{\wedge}}} [Z] \overset{A}{\bar{\bar{\wedge}}} P_{z+a}$ with $z = 0, x, y, \infty$. Hence $P_\infty P_0 P_x P_y \bar{\wedge} P_\infty P_0 P_{x+a} P_{y+a}$, where $P_a(\neq \infty)$ is a fixed point of l.

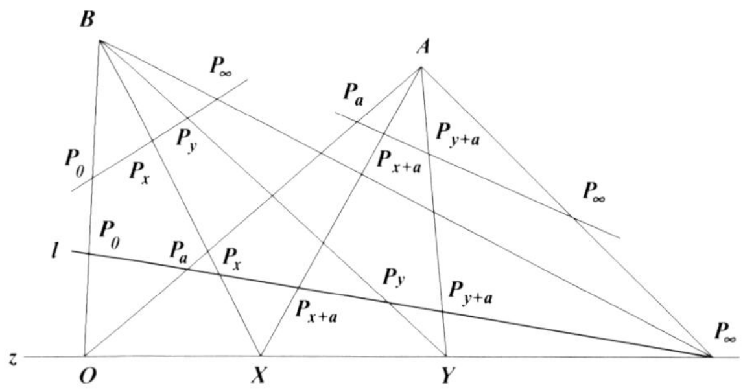

Figure 3.22: *Diagram to show that the operation of addition is projective*

The Euclidean equivalent diagram, Figure 3.21(b) shows the initial point of the vector $\overrightarrow{P_0P_y}$ brought into coincidence with the final point of $\overrightarrow{P_0P_x}$, such that ordinary vector addition yields $\overrightarrow{P_0P_x} + \overrightarrow{P_0P_y} = \overrightarrow{P_0P_{xy}}$, corresponding to the transformation:

(i) $x' = x + a \ (a \neq \infty)$

for any point on l.

Multiplication

In any plane through l let l_0, l_1, l_∞ be three lines through P_0, P_1, P_∞ respectively. Let l_1 meet l_0 and l_∞ in A and B respectively. Let P_x and P_y be any

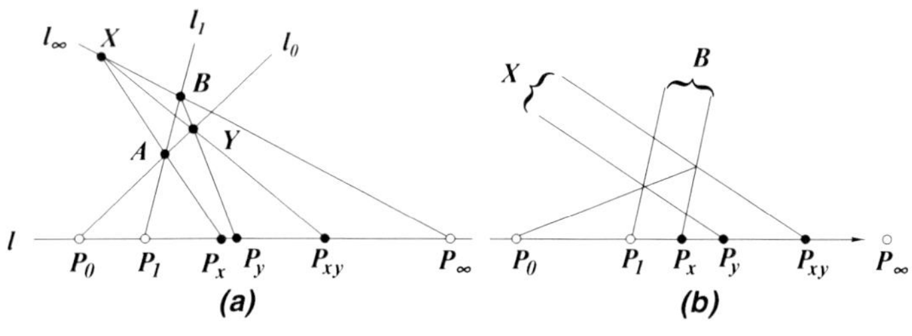

Figure 3.23: *Multiplication defined*

two points of l_1 and let P_xA and P_yB meet l_∞ and l_0 in X and Y respectively. The point P_{xy} in which XY meets l is called the *product* of P_x by P_y $(P_x \cdot P_y = P_{xy})$ in the scale $P_0P_1P_\infty$. The operation to form the product is known as multiplication. P_x and P_y are the *factors* of P_{xy}. As before, since $AXBY$ is a quadrangle, $Q(P_0P_xP_1, P_\infty P_yP_{xy})$ ensures that $P_x \cdot P_y = P_{xy}$.

The pairs of similar triangles (P_0AP_1, P_0YP_y) and (P_0AP_x, P_0YP_{xy}) in the Euclidean diagram, Figure 3.23(b) imply:

$$\frac{P_0P_1}{P_0P_y} = \frac{P_0A}{P_0Y} = \frac{P_0P_x}{P_0P_{xy}}.$$

Setting $P_0P_1 = 1$, $P_0P_x \cdot P_0P_x = P_0P_{xy}$.

The perspectivities $[P_x] \overset{A}{\doublebarwedge} [X] \overset{Y}{\doublebarwedge} [P_{xy}]$ and $[P_y] \overset{B}{\doublebarwedge} [Y] \overset{X}{\doublebarwedge} [P_{xy}]$ which define multiplication, imply the projectivities: $P_\infty P_0P_1P_x \barwedge P_\infty P_0P_yP_{xy}$ and $P_\infty P_0P_1P_y \barwedge P_\infty P_0P_xP_{xy}$.

Figure 3.24 shows that the perspectivities $P_1 \overset{R}{\doublebarwedge} A \overset{Q}{\doublebarwedge} P_a$, $P_x \overset{R}{\doublebarwedge} X \overset{Q}{\doublebarwedge} P_{xa}$, which correspond to $P_\infty P_0P_1P_x \barwedge P_\infty P_0P_aP_{xa}$, define multiplication as a projectivity.

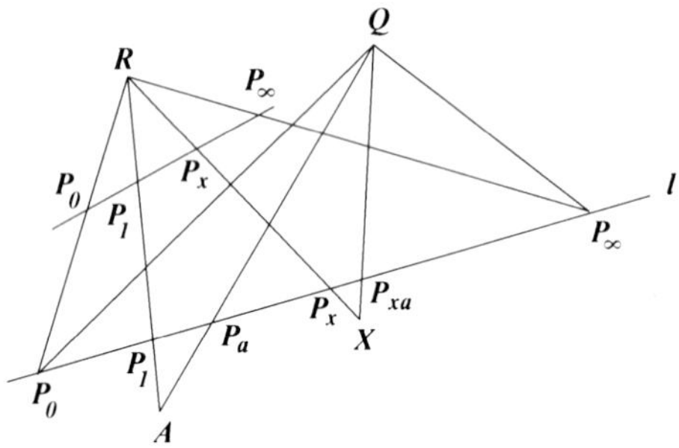

Figure 3.24: *Multiplication as a projectivity*

Symbolically:

(ii) $x' = ax$ $(a \neq 0)$

Involution

As in the construction of a product, let xA meet l_∞ in Figure 3.25.

If Y is on $1X$ and l_0, the line BY on l determines x', such that $xx' = 1$, by definition. Hence $[x] \overset{A}{\doublebarwedge} [X] \overset{1}{\doublebarwedge} [Y] \overset{B}{\doublebarwedge} [x']$, defining the involution:

$(iii) x' = \frac{1}{x}$

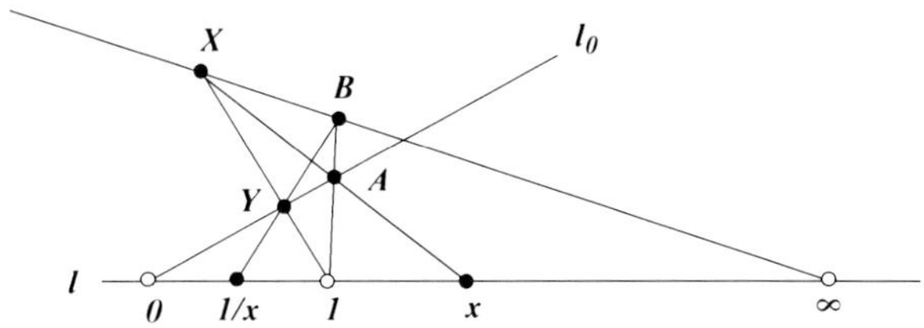

Figure 3.25: *Involution defined as a projectivity*

Projective Transformation

The combination of projectivities (*i*), (*ii*) and (*iii*) yields a more general projective transformation that will be shown to have the form (3.5)

$$x' = \frac{ax+b}{cx+d}.$$

As $x \to \infty$, $x' \to a/c$, the point that corresponds to the point ∞ by (3.5). This equation only applies if the matrix $\begin{pmatrix} a & b \\ c & d \end{pmatrix} \neq 0$. Otherwise, $b = ad/c$, such that

$$x' = \frac{ax + \frac{ad}{c}}{cx+d} = a\frac{(cx+d)}{(cx+d)} = \text{constant}.$$

The transformation (3.5) can now be shown to incorporate all of the features:

Multiplication	:	$x_1 = cx$
Addition	:	$x_2 = x_1 + d$
Involution	:	$x_3 = 1/x_2$
Transformation	:	$x_4 = (b - ad/c)x_3$
Point at ∞	:	$x' = x_4 + a/c$

$$\begin{aligned}
x' &= (b - ad/c)x_3 + a/c = (b - ad/c)(1/x_2) + a/c \\
&= (b - ad/c)/(x_1 + d) + a/c = (b - ad/c)/(cx + d) + a/c \\
&= \frac{1}{c}\left(\frac{bc - ad}{cx+d}\right) + \frac{a}{c} = \frac{(1/c)(bc - ad) + (a/c)(cx+d)}{cx+d} \\
&= (1/c)(bc - ad + acx + ad)/(cx + d) \\
&= \frac{b + ax}{cx + d}
\end{aligned}$$

According to the fundamental theorem a projectivity is completely determined once three pairs of homologies are assigned. Suppose that in a given projectivity the points $0, 1, \infty$ are transformed into the points p, q, r respectively. The transformation,

$$x' = \frac{r(q-p) + p(r-q)}{(q-p)x + (r-q)}$$

with the same form as (3.5), has this property:

$$\begin{array}{lll} \text{For} \quad x = 0 & , & x' = p \\ \phantom{\text{For} \quad} x = 1 & , & x' = (rq - rp + pr - pq)/(q - p + r - q) = q \\ \phantom{\text{For} \quad} x = \infty\,(a/c) & , & x' = r \end{array}$$

3.3 Complex Geometry

The problem of finding the fixed points in a projective mapping of a line onto itself reduces to solution of the quadratic equation

$$cx^2 + (d-a)x - b = 0 \tag{3.6}$$

obtained by equating the coordinate of a general point: $x' = (ax+b)/(cx+d)$ to that of its image: $x = (dx' - b)/(a - cx')$.

A quadratic equation with real coefficients has two roots which depend on the value of the discriminant Δ; they are real and unequal if $\Delta > 0$, real and equal if $\Delta = 0$ and complex if $\Delta < 0$.

An equivalent geometrical condition stipulates that every line in a plane meets every conic in two points, which may coincide if the line is tangent to the conic. If the points are not visible they are imaginary, but the equation makes it possible to find definite complex coordinates for such points. In the same way, every projective mapping of a line onto itself has two fixed points that may coincide.

In equation (3.6) the discriminant $\Delta = (d-a)^2 + 4bc$. The roots are equal if $\Delta = 0$, *i.e.* $d = a$, b and/or $c = 0$. Hence all $x' = x$ and, either $x' = x + b/a$ or $x' = (a/c)x$. This is simply the identity mapping with all points double. The roots are complex if $\Delta < 0$, *i.e.* $d = a$, $b = -c$, $\Delta = -4c^2$ and $x = \pm 2ic/2c = \pm i$. The fixed lines of this mapping are $x = iy$ and $x = -iy$, as shown in the Argand diagram, Figure 3.26.

The fixed lines, known as *isotropic* or *minimal* lines have some extraordinary properties. The distance between two points, z_1 and z_2 on such a fixed line is

$$\begin{aligned} d &= \sqrt{(x_1 - x_2)^2 + (y_1 - y_2)^2} \\ &= \sqrt{-(y_1 - y_2)^2 + (y_1 - y_2)^2} \quad , \quad \text{for } x = iy \\ &= 0 \end{aligned}$$

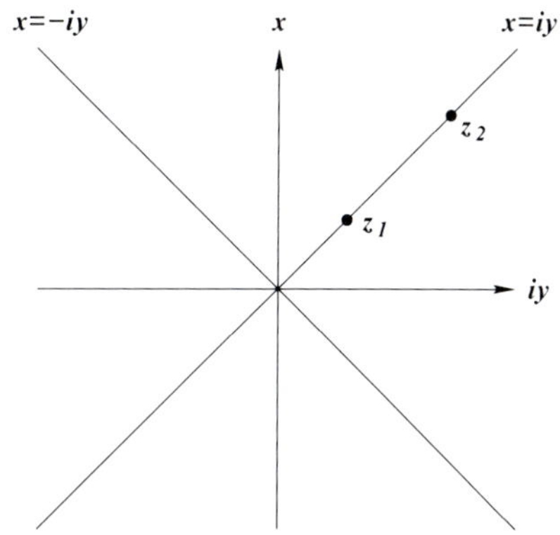

Figure 3.26: *Isotropic lines on an Argand diagram*

It follows that any isotropic dot product $\boldsymbol{z} \cdot \boldsymbol{z} = z^2 \cos\theta = 0$, which means that the isotropic line is perpendicular to itself: $\theta = \pi/2$.

The double points of the involution are called *circular* points, because every circle in the plane cuts the line at infinity in these same points. A circle of radius zero, which in Euclidean geometry consists only of its centre, in complex geometry consists of the two isotropic lines through the centre.

Figure 3.26 is of special interest in the theory of special relativity (Jennings, 1994) where iy is interpreted as the time axis in four-dimensional Minkowski space. The isotropic lines now define a *time cone* and the Lorentz transformation is equivalent to a complex rotation.

3.4 Topology

The only common invariant in geometry is the straight line. In no case does a conic belong to the same equivalence class as a polygon. However, on proceeding from projective geometry to topology even this invariant is abandoned. The new type of transformation that defines this equivalence in topology is known as an *elastic deformation.*

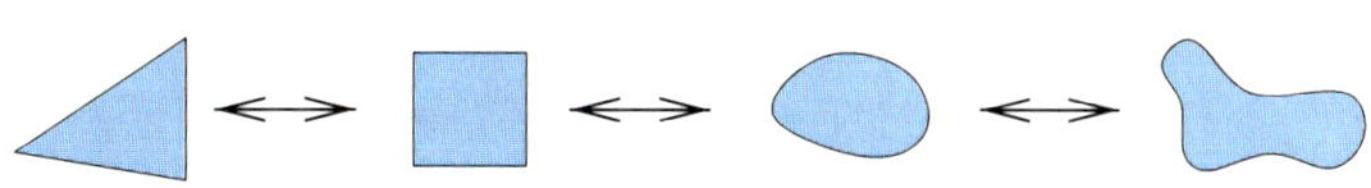

Even the deformation of a plane closed curve into a knot, which is impossible to perform in three dimensional space, is topologically allowed (Flegg, 1974).

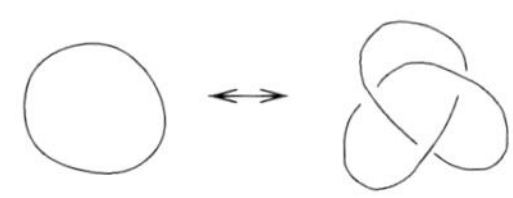

It requires cutting and rejoining of the curve.

To understand the topology of the projective plane $\boldsymbol{P}^2$, we start with the two-dimensional form known as a Möbius band.

Figure 3.27: *Making a Möbius strip*

A paper model is readily constructed by the sequence of operations shown in Figure 3.27 – twisting a strip of paper through 180° before joining the ends.

Whereas all common surfaces are two-sided, the surface of the Möbius band is one-sided. Because of the half-twist one side of the original strip has been joined to the other side, and the resulting surface is one-sided. It is now possible to draw a line from any point to any other point originally on the other side of the paper without crossing a boundary curve, or penetrating the surface in any way. In fact, a Möbius band has only one boundary curve, since another effect of the half-twist is that the original opposite long edges of the paper strip have been joined so as to form a continuous curve topologically equivalent to a circle.

Closely allied to the property of one-sidedness is the property of *non-orienrability*. Starting from any point P in the surface and following the path shown by stippled and dotted lines in Figure 3.28, one complete rotation ends in the point P' on the opposite side of the paper, but still in the same surface. Topologically P and P' are the same point, but they differ in orientation. To understand this, imagine that P is surrounded by a circle that rotates in a given sense, with angular momentum vector directed perpendicular to the circle of motion. If this vector is transplanted continuously from P to P', the

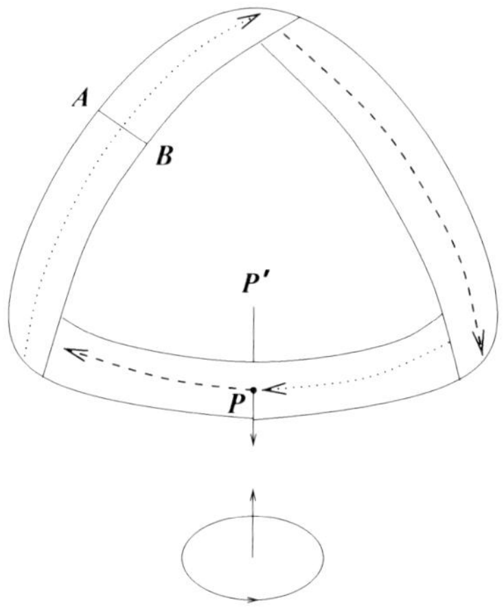

Figure 3.28: *Non-orientable surface of a Möbius band*

sense of rotation and the direction of the vector will be reversed on reaching P'.

The transformation from P to P' is therefore not an identity transformation, but rather an involution, with P and P' as conjugate points. The identity transformation corresponds to a double rotation of 2π along the Möbius surface. The two sides of the paper corresponds to a *double covering* of the non-orientable topological Möbius surface.

In Figure 3.28 let the line AB represent the position at which the two ends of the paper were joined to form the Möbius band. Symbolically this is equivalent to the plane diagram:

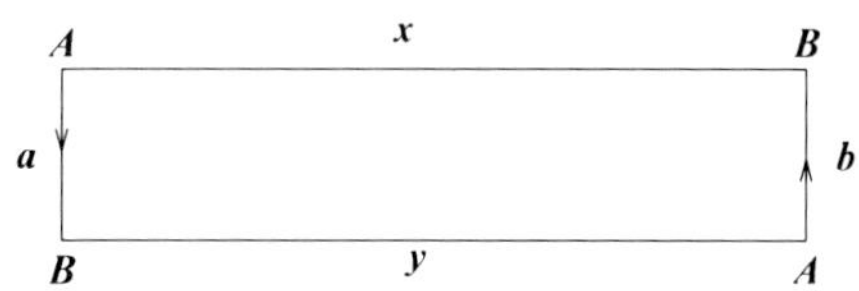

The symbols x and y indicate that the surface remains open perpendicular to this edge. Closure of the Möbius surface can be done in one of two ways, according to the plane diagrams, (a) and (b) shown overleaf.

Neither of these operations is possible in three dimensional space, but feasible in four. The first of these (a), where all vertices are joined to form a closed surface, is known as a *Klein bottle*. It can be constructed by a

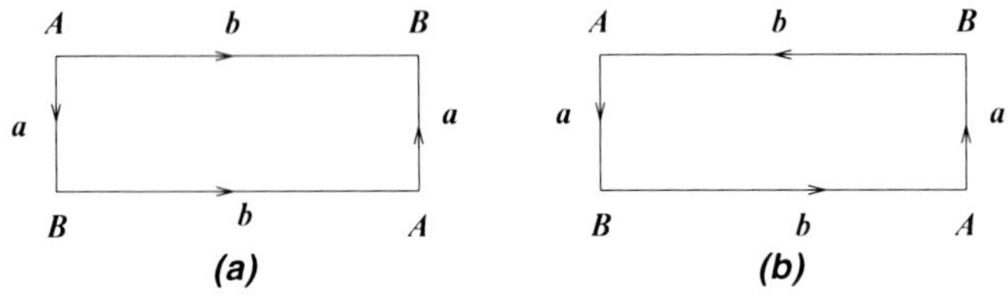

glassblower's trick that allows two surfaces to interpenetrate, resulting in a bottle without distinction between inside and outside. By analogy, the Möbius band, although a two-dimensional surface, cannot be physically constructed in two dimensions only, because of the half-twist.

The second possibility b in which opposite sides of the plane rectangle are joined together, always in the opposite sense, leaving two vertices distinct, is equivalent to a real projective plane. Recall that the real projective plane is obtained by adding a line at infinity to the Euclidean plane.

It follows from the mapping of the points on a sphere to a tangential plane using *central* or *gnomonic* projection through the centre O of the sphere as

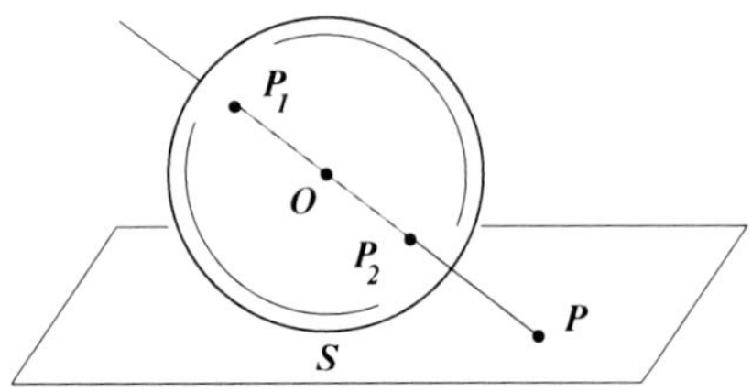

Figure 3.29: *Gnomonic projection*

in Figure 3.29. The tangential plane is considered to touch the sphere at the south pole S. Antipodal points P_1 and P_2 are mapped to the single point P in the plane by extending the diameter P_1OP_2 so as to meet the plane at P. In this way every great circle of the sphere is mapped to a line in the tangential plane with the exception of the equator, which is defined by the plane through O, parallel to the tangential plane. In order to allow for the mapping of the equator, a line at infinity is added to the tangential Euclidean plane. Points at infinity on the line represent pairs of antipodal points on the equator. Thus, all the lines parallel to a given line contain the same point at infinity, but lines in different directions have different points at infinity, all lying on the same line at infinity. This extended plane is the real projective plane. Any straight line through a given point P in ordinary three-dimensional Euclidean space is a point of the real projective plane.

Another way of representing the real projective plane is to project each point of a hemisphere on to the plane of its equator by a line perpendicular to this plane as depicted in Figure 3.30. This projection implies a one-to-

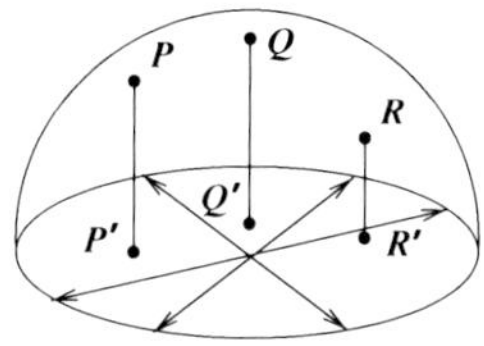

Figure 3.30: *Closing a hemisphere into a projective plane as shown by two-headed arrows*

one correspondence between the points of the hemisphere and the points inside and on the circle. If each pair of diametrically opposed points on the circumference is now made to coincide, a representation of the real projective plane results. Since a circle and a rectangle are topologically equivalent this identification of points is identical to that of plane diagram *b*.

The real projective plane may also be constituted from a Möbius band and a disc. The boundary of a Möbius band is a closed curve, topologically equivalent to a circle. It can therefore be imagined attached by its boundary to the boundary of a disc so as to form a closed surface, the real projective plane. A Möbius band may therefore be thought of as the real projective plane with a disc cut out of it.

The real projective plane, like a Möbius band, is one-sided and non-orientable. Like the Möbius band, which cannot be embedded in two-dimensional space, the deformations needed to produce a real projective plane cannot be performed in ordinary three-dimensional Euclidean space. Quoting Flegg (1974) the merit of non-Euclidean geometries is that they:

> "... provided exactly the release from Euclidean thraldom that was needed. In particular, it had been universally accepted for many centuries not only that Euclidean geometry was founded upon an unshakable axiomatic basis but also that it uniquely represented the real world in which all physical problems were assumed to arise.
>
> ... it was the realization that some of these geometries applied to practical and easily visualizable situations, such as geometry in the surface of a sphere, that set men's minds free and led to

rapidly developing investigations into the nature of spaces, which continue until the present day".

It is indeed the insight gained through non-Euclidean geometry, more than anything else, that inspired this venture, which is aimed at a better understanding of those observations in the physical sciences and astronomy that cannot be reconciled with standard cosmology.

3.4.1 Connectivity

A topological property of cosmological importance is the connectivity of surfaces. A one-sided surface is connected if it is possible to travel continuously from any point of the surface to each and every point of the surface. A two-sided surface is connected if its sides, taken separately, are both connected.

A surface is simply-connected if every continuous non-self-intersecting closed curve on it may be continuously contracted on the surface into a point. Any surface which is homeomorphic to a disc or a sphere is simply connected.

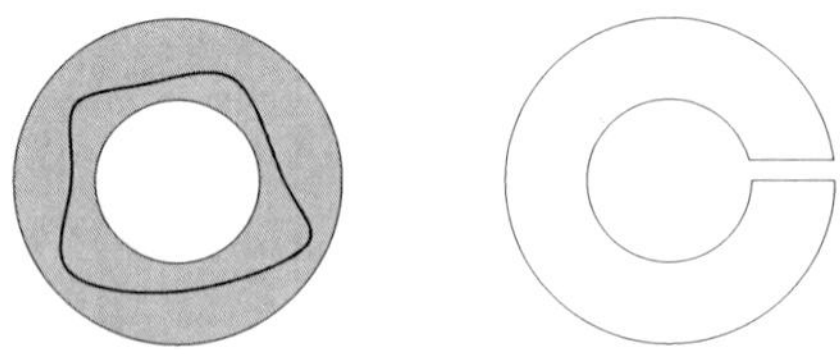

The closed curve shown here on the annulus shows that it is not simply connected. The annulus, cut as shown on the right, becomes simply connected again. Such a surface that requires a single cut to become homeomorphic to a disc is said to be doubly connected. Each additional hole in a disc increases the connectivity by one. In general an *n-tuply* connected surface needs $n-1$ cuts. Of particular importance is a Möbius strip (Figure 3.28) that needs a single cut and a twist to be homeomorphic to a disc. An n-tuply connected surface has *rank* $n-1$.

An important property of multiply-connected surfaces is that several topologically different routes between two points in the surface are possible, which means that multiple images of an object may occur to a single observer.

3.4.2 Fixed Points

Another important property of surfaces is demonstrated by the rotation of a disc in its own plane.

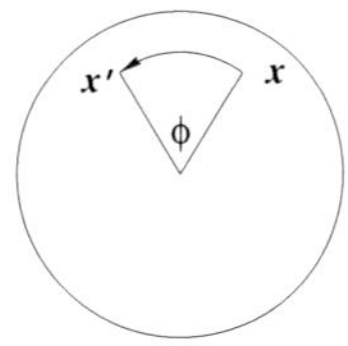

Suppose that the angle of rotation is ϕ. Each point of the disc x is seen to be mapped to some unique point, x', which is the image of no other point. There is only one point, at the centre, that maps to itself for a rotation of $\phi \neq 2\pi n$. For a similar rotation of an annulus, or any surface, not homeomorphic to a disc, the *fixed-point theorem* states that in this case no fixed point occurs. Such surfaces lack a special point. They have the alternative property that hair on such a surface can all be brushed to lie in the same direction, unlike the hair on a disc, a sphere, or a human head which develops a crown. This is a striking property of a Möbius band, showing that all points in the surface are quivalent and any of these can be considered to be the central point.

3.5 Golden Geometry

One of the topics addressed by Euclid in the Elements is the *golden ratio*.

The symbols Φ and τ occur indiscriminately in the literature to represent the golden ratio and/or its reciprocal. We shall follow the convention defined below.

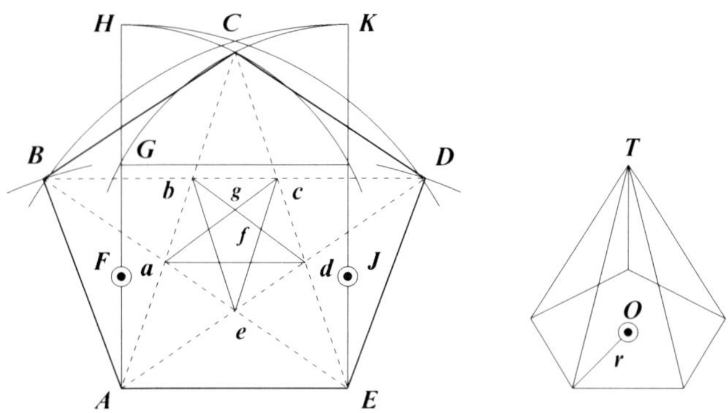

Figure 3.31: *Euclid's construction of the golden ratio and a regular pentagon*

Figure 3.31 shows Euclid's construction of a regular pentagon, starting from a square with unit side AG. Unit circles centred at F and J intersect the extended sides of the square at H and K and each other at C. It is noted that the ratios HG/AG and AG/AH are equal. Let $AH = \Phi$, such that

$$\Phi - 1 = \frac{1}{\Phi} \quad i.e. \quad \Phi^2 - \Phi - 1 = 0 \tag{3.7}$$

and
$$\Phi = \frac{1 \pm \sqrt{1+4}}{2} = (1 \pm \sqrt{5})/2 = 1.61803\ldots$$

The ratio $AG/AH = 1/\Phi$ defines the golden ratio $\tau = 1/\Phi = 0.61803....$

To complete the construction circles of radius Φ centred at A and E intersect the unit circles centred at E and A respectively, in points D and B. Point C, unit distance from B and D completes the pentagon $ABCDE$.

By construction each diagonal of the pentagon with unit sides measures Φ. The five diagonals define a pentagram with an inscribed pentagon $abcde$. Ratios such as Aa/AC, ab/Ab are again in golden ratio. Taking R and r as the radii of the circumcircles of the two pentagrams and $ab = 1$ as unit length, the ratio $R/r = \Phi$.

By folding $\triangle Aae$, and the other four corresponding triangles, about ae, *etc.*, so that $ABCDE$ meet in T, as shown, a pyramid of height $OT = h$ is formed, such that $h/r = \Phi$.

3.5.1 The Fibonacci Series

The diagonals of the inscribed pentagon define a smaller self-similar pentagram. From the defining equation (3.7) it follows that any power of Φ can be written as the sum of smaller powers, *e.g.*

$$\begin{aligned}
\Phi^5 &= \Phi^4 + \Phi^3 \\
\Phi^6 &= \Phi^5 + \Phi^4 \\
&= 2\Phi^4 + \Phi^3 \\
&= 2(\Phi^3 + \Phi^2) + (\Phi^2 + \Phi) \\
&= 2(2\Phi^2 + \Phi) + (2\Phi + 1) \\
&= 2(3\Phi + 2) + 2\Phi + 1 \\
&= 8\Phi + 5
\end{aligned}$$

The two coefficients are successive terms in the Fibonacci series:

$$0, 1, 1, 2, 3, 5, 8, 13, \ldots,$$

in which each term is the sum of the two preceeding terms, *i.e.*:

$$F_{n+1} = F_n + F_{n-1} \quad , \quad n > 0\,,\, F_1 = 1.$$

A general power of Φ therefore becomes

$$\Phi^n = F_n\Phi + F_{n-1}\,.$$

The ratio between successive Fibonacci numbers converges to the golden ratio:

$$\lim_{n\to\infty}\frac{F_{n+1}}{F_n} = \Phi\,.$$

3.5.2 Self-similarity

The distances CE, Cd, Cc, cd, cf, fg, *etc.*, are in geometrical progression, equal to

$\Phi\,,\,1\,,\,\Phi^{-1}(\tau)\,,\,\Phi^{-2}(\tau^2)\,,\,\tau^3\,,\,\tau^4\,,\,etc.$

The construction of smaller or larger pentagrams around the central pentagon can be continued indefinitely to define an infinite geometrical series based on τ:

$$S = \{\Phi^n\,,\,n = -\infty,\infty\}$$

and the corresponding infinite structure consisting of self-similar pentagrams.

In a sense, as affine geometry is based on triangles and projective geometry on quadrangles, so is golden geometry based on the pentagon. As all of these geometries are conspicuously evident in Nature cosmology should not be based only on the first.

3.5.3 The Golden Spiral

A rectangle, the sides of which are in golden ratio, is called a golden rectangle. A golden rhombus has perpendicular diagonals with a $1:\tau$ ratio of lengths. Its acute apical angle is $2\tan^{-1}\tau = 63.4°$.

Each golden rectangle consists of a square and another, smaller, golden rectangle. In different terminology the square is a gnomon to the golden rectangle. As shown in Figure 3.32 this relationship continues indefinitely, and converges to the point where the diagonals of the first two golden rectangles intersect at right angles. The angle, $\alpha = \tan^{-1}\tau$. By inscribing circular segments into each of the squares, as shown, the resulting *golden spiral* is virtually identical to the logarithmic spiral given by the equation $r = a\Phi^{2\theta/\pi}$ with constant a, which transforms into the classical form

$$r = ae^{\theta\cot\varphi} \quad \text{with} \quad \varphi = 72.97°\,.$$

The independent variable angle θ may have any value so that the curve is unlimited in length.

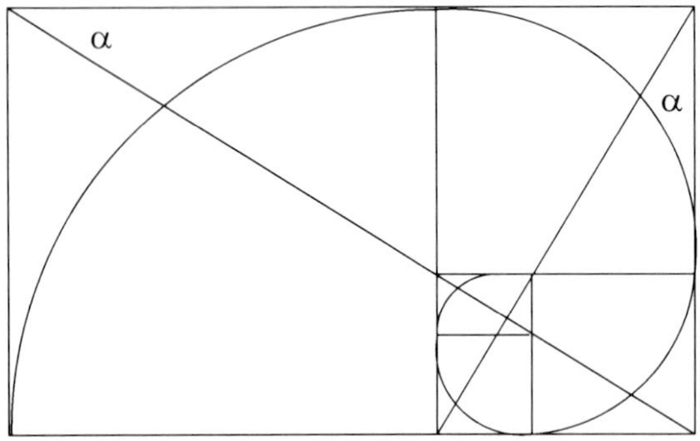

Figure 3.32: *A golden spiral*

3.5.4 Platonic Solids

Kepler's obsession with the Platonic solids probably relates to the close connection of these bodies with the golden ratio.

The five solids are distinguished by the property of being bounded by geometrically congruent polygons. Each polyhedron is characterized by the Schäffli symbol $\{p, g\}$, which means that it has p-gonal faces, g of which meet at each vertex. The five solids shown in Figure 3.33 are all closely related. The clearest connection

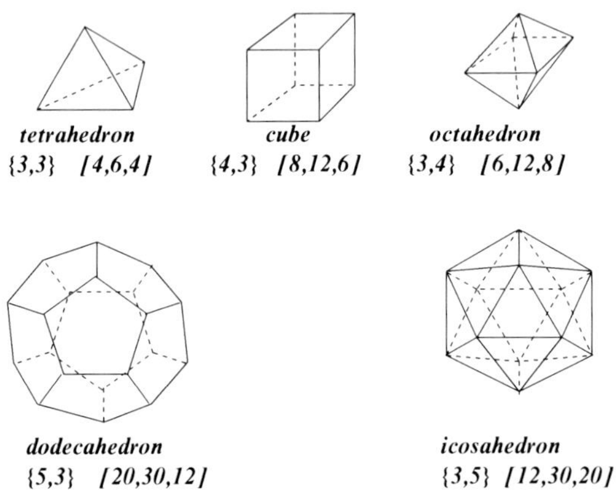

Figure 3.33: *The Platonic solids. Euler numbers [V,E,F] denote the numbers of vertices, edges and faces*

is between a tetrahedron and a cube. Each tetrahedral edge coincides with the face diagonal of a cube.

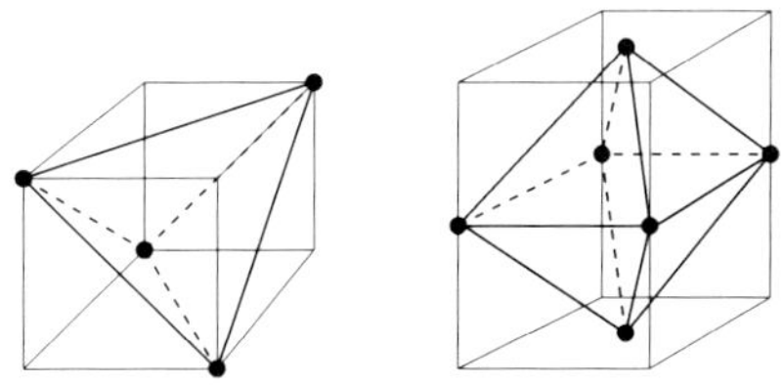

The cube has a three-fold axis along each body diagonal. Grinding down the vertex of a solid cube along the three-fold axis exposes a new triangular face. Grinding away from all vertices until the triangles meet, an octahedron is produced. A similar, inverse operation performed on an octadedron produces a cube. The two polyhedra are said to be reciprocal. The dodecahedron and the icosahedron are reciprocal in the same sense.

Because of their elements of five-fold symmetry the two large polyhedra have a close relationship with the golden ratio. For instance, three mutually perpendicular and interpenetrating golden rectangles define the vertices of an icosahedron.

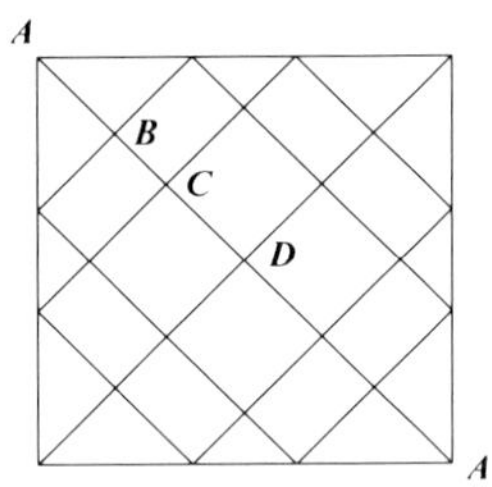

In the case of a cube it is noted that two perpendicular golden rectangles can be inscribed in a square such that each vertex divides a side of the square in the ratio $1 : \tau$. The edge of an inscribed tetrahedron (AA) is divided by the rectangles in the ratios $AC/AD = BD/AD = \tau$.

A circle, inscribed in the triangular face of the tetrahedron intersects the perpendicular (ADA) in the points F of Figure 3.34.

A circle of radius $\tau/2$, with respect to $AD = 1$, centred at F, intersects the edges in the points B and C, such that BBE defines a golden triangle and $BCECB$ is a regular pentagon.

In conjunction with the divine ratio, Kepler saw a relationship between the five regular solids and a sphere, with the six planetary orbits and the Pythagorean music of the spheres. This relationship is known as *squaring the circle* and represents the design principle of Egyptian pyramids. His original model of the universe was based on a nesting of the Platonic solids such that the vertices of an inner figure just touch the surface of the next solid. Starting from the sun at the centre it was surrounded by an inner sphere, representing the orbit of Mercury, enclosed by an octahedron, within an icosahedron, dodecahedron, tetrahedron and finally a cube, surrounded by an outer sphere for Saturn. The radii of the six spheres provided a rough

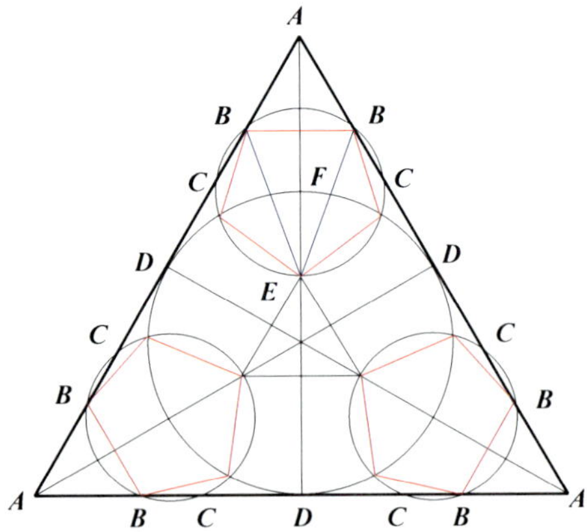

Figure 3.34: *The reationship between tetrahedron and* τ

estimate of the six planetary orbits, but the model was eventually abandoned in favour of Kepler's own discovery of elliptic rather than spherical planetary orbits.

3.6 Differential Geometry

Recognition of non-Euclidean geometries creates the new problem of describing the position and motion of finite objects in curved space, a problem which does not occur in Euclidean spaces that extend uniformly to infinity. An object such as the distance between two points,

$$AB = \sqrt{(x_2 - x_1)^2 + (y_2 - y_1)^2}$$

is independent of its position in the two-dimensional plane, but not in a curved surface. In technical terms the direction and magnitude of a vector in a curved surface are not Euclidean invariants, but functions of the coordinates. The ideal solution of the problem is to find a *covariant* description of an object, independent of the coordinate system.

This problem was addressed by Gauss for two-dimensional surfaces in Euclidean space and later extended by Riemann to general n-dimensional non-Euclidean spaces. The procedure is of interest here as it provides the facility to investigate the gravitational field in general relativity. The vital assumption is that in the limit of an infinitesimally small object simple Euclidean geometry would apply, suggesting that the methods of infinitesimal calculus, which deals with limits, should be followed.

The concept of the derivative of a function in calculus is essentially the same as the tangent line or slope of a curve, and the integral of a function can be geometrically interpreted as the area under a curve.

According to the ordinary meaning of partial differentiation, the partial derivatives of a position vector

$$\boldsymbol{r} = \boldsymbol{i}x + \boldsymbol{j}y + \boldsymbol{k}z$$

are the unit vectors along the Cartesian axes:

$$\frac{\partial \boldsymbol{r}}{\partial x} = \boldsymbol{i} \quad , \quad \frac{\partial \boldsymbol{r}}{\partial y} = \boldsymbol{j} \quad , \quad \frac{\partial \boldsymbol{r}}{\partial z} = \boldsymbol{k}.$$

The differential of $\boldsymbol{r}$, representing displacement in any given direction, is given by the scalar product

$$(ds)^2 = |d\boldsymbol{r}|^2 = d\boldsymbol{r} \cdot d\boldsymbol{r} = (dx)^2 + (dy)^2 + (dz)^2. \tag{3.8}$$

Where there is no possible confusion the parentheses will henceforth be deleted in expressions of this type. Also

$$\frac{\partial \boldsymbol{r}}{\partial u^\alpha} = \boldsymbol{i}x_\alpha + \boldsymbol{j}y_\alpha + \boldsymbol{k}z_\alpha$$

and

$$d\boldsymbol{r} = \boldsymbol{r_1}du^1 + \boldsymbol{r_2}du^2 + \boldsymbol{r_3}du^3 = \sum \boldsymbol{r}_\alpha du^\alpha. \tag{3.9}$$

The second problem is to find a description of geometrical objects that remains invariant under a change of coordinate system. The mathematical solution is to define such an object as a *tensor* ; a concept which finds wide application in the analysis of anisotropic media.

Curvature

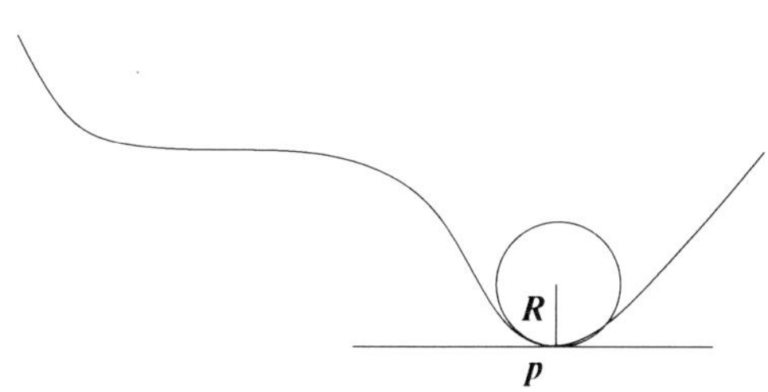

The differential that characterizes non-Euclidean space is known as *curvature* (Lee, 1997). In two dimensions it describes how a smooth curve deviates from linearity. The curvature, which varies from point to point, is specified in terms of the *osculating circle* of radius R and centred on the perpendicular to the tangent at p, and which follows the curve in the vicinity of p. On an infinitesimal scale each point on a curve has a unique osculating circle.

A straight line has zero curvature and is considered to be its own osculating circle of infinite radius. Curvature in general is therefore defined in terms of the radius of the osculating circle as $K = 1/R$. The larger the curvature, the smaller is the osculating circle, and therefore the faster the curve is turning. To distinguish between positive and negative curvature a unit vector is defined normal to the curve. A curve of positive curvature turns toward the unit vector; if it turns away the curvature has a negative sign.

In the case of a two-dimensional surface, embedded in $\mathbb{R}^3$ the curvature at a point is described by two numbers, called the principal curvatures. As an example, consider a regular cylinder. The two principal curvatures at p are clearly $\kappa_1 = 0$ and $\kappa_2 = 1/R$. More complicated expressions are obtained by different choice of principal directions. We give without proof Gauss's *Theorema Egregium* which states that irrespective of direction, the product $K = \kappa_1\kappa_2$ on a curved two-dimensional surface (quadric) is a constant. This means that the Gaussian curvature is an intrinsic property of the surface itself[3].

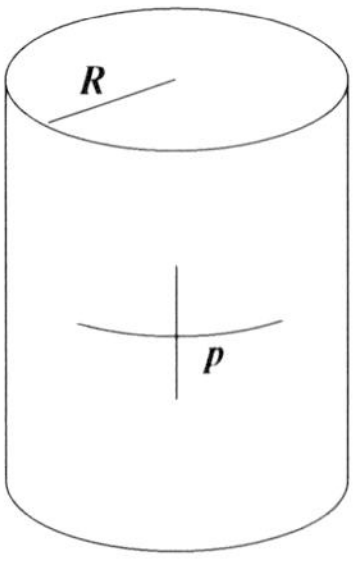

Another simple example is a sphere of radius R. The principal curvatures at any point $K = \pm 1/R$ and the Gaussian curvature, $\kappa_1\kappa_2 = 1/R^2$, which is a positive constant. The only other quadric of constant curvature is the projective plane. The curvature of any dome-shaped surface, although not necessarily constant, is always positive. On the other hand, the curvature of a saddle-shaped surface is negative, because the principal curvatures are of opposite sign. The surface of constant negative curvature, $K = -1/R^2 = 1$, in the special case where

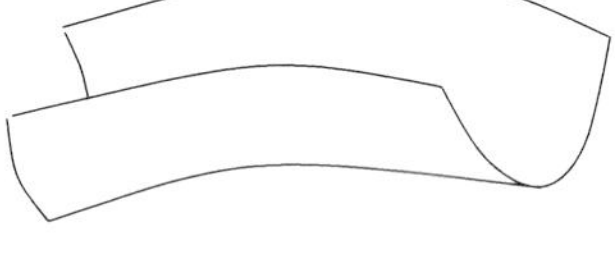

[3]Notice how a flat sheet of paper can be rolled into a cylinder or a cone ($K = 0$) without getting crumpled.

$R = 1$, is known as a *hyperbolic plane*. The same principle applies for the definition of curvature in high-dimensional surfaces. It relies on the idea of *geodesics* – the shortest curves between nearby points – to be discussed next.

3.6.1 Tensor Analysis

In an isotropic medium, vectors such as stress $\boldsymbol{S}$ and strain $\boldsymbol{X}$ are related by vector equations such as, $\boldsymbol{S} = k\boldsymbol{X}$, where $\boldsymbol{S}$ and $\boldsymbol{X}$ have the same direction. If the medium is not isotropic the use of vectors to describe the response may be too restrictive and the scalar k may need to be replaced by a more general operator, capable of changing not only the magnitude of the vector $\boldsymbol{X}$, but also its direction. Such a construct is called a *tensor*.

An important purpose of tensor analysis is to describe any physical or geometrical quantity in a form that remains invariant under a change of coordinate system. The simplest type of invariant is a scalar. The square of the line element ds of a space is an example of a scalar, or a *tensor of rank zero*.

In a space of ν dimensions two coordinate systems may be defined in such a way that the same point has coordinates $(x^1, x^2, \ldots, x^\nu)$ and $(\bar{x}^1, \bar{x}^2, \ldots, \bar{x}^\nu)$ in terms of the two systems respectively. The two coordinate systems are related in such a way that a transformation can be effected from one to the other, *i.e.*

$$\bar{x}^j = f^j(x^1, x^2, \ldots, x^\nu) \quad ; \quad x^j = h^j(\bar{x}^1, \bar{x}^2, \ldots, \bar{x}^\nu), \quad j = 1, \nu$$

Suppose that an infinitesimal displacement moves point A (coordinates x^i) to position B (coordinates $x^i + dx^i$). To describe the same displacement in the other coordinate system, it is necessary to differentiate the expression for $\bar{x}^i$, *i.e.*

$$d\bar{x}^i = \sum_{j=1}^{\nu} \frac{\partial h^i}{\partial x^j} dx^j = \sum_{j=1}^{\nu} \frac{\partial \bar{x}^i}{\partial x^j} dx^j, \quad j = 1, 2, \ldots, \nu$$

Any set of quantities A^i that transform in this way, *i.e.*

$$\bar{A}^i = \sum_{j=1}^{\nu} \frac{\partial \bar{x}^i}{\partial x^j} A^j$$

are the contravariant elements of a vector, or a *tensor of first rank*.

To simplify the notation, it is customary to omit the summation sign and sum over indices which are repeated on the same side of the equation. An index which is not repeated is understood to take successively the values

$1, 2, \ldots, \nu$, so that there are altogether ν different equations. With these conventions the transformation equations become

$$\bar{A}^i = \frac{\partial \bar{x}^i}{\partial x^j} A^j\,.$$

Consider a function (or scalar field) $\phi(x^i)$ of the point M (coordinates x^i) and defined in the neighbourhood of M. Being a function of a point, the value of ϕ does not change when described in a different coordinate system. By the rules of differentiation

$$\frac{\partial \phi}{\partial \bar{x}^i} = \sum_{j=1}^{\nu} \frac{\partial x^j}{\partial \bar{x}^i} \frac{\partial \phi}{\partial x^j}.$$

It follows that the quantities $A_i = \partial\phi/\partial x^j$ (*e.g.* gradients) transform under a change of coordinate system from x^i to $\bar{x}^i$ like

$$\bar{A}_i = \frac{\partial x^j}{\partial \bar{x}^i} A_j.$$

Any set of quantities that transform according to this prescription are known as the covariant components of a vector, and represented by subscripted indices[4].

These ideas may be extended to define tensors of any rank. There are three varieties of second rank tensors, defined by the transformations

$$\begin{aligned}
\bar{A}^{mn} &= \frac{\partial \bar{x}^m}{\partial x^i} \frac{\partial \bar{x}^n}{\partial x^j} A^{ij} \\
\bar{A}_{mn} &= \frac{\partial x^i}{\partial \bar{x}^m} \frac{\partial x^j}{\partial \bar{x}^n} A_{ij} \\
\bar{A}^m_n &= \frac{\partial \bar{x}^m}{\partial x^i} \frac{\partial x^j}{\partial \bar{x}^n} A^i_j
\end{aligned}$$

They are called contravariant, covariant and mixed tensors, respectively.

The process of reducing the rank of a tensor by summing over a pair of contravariant and covariant indices is known as *contraction*, *e.g.*

$$B_{np} = A^m_{npm}\,.$$

[4]Although it is customary to refer to covariant and contravariant vectors, this may be misleading. Any vector can be described in terms of its contravariant or its covariant components with equal validity. There is no reason other than numerical simplicity for the preference of one set of components over the other.

It always reduces the rank of a mixed tensor by two.

Tensors are seen to be formulated in terms of the dual bases with unit vectors $\boldsymbol{r_1}$, $\boldsymbol{r_2}$, $\boldsymbol{r_3}$ and $\boldsymbol{r^1}$, $\boldsymbol{r^2}$, $\boldsymbol{r^3}$ respectively, and related by the equation

$$\boldsymbol{r^\alpha} \cdot \boldsymbol{r_\beta} = \delta^\alpha_\beta$$

where the *Kronecker delta*, $\delta^\alpha_\beta = \begin{cases} 1, & \alpha = \beta \\ 0, & \alpha \neq \beta. \end{cases}$

It follows that $\boldsymbol{r^1}$ is perpendicular to the plane $\boldsymbol{r_2 r_3}$, and of length such that $\boldsymbol{r^1} \cdot \boldsymbol{r_1} = 1$; similarly for $\boldsymbol{r^2}$ and $\boldsymbol{r^3}$. A familiar example of such dual bases is found in crystallographic practice as direct and reciprocal lattices.

The Fundamental Tensor

Any vector $\boldsymbol{u}$ may be expressed as

$$\sum u_\alpha \frac{\partial \boldsymbol{r}}{\partial u^\alpha} \equiv u_1 \frac{\partial \boldsymbol{r}}{\partial u^1} + u_2 \frac{\partial \boldsymbol{r}}{\partial u^2} + u_3 \frac{\partial \boldsymbol{r}}{\partial u^3}$$

or as $\quad \sum u^\alpha \frac{d\boldsymbol{r}}{\partial u_\alpha}$.

The covariant components u_α are the inner products $\boldsymbol{u} \cdot \frac{\partial \boldsymbol{r}}{\partial u^\alpha}$ and $\boldsymbol{u} \cdot \frac{\partial \boldsymbol{r}}{\partial u_\alpha}$ as in:

$$\boldsymbol{u} \cdot \frac{\partial \boldsymbol{r}}{\partial u^\beta} = \sum u_\alpha \frac{\partial \boldsymbol{r}}{\partial u_\alpha} \cdot \frac{\partial \boldsymbol{r}}{\partial u_\beta} = \sum u_\alpha \delta^\alpha_\beta = u_\beta\,,$$

$$\boldsymbol{u} \cdot \frac{\partial \boldsymbol{r}}{\partial u_\beta} = \sum u^\alpha \frac{\partial \boldsymbol{r}}{\partial u^\alpha} \cdot \frac{\partial \boldsymbol{r}}{\partial u^\beta} = \sum u^\alpha \delta^\beta_\alpha = u^\beta\,.$$

In an expression such as

$$\sum u^\alpha \frac{\partial \boldsymbol{r}}{\partial u_\alpha} \cdot \frac{\partial \boldsymbol{r}}{\partial u^\beta}$$

summation is understood taken over the index α that appears twice (up and down) and not over β, which only appears once.

In the special case where $\boldsymbol{u} = \partial \boldsymbol{r}/\partial u^\alpha$ or $\partial \boldsymbol{r}/\partial u_\alpha$, the components

$$u_\beta = \frac{\partial \boldsymbol{r}}{\partial u^\alpha} \cdot \frac{\partial \boldsymbol{r}}{\partial u_\beta} \equiv g_{\alpha\beta} \quad \text{or} \quad u^\beta = \frac{\partial \boldsymbol{r}}{\partial u_\alpha} \cdot \frac{\partial \boldsymbol{r}}{\partial u^\beta} \equiv g^{\alpha\beta} \quad ,$$

so that

$$\frac{\partial \boldsymbol{r}}{\partial u^\alpha} = \sum g_{\alpha\beta} \frac{\partial \boldsymbol{r}}{\partial u^\beta} \quad , \quad \frac{\partial \boldsymbol{r}}{\partial u_\alpha} = \sum g^{\alpha\beta} \frac{\partial \boldsymbol{r}}{\partial u^\beta}\,.$$

The components of a vector may be transformed from one base to the other by

$$u_\alpha = \sum g_{\alpha\beta} u^\beta \quad \text{or} \quad u^\alpha = \sum g^{\alpha\beta} u_\beta \,.$$

An element of arc (3.8) in the direction of a given displacement $\boldsymbol{r}$ now appears as

$$\begin{aligned} ds^2 &= d\boldsymbol{r} \cdot d\boldsymbol{r} = \sum \frac{\partial \boldsymbol{r}}{\partial u^\alpha} du^\alpha \cdot \sum \frac{\partial \boldsymbol{r}}{\partial u^\beta} du^\beta \\ &= \sum \sum g_{\alpha\beta} du^\alpha du^\beta \\ &= g_{11}(du^1)^2 + g_{22}(du^2)^2 + g_{33}(du^3)^2 \\ &+ 2g_{23} du^2 du^3 + 2g_{31} du^3 du^1 + 2g_{12} du^1 du^2 \,. \end{aligned}$$

In the special case when u_1, u_2, u_3 are x, y, z this result reduces to (3.8).

The tensor g with components $g_{\alpha\beta}$ or $g^{\alpha\beta}$ is known as the metric tensor or the fundamental tensor. In terms of g the length of any vector is defined by any of the following,

$$|x|^2 = g_{ij} x^i x^j = g^{ij} x_i x_j = x^i x_j \,,$$

as before.

Geodesics

By the law of inertia a projectile in uniform motion moves on a straight line in Euclidean space. In a curved space there should be a comparable preferred trajectory which follows a *geodesic*. This is the principle that Einstein used in developing the general relativistic theory of the gravitational field. It is not immediately obvious how objects with different initial velocities should follow the same geodesic when starting in the same direction from the same point. The geodesic, after all, is a property of only space variables and direction, whereas velocity is a function of time. A reduction of gravitational theory to geometry can therefore only succeed if space and time coordinates are inextricably connected and transform together as in the Lorentz–Minkowski space of special relativity.

We agree that motion in the gravitational field will be geodesic only in curved Minkowski space, noting that this principle is conveniently ignored in virtually all cosmological models which claim a relativistic basis.

An important device that features in the calculation of geodesics is known as the Christoffel symbols, which occur as coefficients that describe the covariant transplantation of a vector $\boldsymbol{r}^i$ at the point x^m to $\boldsymbol{r}^j = \boldsymbol{r}^i + d\boldsymbol{r}^i$ at $x^m + dx^m$:

$$d\boldsymbol{r}^i = \Gamma^i_{mj} dx^m \boldsymbol{r}^j \,.$$

The coefficients are of the form

$$\Gamma^i_{mj} = \frac{\partial^2 x^i}{\partial x^k \partial x^l}\frac{\partial x^l}{\partial x^m}\frac{\partial x^k}{\partial x^j} \, .$$

Using this formula it can be shown that

$$\begin{aligned} \frac{\partial g_{ij}}{\partial u^k} &= \frac{\partial}{\partial u^k}(\boldsymbol{r}_i \cdot \boldsymbol{r}_j) \\ &= \Gamma_{ijk} + \Gamma_{jki} \end{aligned}$$

and hence

$$\Gamma_{ijk} = \frac{1}{2}\left(\frac{\partial g_{ij}}{\partial x^k} + \frac{\partial g_{jk}}{\partial x^i} - \frac{\partial g_{ik}}{\partial x^j}\right) \quad , \quad \Gamma^j_{ik} = \sum_i g^{jl}\Gamma_{ijk} \, .$$

The alternative notation $-\left\{ {i \atop jk} \right\} = \Gamma^i_{jk}$ is often used to define a parallel displacement of vectors along curves.

By defining the geodesic as a curve in a surface for which the tangent vector remains constant it is found to obey the equations

$$\frac{d^2x^i}{ds^2} + \left\{ {i \atop \alpha\beta} \right\} \frac{dx^\alpha}{ds}\frac{dx^\beta}{ds} = 0 \, . \tag{3.10}$$

3.6.2 Riemannian Geometry

Differential geometry of n-dimensional non-Euclidean space was developed by Riemann and is best known in its four-dimensional form that provided the basis of the general theory of relativity. Elementary examples of Riemann spaces include Euclidean space, spherical surfaces and hyperbolic spaces.

The measurable distance between points infinitely close together in three-dimensional Euclidean space is specified as

$$ds^2 = dx^2 + dy^2 + dz^2 \, ,$$

which formalism can be extended to any number of dimensions. In particular, where the special theory of relativity requires that a time dimension be included, on the same footing as the three space dimensions, in order to give a correct discription of events, separated in both space and time:

$$ds^2 = c^2dt^2 - dx^2 - dy^2 - dz^2 \, .$$

This space-time continuum differs from Euclidean space in that ds may be either positive or negative to differentiate between time-like and space-like line elements. The boundary between the two resulting regions of space-time corresponds to the so-called *light cone*, which starts at each point in space, and has $ds^2 = 0$. The general theory of relativity indicates that, in order to accommodate accelerated systems or a gravitational field it is further necessary to consider the space-time continuum as being non-Euclidean. This means that the simple differential geometry of Euclidean space is no longer adequate to allow for the possibility of curved space-time.

The effect of curvature is that the simple differential form can only be valid in the immediate neighbourhood of any point. For a finite region of a curved surface ds is not invariant. It depends on the position in the surface and takes the form

$$ds^2 = \sum_{i,k} g_{ik} dx^i dx^k \ ,$$

which allows the definition of the arc length of a curve $x^i = x^i(t)$ as the integral

$$s = \int_{t_0}^{t} \sqrt{\left| \sum_{i,k} g_{ik} \frac{dx^i}{dt} \frac{dx^k}{dt} \right|} dt$$

Geodesics are the curves that trace the shortest distance between two points, sufficiently close to each other. An n-dimensional domain has the volume given by the integral

$$\int \sqrt{|g|} dx^1 \dots dx^n$$

in which $|g|$ is the determinant of g_{ik}. A geodesic is either a curve of zero length or one along which the unit tangent vector is parallel. In the four-dimensional Riemannian space of general relativity the components of the fundamental tensor g_{ik} define the gravitational potential. Light rays are the null curves $ds^2 = 0$ and the trajectories of particles in the field are the geodesics.

Attempts to accommodate the electromagnetic field in addition to the gravitational lead to a further generalization of the Riemannian geometry, for instance to the five-dimensional Kaluza metric or projective relativity of Veblen.

Riemannian coordinate sytems related by continuous transformation, such as

$$\bar{x}^j \leftrightarrow x^i \quad , \quad i, j, = 0, 3$$

and for which the derivatives $\partial \bar{x}^i / \partial x^i$ are assumed to be continuous, are not extended over all space. The coordinate system x^i will, in general, be defined only in a part Σ of the space considered, and $\bar{x}^j$ only in a different part $\bar{\Sigma}$ of the space. The transformation is assumed to hold only in a common part Σ_c of Σ and $\bar{\Sigma}$. Each coordinate neighbourhood may be conceived of as a local map of the space which overlaps with neighbouring regions in a way that finally covers all space. A space which allows such a covering by local maps and admits a Riemannian metric is called a Riemannian manifold. A homogeneous Riemann manifold looks geometrically the same at any point while an isotropic one looks the same in every direction.

Since Riemannian manifolds are not defined as embedded in Euclidean space Gaussian methods cannot be used to define their curvature. An alternative definition is possible in terms of geodesics. Given any point p in the manifold $p \in M$, and any vector V tangent to M at p, there is a unique geodesic starting at p with initial tangent vector V. It is possible by a complicated procedure to derive an approximate curvature at p by choosing a two-dimensional subspace S^{II} that contains a set of geodesics whose initial tangent vectors lie in S^{II}. The Gaussian curvature of S^{II} can then be computed from its Riemannian metric. There is still no general method for the rigorous calculation of curvatures of high-dimensional surfaces.

3.7 References

Coxeter, H.S.M. (1998): *Introduction to Geometry*, 2nd ed., Wiley, N.Y.

Flegg, H.G. (1974): *From Geometry to Topology*, Reprinted 2001, Dover, Mineola, NY.

Jennings, G.A. (1994): *Modern Geometry with Applications*, Springer, N.Y.

Lee, J.M. (1997): *Riemannian Manifolds*, Springer, N.Y.

Veblen, O. & J.W. Young (1910): *Projective Geometry*, Vol.I, 1910; Vol.II (O. Veblen sole author), 1918, Ginn and Co., Boston.

Chapter 4

Physical Evidence

Cosmology is notorious for its reliance on conviction rather than data and the consequent selective use of observational evidence. The process is aggravated by distortion of the evidence by half-scholars with a limited understanding of information received at second hand. Science operates differently and strives to reduce all observed material to a limited set of fundamentals, using Occam's razor.

Because of these radically different cultures there is no incorporation of science into cosmology. An alternative is to objectively examine all scientific facts of relevance, leading to an improved understanding of the cosmic whole. The end product, as in all of science, will be riddled with uncertainty and open to re-interpretation. What is lost in dogmatic certainty will be compensated by a falsifiable model that grows with new scientific insight.

This methodology inspired the chapter on world geometry, based on the assumption that the cosmos is best described by the most consistent version of geometry available at the time. It enables the study of subtleties that cannot be revealed by a more primitive version and anticipates better comprehension of the world through geometries of the future. In the same spirit the present chapter explores the theories of physics and the cosmic picture that they reveal. The next chapter will examine chemical theory in the same way.

The two fundamental theories of relativity and the quantum still await reduction to a common basis, the full implication of which would doubtlessly be a radically different future reading of the cosmos.

J.C.A. Boeyens, *Chemical Cosmology*, DOI 10.1007/978-90-481-3828-9_4,

4.1 Special Relativity

Special relativity was the first theory to provide direct evidence that the physical universe is four-dimensional. Although this happened a century ago the amazing reality is that leading cosmologists still make the assumption of a large-scale three-dimensional flat universe. The theory of special relativity was the first to augment the classical theory of mechanics by taking the electromagnetic field into account. Extended into a general theory it opened up the possibility of a unified electromagnetic and gravitational field, which seemed to require yet another extra dimension for space-time.

The inadequacy of the mechanical model of light first became apparent when the electromagnetic equation of motion was seen to violate Galileo's principle of relative motion. As derived by Maxwell, electromagnetic motion is described by a wave equation:

$$\nabla^2\Phi = (\epsilon_0\mu_0)\frac{\partial^2\Phi}{\partial t^2} \tag{4.1}$$

In one space dimension, putting $\sqrt{\epsilon_0\mu_0} = 1/c$:

$$\frac{\partial^2\Phi}{\partial x^2} = \frac{1}{c^2}\frac{\partial^2\Phi}{\partial t^2}$$

The constants ϵ_0 and μ_0 are the electric permittivity and the magnetic permeability of free space. The constant product c has the dimensions of a velocity, which implies that the electromagnetic field is transmitted at constant speed through the vacuum. It means that a light beam, considered an electromagnetic disturbance, moves through the vacuum at constant speed, irrespective of the motion of the light source. Whereas ϵ_0 and μ_0 are constant properties of space they are independent of any source. The disturbance carries momentum and energy, but by a mechanism that seems to be inconsistent with the accepted ideas of classical dynamics.

The familiar Galilean law of relative motion dictates that a stationary observer measures the position of an object in relative motion, at constant speed v, to change by an amount vt during time t. In the moving frame of reference, where the position z' remains constant, the relative motion is described correctly by:

$$z' = z - vt \tag{4.2}$$

For constant relative velocity the first derivative of (4.2) is

$$\frac{dz'}{dt} = \frac{dz}{dt} - v$$

and another differentiation gives:

$$\frac{d^2z'}{dt^2} = a' = \frac{d^2z}{dt^2} = a$$

so that the acceleration is the same in both systems and the Newtonian equation of motion remains unaffected: $F = ma$.

However, a transformation such as (4.2) completely destroys equation (4.1). In a relatively moving frame it takes the form:

$$(1 - v^2/c^2)\frac{\partial^2\Phi}{\partial x'^2} + \frac{2v}{c^2}\frac{\partial^2\Phi}{\partial z'\partial x} = \frac{1}{c^2}\frac{\partial^2\Phi}{\partial t'^2}$$

which approaches (4.1) only for $c >> v$.

To ensure that the equation of motion remains the same at all relative velocities (4.2) is replaced by the *Lorentzian* transformation:

$$\begin{aligned} x' &= x \\ y' &= y \\ z' &= \frac{z - vt}{\sqrt{1 - v^2/c^2}} \\ t' &= \frac{t - vz/c^2}{\sqrt{1 - v^2/c^2}} \end{aligned}$$

In the limit of small relative velocities these equations reduce to the Galilean transformation.

A spherical wavefront that propagates with velocity c is described by the equation $x^2 + y^2 + x^2 = c^2t^2$. Seeing that the Lorentz transformation mixes time and space coordinates it is convenient to formulate it in an orthogonal four-dimensional coordinate system, with $x = x_1$, $y = x_2$, $z = x_3$, $ict = x_4$. In this, socalled *Minkowski space*, define $\beta = v/c$, $\gamma = 1/\sqrt{1 - \beta^2}$, such that

$$\begin{aligned} x'_1 &= x_1 \\ x'_2 &= x_2 \\ x'_3 &= \gamma x_3 + i\beta\gamma x_4 \\ x'_4 &= -i\beta\gamma x_3 + \gamma x_4 \end{aligned}$$

This transformation in matrix notation becomes:

$$\boldsymbol{x'} = \boldsymbol{x} \times \begin{pmatrix} 1 & 0 & 0 & 0 \\ 0 & 1 & 0 & 0 \\ 0 & 0 & \gamma & i\beta\gamma \\ 0 & 0 & -i\beta\gamma & \gamma \end{pmatrix} \tag{4.3}$$

The 2×2 submatrix of x_3 and x_4 resembles the rotation matrix, equation (3.3) for rotation about one axis in a three-dimensional coordinate system. Correspondingly the transformation (4.3) can be said to be a rotation in the x_3x_4 plane of four-dimensional Minkowski space, through an imaginary angle ϕ, such that

$$\cos\phi = \gamma = 1/\sqrt{1-\beta^2} \quad , \quad \sin\phi = i\beta\gamma = i\beta/\sqrt{1-\beta^2}\,.$$

In terms of the real angle $\psi = -i\phi$, the submatrix becomes

$$\begin{pmatrix} \cosh\psi & i\sinh\psi \\ -i\sinh\psi & \cosh\psi \end{pmatrix}.$$

Two successive rotations of ϕ_1 followed by ϕ_2 amounts to a total rotation of $\Phi = \phi_1 + \phi_2$. Noting that $\tan\phi = \sin\phi/\cos\phi = i\beta$ it follows that:

$$\tan\Phi = \tan(\phi_1+\phi_2) = \frac{\tan\phi_1 + \tan\phi_2}{1-\tan\phi_1\tan\phi_2}\,.$$

Two successive Lorentz transformations with relative speeds β_1 and β_2 therefore correspond to a single transformation with relative speed

$$\beta = \frac{\beta_1+\beta_2}{1+\beta_1\beta_2}\,.$$

This is the well-known Einstein addition law for parallel velocities. It shows that, no matter how close β_1 and β_2 approach unity ($v = c$), their sum β can never exceed unity.

Consider a source that emits light of velocity c and which moves relative to an observer with velocity v. The observed speed of the light ray, according to the addition formula follows from:

$$\begin{aligned} \frac{v_{obs}}{c} &= \left(\frac{v+c}{c}\right)\cdot\frac{1}{1+vc/c^2} \\ v_{obs} &= \frac{v+c}{(c+v)/c} = c \end{aligned}$$

Vectors in Minkowski space

The revolutionary feature of the Lorentz equations is that in order to perform a coordinate transformation between relatively moving frames of reference a complex time coordinate must be taken into account. This transformation takes the form of a complex rotation in a four-dimensional pseudo-Euclidean

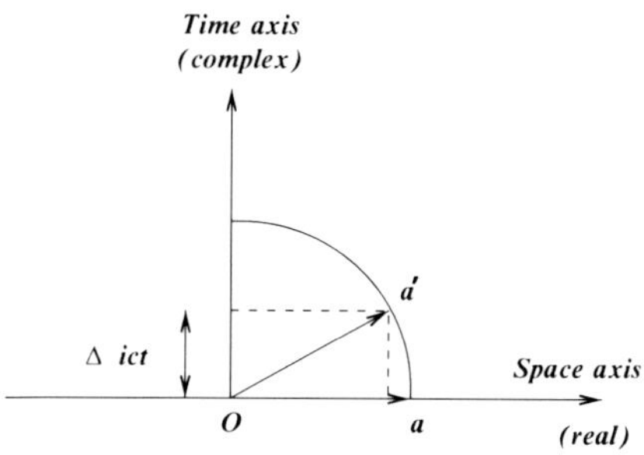

or Minkowski space. The amount of rotation is related to the relative velocity of the observers.

A three-dimensional vector in familiar Euclidean space is represented by $\overrightarrow{Oa}$. Rotation in three-dimensional Euclidean space leaves the length

$$|\overrightarrow{Oa'}| = \sqrt{\Delta x^2 + \Delta y^2 + \Delta z^2} = r$$

invariant. However, on rotation into complex space the real part of the transformed vector is no longer invariant, as the vector acquires an extra component, such that

$$|\overrightarrow{Oa'}| = \sqrt{\Delta x^2 + \Delta y^2 + \Delta z^2 + \Delta(ict)^2} = |\overrightarrow{Oa}|\,.$$

The real (space) part therefore contracts by Δr, which is compensated by an increase Δict. Since c is constant this effect is observed as a *time dilation*, Δt. Clocks in relatively moving frames do not remain synchronized.

Thus, when an object moves by an amount $dx + dy + dz$ in time dt with respect to a stationary observer, the time as measured by an observer moving with the object, is $d\tau$, where

$$(d\tau)^2 = dt^2 - (1/c^2)(dx^2 + dy^2 + dz^2)$$

and $d\tau$ is known as the *proper time* for the observer in relative motion. While the velocities, such as $dz/d\tau$ remain small compared to c, there is little difference between dt and $d\tau$. However, when the relative velocity of two observers is comparable to the speed of light, their time directions are measurably not parallel.

Another consequence of time dilation is that the concept of simultaneity depends on the frame of reference. Suppose two events occur at the same time t at two points z_1 and z_2 in system S. Observed in a relatively moving reference frame S' these events occur at times

$$t'_1 = \frac{t - (v/c^2)z_1}{\sqrt{1 - v^2/c^2}} \quad \text{and} \quad t'_2 = \frac{t - (v/c^2)z_2}{\sqrt{1 - v^2/c^2}} .$$

Obviously $t'_1 \neq t'_2$, unless $z_1 = z_2$. This inability to define absolute simultaneity also prevents the definition of an absolute universal time, as assumed in standard cosmology.

The path followed by an object in *space-time* is called its *world line* and the distance along it is seen to measure its proper time:

$$(icd\tau)^2 = dx^2 + dy^2 + dz^2 + (icdt)^2 .$$

Since the apparent size of objects changes with the relative velocity of the observer, the apparent density of matter is also not an invariant under Lorentz transformation. It is inferred that the mass of a body is not a four-dimensional invariant. This is confirmed by the conservation of momentum as discussed below.

The vector dx_μ, ($\mu = 1,4$, or $\mu = 0,3$ in alternative notation) represents the change in the four-dimensional position vector of a particle in differential motion along its world line. The absolute magnitude of the 4-vector, as for any 3-vector, is described by the dot product with itself and defines an invariant *world scalar*[1],

$$ds^2 = -\frac{1}{c^2}\sum_{\mu=1}^{4} dx_\mu \cdot dx_\mu \equiv -\frac{1}{c^2}dx_\mu dx_\mu .$$

In expanded form:

$$ds^2 = -\frac{1}{c^2}\left[dx^2 + dy^2 + dz^2 - c^2dt^2\right]$$

or

$$ds = dt\sqrt{1 - \frac{1}{c^2dt^2}\left[dx^2 + dy^2 + dz^2\right]}$$

which is equivalent to writing $dt = d\tau/\sqrt{1-\beta^2}$ for proper time $d\tau$.

Exactly like the proper time all other vectors such as the momentum or force vectors have one time-like and three space-like components. The 4-velocity

$$\boldsymbol{u}_\nu = \frac{dx_\nu}{d\tau} = \frac{dx_\nu}{\sqrt{1-\beta^2}dt}$$

[1]The Einstein summation convention applies: If an index occurs twice in one term of an expression, it is always to be summed unless the contrary is expressly stated.

has space-like components $u_i = v_i/\sqrt{1-\beta^2}$, $i = 1,3$ and time component $u_4 = ic/\sqrt{1-\beta^2}$. The world velocity

$$\boldsymbol{u}_\nu \cdot \boldsymbol{u}_\nu = \frac{v^2 - c^2}{1-\beta^2} = -c^2 \,.$$

To an observer moving with a particle at constant velocity v in the z-direction the mass of the particle is m_0 and the proper time is τ. With respect to a stationary observer the particle travels a distance dz, where $d\tau^2 = dt^2 - (dz/c)^2$. To ensure that momentum is conserved it is necessary to define the components of the momentum as:

$$p_z = m_0 \frac{dx}{dt} = \frac{m_0 v}{\sqrt{1 - v^2/c^2}} \; ; \; p_t = m_0 c \frac{dt}{d\tau} = \frac{m_0 c}{\sqrt{1 - v^2/c^2}} \; ; \; v = \frac{dz}{dt} \,.$$

The invariant square of the momentum

$$p^2 = p_z^2 - p_t^2 = \frac{(m_0 v)^2}{1 - v^2/c^2} - \frac{(m_0 c)^2}{1 - v^2/c^2} = \frac{m_0^2(v^2 - c^2)}{1 - v^2/c^2} = -(m_0 c)^2 \,.$$

The momentum four-vector as measured by a stationary observer, for a particle moving with relative velocity v is

$$p_{x_i} = \frac{m_0 (dx_i/dt)}{\sqrt{1 - v^2/c^2}} \,, \; i = 1,3, \quad p_t = \frac{m_0 c}{\sqrt{1 - v^2/c^2}} \,, \; v^2 = \sum \left(\frac{dx_i}{dt}\right)^2 \,.$$

The time component of the momentum is proportional to the total energy of the particle:

$$E = c p_t = \frac{m_0 c^2}{\sqrt{1 - v^2/c^2}} = mc^2$$

where $m = m_0/\sqrt{1 - v^2/c^2}$ is the relativistic mass and m_0 is the *rest* mass.

Electromagnetism

Electromagnetic potential in relativistic notation is a Minkowski 4-vector:

$$\boldsymbol{V} = (A_x, A_y, A_z, i\phi/c)$$

where $\boldsymbol{A}$ is a space-like vector and ϕ a time-like potential. The 4-gradient of $\boldsymbol{V}$, *i.e.* $\Box \cdot \boldsymbol{V}$, where

$$\Box = \frac{\partial}{\partial x} + \frac{\partial}{\partial y} + \frac{\partial}{\partial z} + \frac{1}{c^2}\frac{\partial}{\partial t}$$

follows as

$$\square \cdot \boldsymbol{V} = \boldsymbol{\nabla} \cdot \boldsymbol{A} + \frac{1}{c^2}\frac{\partial \phi}{\partial t} .$$

This is recognized as the equation that relates the magnetic vector potential to the electrostatic scalar potential:

$$\boldsymbol{\nabla} \cdot \boldsymbol{A} + \mu\epsilon \frac{\partial \phi}{\partial t} = 0 .$$

The electromagnetic field is defined by Maxwell's four equations in vector notation:

$$\boldsymbol{\nabla} \times \boldsymbol{E} = -\frac{\partial \boldsymbol{B}}{\partial t} \tag{4.4}$$
$$\boldsymbol{\nabla} \cdot \boldsymbol{B} = 0 \tag{4.5}$$
$$\boldsymbol{\nabla} \times \boldsymbol{B} - \frac{1}{c^2}\frac{\partial \boldsymbol{E}}{\partial t} = \mu_0 \boldsymbol{J} \tag{4.6}$$
$$\boldsymbol{\nabla} \cdot \boldsymbol{E} = \frac{\rho}{\epsilon_0} \tag{4.7}$$

In free space the net charge density vanishes, hence $\mu_0 \boldsymbol{J} = 0$ and $\rho/\epsilon_0 = 0$. To show their relationship to the theory of relativity these equations are reformulated in tensor notation by pairwise combination, setting $x_4 = ict$. Noting that:

$$\boldsymbol{\nabla} \times \boldsymbol{a} = \text{curl } \boldsymbol{a} = \boldsymbol{i}\left(\frac{\partial a_z}{\partial y} - \frac{\partial a_y}{\partial z}\right) + \boldsymbol{j}\left(\frac{\partial a_x}{\partial z} - \frac{\partial a_z}{\partial x}\right) + \boldsymbol{k}\left(\frac{\partial a_y}{\partial x} - \frac{\partial a_x}{\partial z}\right)$$

$$\boldsymbol{\nabla} \cdot \boldsymbol{a} = \text{div } \boldsymbol{a} = \frac{\partial a_x}{\partial x} + \frac{\partial a_y}{\partial y} + \frac{\partial a_z}{\partial z}$$

equations (4.4) and (4.5) in the expanded form of 4-vectors are:

$$\frac{\partial E_3}{\partial x_2} - \frac{\partial E_2}{\partial x_3} + ic\frac{\partial B_1}{\partial t} = 0$$

$$\frac{\partial E_1}{\partial x_3} - \frac{\partial E_3}{\partial x_1} + ic\frac{\partial B_2}{\partial t} = 0$$

$$\frac{\partial E_2}{\partial x_1} - \frac{\partial E_1}{\partial x_3} + ic\frac{\partial B_3}{\partial t} = 0$$

$$\frac{\partial B_1}{\partial x_1} + \frac{\partial B_2}{\partial x_2} + \frac{\partial B_3}{\partial x_3} = 0$$

readily rearranged into:

$$\left(\frac{\partial}{\partial x_\mu}\right)(G_{\mu\nu}) = O \tag{4.8}$$

where the matrix

$$G_{\mu\nu} = \begin{pmatrix} 0 & -\frac{i}{c}E_3 & \frac{i}{c}E_2 & B_1 \\ \frac{i}{c}E_3 & 0 & -\frac{i}{c}E_1 & B_2 \\ -\frac{i}{c}E_2 & \frac{i}{c}E_1 & 0 & B_3 \\ -B_1 & -B_2 & -B_3 & 0 \end{pmatrix}$$

defines an antisymmetric tensor.

Alternatively, starting from (4.6) and (4.7) a second set of equations

$$\left(\frac{\partial}{\partial x_\mu}\right)(F_{\mu\nu}) = \{\mu_0 J^\mu\} \tag{4.9}$$

in which $i\rho/c\epsilon_0 = \mu_0 J^4$, and

$$F_{\mu\nu} = \begin{pmatrix} 0 & B_3 & -B_2 & -\frac{i}{c}E_1 \\ -B_3 & 0 & B_1 & -\frac{i}{c}E_2 \\ B_2 & B_1 & 0 & -\frac{i}{c}E_3 \\ \frac{i}{c}E_1 & \frac{i}{c}E_2 & \frac{i}{c}E_3 & 0 \end{pmatrix}$$

is obtained.

Despite the similarity of the magnetic and electric tensors, equations (4.8) and (4.9) show an important difference which has led to many futile attempts to fully symmetrize the system by the introduction of magnetic monopoles. However, an equally pleasing symmetry already exists in that a positron source defines a conjugate tensor $F^*_{\mu\nu}$, the mirror image of $F_{\mu\nu}$. The two conjugate fields generate equivalent magnetic fields and the overall symmetry: $F_{\mu\nu}|G_{\mu\nu}|F^*_{\mu\nu}$.

In tensor formalism the difference between electric and magnetic fields practically disappears. What one observer interprets as an electric process another may regard as magnetic, although the actual particle motions that they predict will be identical. It can be shown (Schwarz, 1972) that the two tensors are related through a completely antisymmetrical tensor of fourth rank

$$G_{\mu\nu} = \tfrac{i}{2c}\varepsilon_{\mu\nu\lambda\rho}F^{\lambda\rho}\,.$$

It would be mistaken to assume that this line of reasoning may be inverted in order to derive Maxwell's equations from the axioms of special relativity.

The electromagnetic field, in this context, is an emergent property – although it implies relativity, special relativity does not imply electromagnetism.

4.2 General Relativity

The quadratic invariant of special relativity

$$ds^2 = dx^2 + dy^2 + dz^2 - (cdt)^2$$

is a special case of the more general expression for non-Euclidean Riemann geometry,

$$ds^2 = \sum_{\mu,\nu=1}^{4} \eta_{\mu\nu} dx^\nu dx^\mu$$

with $x^1 = x$, $x^2 = y$, $x^3 = z$, $x^4 = ict$ and the metric tensor

$$\eta = \begin{pmatrix} 1 & 0 & 0 & 0 \\ 0 & 1 & 0 & 0 \\ 0 & 0 & 1 & 0 \\ 0 & 0 & 0 & -c^2 \end{pmatrix}$$

As the presence of gravity (mass) imparts a variable curvature on space the metric tensor is no longer constant[2]. As the summation extends over all values of ν and μ, the sum consists of 4×4 terms, of which 12 are equal in pairs, hence 10 independent functions. The motion of a free material point in this field will follow a curvilinear non-uniform path.

The curved space may be considered as covered by local metric neighbourhoods and regions of overlap where the transformation law of general relativity applies. It is called a Riemannian *manifold*. Vectors transplanted to neighbouring points may change their orientation and the intuitive idea of a straight line must be replaced by that of a *geodesic*, which may be regarded as the shortest or straightest curve in that region. A manifold on which vector transplantation is described by a law of the form

$$d\xi^i = \Gamma^i_{mj} dx^m \xi^j \tag{4.10}$$

[2]Any symmetric matrix can be transformed into a form with only 0, 1 and -1 as diagonal elements, said to define the *signature* of the matrix. The Lorentz transformation in the form $(cd\tau)^2 = x_0^2 - x_1^2 - x_2^2 - x_3^2$ defines the signature (1,-1,-1,-1), known as a hyperbolic metric – not to be confused with hyperbolic geometry.

(in which the summation rule operates) is called an *affine* (or linear) space and the Γ coefficients are called the affine connections. To ensure that the length of a vector is not affected by the transplantation the metric requirement that the scalar product of two vectors be invariant is retained. This condition defines an unique connection, compatible with a given metric tensor g, and its components in any coordinate system (x) are given by

$$\Gamma^i_{jk} = -\tfrac{1}{2}g^{im}\left(\frac{\partial g_{mk}}{\partial x^i} + \frac{\partial g_{mj}}{\partial x^k} - \frac{\partial g_{jk}}{\partial x^m}\right).$$

The defining equations for a geodesic in Riemann space become

$$\frac{d^2x^i}{ds^2} - \Gamma^i_{\alpha\beta}\frac{dx^\alpha}{ds}\frac{x^\beta}{ds} = 0$$

where s is the arc length of the geodesic.

An affine manifold is said to be *flat* or *Euclidean* at a point p, if a coordinate system in which the functions Γ^i_{jk} all vanish, can be found around p. For a cartesian system the geodesics become

$$\frac{d^2x^i}{ds^2} = 0, \quad \textit{i.e.}\ \text{straight lines.}$$

To specify the directions of two different vectors at nearby points it is necessary to define *tangent* vectors at these points. Stated in different terms, at each point of space-time, known as the *contact point*, there is an associated tangent Minkowski space. The theory of these spaces together with the underlying space becomes a Riemannian geometry if a Euclidean metric is introduced in each tangent space by means of a differential quadratic form[3].

Important tangent spaces are the light cones. In Minkowski space these all have the same shape and orientation, but not in general relativistic curved space, where the light cones may tilt, expand or contract as one moves from one tangent space to another.

Our immediate objective here is an equation that relates a geometrical object representing the curvature of space-time to a geometrical object representing the source of the gravitational field. The condition that all affine connections must vanish at a Euclidean point, defines a tensor

$$R^\alpha_{\mu\beta\nu} = \frac{\partial\Gamma^\alpha_{\mu\nu}}{\partial x_\beta} - \frac{\partial\Gamma^\alpha_{\mu\beta}}{\partial x_\nu} + \Gamma^\gamma_{\mu\nu}\Gamma^\alpha_{\gamma\beta} - \Gamma^\gamma_{\mu\beta}\Gamma^\alpha_{\gamma\nu}$$

[3]The tangent spaces can be converted into projective spaces by introducing points at infinity, in the usual manner, in each of them. The projective spaces can be studied analytically by means of homogeneous coordinates X^0, X^1, X^2, X^3, X^4 and projective tensors.

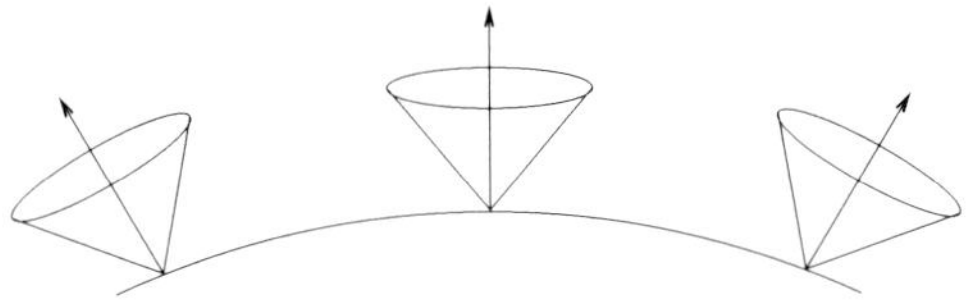

known as the *Riemann curvature tensor*, which can be contracted to the symmetrical tensor $R_{\mu\nu}$ with 10 independent components. Since this tensor has non-zero divergence it cannot feature in a conservation law (of mass and energy), but this is achieved by definition of the *Ricci tensor*

$$G_{\mu\nu} = R_{\mu\nu} - \tfrac{1}{2}\, g_{\mu\nu} R$$

where the doubly contracted tensor R is called the Riemann scalar. The components $g_{\mu\nu}$ of the fundamental tensor define the gravitational potential. Light rays are null curves $ds^2 = 0$ and the trajectories of particles in uniform motion are the geodesics. The geometry of space is not known *a priori*, but depends on the distribution of matter.

The Ricci tensor that represents the geometry of space is next equated with the so-called energy-momentum (stress) tensor of the matter field that defines the influence of matter and field energy

$$T^{\mu\nu} = \rho_0(x) u^\mu(x) u^\nu(x)$$

in terms of a scalar density field ρ_0 and a four-vector field of flow.

This procedure leads to the Einstein gravitational field equations, one form of which, without cosmological term, is

$$\begin{aligned} G_{\mu\nu} = R_{\mu\nu} - \tfrac{1}{2}\, g_{\mu\nu} R &= -\frac{8\pi\kappa}{c^2} T_{\mu\nu} \\ &= kT_{\mu\nu} \end{aligned} \tag{4.11}$$

where k is Newton's gravitation constant, or equivalently

$$R_{\mu\nu} = -\frac{8\pi\kappa}{c^2}\left(T_{\mu\nu} - \tfrac{1}{2}\, g_{\mu\nu} T\right) \tag{4.12}$$

where the Laue scalar T is the doubly contracted stress tensor.

The important result is the obvious symmetry between $T_{\mu\nu}$ and $R_{\mu\nu}$ as shown in (4.11) and (4.12). Both of these tensors vanish in empty Euclidean space and a reciprocal relationship between them is inferred: The presence of matter causes space to curl up and curvature of space generates matter.

The field equations of general relativity are rarely used without simplifying assumptions. The most common application treats of a mass, sufficiently distant from other masses, so as to move uniformly in a straight line. All applications of special relativity are of this type, in order to stay in Minkowski space-time. A body that moves inertially (or at rest) is thus assumed to have four-dimensionally straight world lines from which they deviate only under acceleration or rotation. The well-known Minkowski diagram of special relativity is a graphical representation of this assumption and therefore refers to a highly idealized situation.

In the real world the stress tensor never vanishes and so requires a non-vanishing curvature tensor under all circumstances. Alternatively, the concept of mass is strictly undefined in flat Minkowski space-time. Any mass point in Minkowski space disperses spontaneously, which means that it has a space-like rather than a time-like world line. In perfect analogy a mass point can be viewed as a local distortion of space-time. In Euclidean space it can be smoothed away without leaving any trace, but not on a curved manifold. Mass generation therefore resembles distortion of a Euclidean cover when spread across a non-Euclidean surface. A given degree of curvature then corresponds to creation of a constant quantity of matter, or a constant measure of misfit between cover and surface, that cannot be smoothed away. Associated with the misfit (mass) a strain field appears in the curved surface.

4.3 Unified Fields

The demonstration in the theory of general relativity that gravitation reflects the geometry of space-time raised the reasonable expectation that all of physics could be reduced to a common geometrical principle.

4.3.1 The Gauge Principle

A pioneering advance was made by Hermann Weyl with his proposal that the parallel displacement of a vector does not necessarily leave its length invariant, but changes it by an amount

$$dl = l\varphi_\mu dx^\mu$$

where the proportionality factors φ_μ are functions of the coordinates. Rearranged into

$$d(\log l) = \varphi_\mu dx^\mu \tag{4.13}$$

i.e.

$$l = l_0 e^{\int \varphi_\mu dx^\mu}$$

where l_0 is the initial value of l.

As the gauge factors φ_μ depend on the path of displacement, equation (4.13) can be integrated only if the circulation vectors of the type

$$F_{\mu\nu} = \frac{\partial \varphi_\mu}{\partial x^\nu} - \frac{\partial \varphi_\nu}{\partial x^\mu} \tag{4.14}$$

should vanish. In the geometry proposed by Weyl these vectors are non-zero, with divergence

$$\frac{\partial F_{\mu\nu}}{\partial x^\rho} + \frac{\partial F_{\nu\rho}}{\partial x^\mu} + \frac{\partial F_{\rho\mu}}{\partial x^\nu} = 0 \tag{4.15}$$

The formal resemblance of (4.14) and (4.15) with Maxwell's equations (4.4) and (4.5) prompted the identification of φ_μ with the electromagnetic 4-potential and the tensor (4.14) as the electromagnetic field. The absence of an electromagnetic field ($F = 0$) is the necessary condition for the validity of general relativity which only accounts for gravitation.

An obvious objection to Weyl's theory is that an atom carried around a closed path in an electromagnetic field would radiate at a different wavelength when reaching the end of the loop. This is refuted by experiment. It was shown by London how to address this problem quantum-mechanically.

Accepting the Bohr postulate of an electronic orbit on a hydrogen atom, stabilized by a balance between mechanical and electrostatic forces,

$$\frac{mv^2}{r} = \frac{e^2}{r^2}$$

one calculates the orbital velocity $v = e/\sqrt{mr}$ and period $\tau = 2\pi r/v$ for an electron at a distance r from the proton. At constant r the Weyl gauge parameters are $\varphi_0 = a/r$, $\varphi_1 \equiv 0$, where a is a dimensionless proportionality constant, and describe the variation of scale as

$$l = l_0 \exp\left(\int_0^{x^0} \varphi_0 dx^0\right) = l_0 e^{\varphi_0 x^0}$$

in terms of the time coordinate x^0. Assuming that the change of scale vanishes for special orbits specified by an appropriate choice of a, it follows that

$$\exp\left(\varphi_0 c\tau\right) = 1, \quad \varphi_0 c\tau = 2\pi i n$$

where n is an arbitrary integer, such that

$$\frac{ac\tau}{r} = ac \cdot \frac{2\pi\sqrt{mr}}{e} = 2\pi i n$$

or $$r = -\frac{n^2e^2}{a^2c^2m} \equiv a_0,$$

for the first Bohr orbit, $$a_0 = \frac{n^2h^2}{4\pi^2e^2m}.$$

Hence $$a^2 = -\frac{4\pi^2e^4}{h^2c^2} \quad \text{and} \quad a = i\alpha,$$

where α is the Sommerfeld fine-structure constant.

By this analysis a complex gauge factor leads directly to the quantization conditions for the hydrogen atom, but complicates the interpretation of the scaling of a real vector. Essentially the same result was obtained earlier by Schrödinger, later identified as describing a phase change, associated with wave motion, rather than a change of scale.

Although Weyl's geometry did not produce the desired unification, the modified theory developed into a convincing definition of the electromagnetic field in wave formalism. The gauge transformation is now formulated as

$$\psi(x) \rightarrow \psi'(x) = e^{i\alpha}\psi(x)$$

with an associated charge q and known as the transformation group $U(1)$. While this α is a constant, the derivative of the field

$$\partial_\mu\psi \rightarrow e^{i\alpha}\partial_\mu\psi$$

transforms like the field itself and the charge density $q|\psi_t|^2$ is invariant under (global) gauge transformation. However, in curved space the gauge factor is an arbitrary function of position, $\alpha(x)$. The derivative of ψ now acquires an extra term and charge is no longer conserved, unless compensated for by the introduction of an additional gauge field which under the (local) transformation

$$A'_\mu(x) \rightarrow A_\mu(x) + \partial_\mu\alpha(x)$$

leaves the electromagnetic field (4.14) invariant. The four-vector A describes the electromagnetic field, or the wave functions of the photon.

The principle of local invariance in a curved Riemannian manifold leads to the appearance of compensating fields. Like the electromagnetic field, which is the compensating field of local phase transformation, the gravitational field may be interpreted as the compensating field of local Lorentz transformations. In modern physics all interactions are understood in terms of theories which combine local gauge invariance with spontaneous symmetry breaking.

Despite its general utility the gauge principle remains an empirical assumption. It clearly identifies all fields as manifestations of space-time configurations, but not characterized more closely in any way. Each field is associated with a specific symmetry group without specification of the underlying geometry. No unified-field theory can be considered complete until the common geometrical basis has been identified.

4.3.2 Kaluza–Klein Theory

Another attempt by Kaluza and Klein to incorporate electromagnetic interaction in the field equations of general relativity was to generate extra degrees of freedom by increasing the number of space-time dimensions from four to five, such that the line element is:

$$ds = \left(\sum \gamma_{\mu\nu} dx^\mu dx^\nu\right)^{\frac{1}{2}} \qquad \mu, \nu = 0, 4. \tag{4.16}$$

Four coordinates, $x^\mu (\mu \neq 4)$, are used to characterize regular space-time and $\gamma_{\mu,\nu}$ is assumed independent of x^4. To explain why the fifth dimension is not observed, space was proposed to have cylindrical symmetry with respect to x^4,

$$\frac{\partial \gamma_{\mu\nu}}{\partial x^4} = 0.$$

On this basis it was possible to obtain equations which agreed with known relativistic equations of the gravitational field and generalized Maxwell equations of electromagnetism. Writing (4.16) in the form

$$ds^2 = \left(dx^4 + \gamma_{i4} dx^i\right)^2 + g_{ik} dx^i dx^k \qquad i, k = 0, 3$$

the g_{ik} are invariant under the transformation (Pauli, 1981)

$$\left(x^4\right)' = x^4 + f\left(x^0, \ldots, x^3\right)$$

while

$$\gamma_{i4}' = \gamma_{i4} - \frac{\partial f}{\partial x^i}. \tag{4.16a}$$

This is analogous to a gauge transformation, suggesting identification of γ_{i4} with the electromagnetic potential φ_i, apart from a proportionality constant. The antisymmetric tensor

$$\frac{\partial \gamma_{k4}}{\partial x^i} - \frac{\partial \gamma_{i4}}{\partial x^k} = f_{ik}$$

is invariant with respect to the gauge transformation (4.16a) and proportional to the electromagnetic field strength, (4.14).

As $\gamma_{\mu\nu}$ is independent of x_4 the equation for the geodesics of the metric,

$$\frac{d}{ds}\left(g_{ik}\frac{dx^k}{ds}\right) - \tfrac{1}{2}\frac{\partial g_{rs}}{\partial x^i}\frac{dx^r}{ds}\frac{dx^s}{ds} = C \cdot f_{ik}\frac{dx^k}{ds}$$

represents the orbit of a charged particle in external gravitational and electromagnetic fields. The integration constant C is proportional to the quotient e/m of charge and mass of the particle.

4.3.3 Einstein's Alternatives

Although Kaluza's theory comes very close to a rigorous unification of gravity and electromagnetism it has a few blemishes as first pointed out by Einstein and Mayer (1931), who nevertheless recognized its merits and attempted a reformulation to eliminate the defects. They rejected Kaluza's model on the grounds that the world is experienced to be four and not five-dimensional; that the assumption of a cylindrically compacted dimension is unnatural; that the theory fails to account for the mass of moving charges; and finally, that the theory offers no physical explanation of the extra coefficient γ_{44}, associated with neither gravitational nor electromagnetic fields. To overcome these problems a four-dimensional Riemannian continuum that incorporates a five-dimensional linear vector space was proposed: a five-component vector is defined at each point of the four-dimensional continuum, with fixed rules of combination and transformation. Unlike Kaluza's theory the Einstein–Mayer theory is not covariant and amounts to little more than a deliberate manipulation of tensors to produce a preferred result. It was not pursued any further and abandoned soon.

Despite his misgivings, Einstein, with Bergmann[4], returned to Kaluza's model by turning Klein's curled-up dimension into a closed macroscopic dimension, invoking quantum-mechanical uncertainty in four dimensions as rationale. One closed geodesic line goes through each point of the five-dimensional space. This geodesic that connects the point to itself intersects itself at a zero angle. The length of such a geodesic is called the *circumference* of the space in five dimensions and this circumference is defined to be everywhere the same.

The model produced nothing beyond Kaluza's original and confirmed the periodicity condition observed by Klein – a mathematical consequence of a closed system. The coordinates they used to describe the closed five-dimensional space are strangely reminiscent of the homogeneous coordinates of four-dimensional projective geometry, but without a point at infinity.

Although Einstein must have been familiar with the concepts of projective relativity, developed by his colleagues Oswald Veblen, Banesh Hoffmann and

[4]See Bergmann (1976) p.272.

others at Princeton, he never mentioned this alternative. A possible clue to his reticence appears in his defence of the Doppler interpretation of galactic redshift (Einstein,1950,p.122).

It was argued that atomic properties, including sharp spectral lines, are not related by "similarity" but by "congruence". Should there be a redshift between galaxies, which remain at a fixed distance apart, "there would exist no metric in the sense of relativity". Metrical properties, by definition, do not feature in projective geometry. It is of interest to note that this argument is not consistent with the analysis of gravitational redshifts in the same volume (p.88). The difference is that the latter argument assumes curved space–time, whereas the former relies on the universal time of the Robertson–Walker metric. We shall return to Einstein's habit of switching between incompatible models as convenient.

Einstein's legendary search for a unified field continued for the rest of his life. What was probably his final serious effort is outlined in Appendix II, Einstein (1950). To increase the number of field variables unsymmetrical affine connections and metric tensors are considered instead of the usual symmetrical structures. Overwhelmed by mathematical complexity and lack of comprehensible geometry and physics, the theory remained unfinished.

Towards the end, in his search for unification, Einstein was even prepared, Michelmore(1963):

> "... to discard the principle of 'no action at a distance' – in other words, to assume that a physical event may directly affect faraway events as well as just its immediate surroundings in space and time".

All this, despite his rejection of non-local quantum correlations and of David Bohm's recognition of the quantum potential.

The searh for a unified theory goes deeper than finding common ground for gravity and electromagnetism. The search is for a single theory that accounts for macroscopic phenomena, the atomic structure of matter and electricity, and for quantum effects.

In an effort to derive particle properties from general relativity Einstein and Rosen (1935) investigated spherically symmetrical solutions of the field equations, including the Schwarzschild solution. On the premise that every field theory should exclude singularities of the field, they found this to be possible provided the physical space is represented by two identical sheets connected by bridges at the position of a singular point of the unmodified metric. The final solution was found only to pertain to massless particles, and quantum phenomena could not be demonstrated *a priori*. However, in the context of projective cosmology the two-sheet concept will be shown to be of fundamental importance.

4.4 Quantum Theory

One of the pillars of standard cosmology is the so-called black-body spectrum of the universal microwave background. It is precisely such an observation that also led to the first recognition of a quantum effect by Max Planck in 1901.

4.4.1 The Seminal Ideas

The experimental observation consists of the analysis of radiation eminating from an orifice in a hollow cavity kept at a fixed temperature. The partition formula by frequency is obtained on integration over allowed modes of vibration,

$$\varrho(\nu) = 8\pi\bar{\varepsilon}\nu^2/c^3$$

where $\bar{\varepsilon}$ is the average energy for a given mode.

In statistical thermodynamics, which treats energy as a continuous function, $\bar{\varepsilon} = kT$ for each mode of the Boltzmann distribution. The radiation density, calculated as

$$\varrho(\nu) = \frac{8\pi kT\nu^2}{c^3}$$

follows the experimental curve only at long wavelength. At short wavelength

$$\lim_{\nu\to\infty} \varrho(\nu) = \infty.$$

This means that an infinite amount of thermal energy needs to be converted into radiant energy to establish an equilibrium at short wavelength.

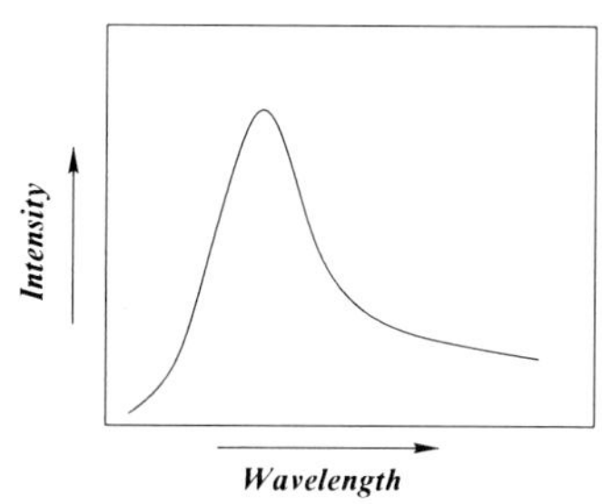

Figure 4.1: *Planck distribution*

To overcome the problem it was necessary to assume that radiant energy occurs as multiples of discrete units, proportional to the frequency, $\varepsilon = h\nu$; the proportionality factor is known as Planck's constant. The Boltzmann factor $\exp -\varepsilon_n/kT$ becomes $\exp(-nh\nu/kT)$ and the energy per degree of freedom is obtained by summation of infinite series rather than integration. The result is the Planck distribution:

$$\varrho(\nu) = \frac{8\pi\nu^3}{c^3}\frac{1}{e^{h\nu/kT} - 1}$$

that corresponds to the so-called black-body distribution, which is observed experimentally and shown in Figure 4.1. Integration of the Planck distribution over all frequencies leads to the observed Stefan–Boltzmann energy density

$$\varrho = \left(\frac{8\pi^5 k^4}{15h^3c^3}\right) T^4 = aT^4 ,$$

as final confirmation of the model.

Solution of the black-body problem had an immediate knock-on effect on two other unsolved problems of classical physics, the photoelectric effect and the interpretation of atomic spectra.

Irradiation of active metal surfaces with ultraviolet light, above a characteristic threshold frequency, regardless of intensity, was known to dislodge a stream of electrons from the metal, with kinetic energies proportional to the frequency of the light. The puzzle was to understand how the energy in a wavefront, assumed to increase with intensity, only became effective above the threshold frequency.

The solution, proposed by Einstein, was that the discrete energy units, identified by Planck, correspond to quanta of light, called *photons*, which interact with electrons in the metal surface during direct collision. This dual wave/particle nature of light inspired de Broglie to postulate a similar behaviour for electrons. Experimental observation of electron diffraction confirmed the wave nature of electrons and firmly established the dual character of all quantum objects as mysterious reality. As the logical picture of an entity, which is wave as well as particle, is hard to swallow, it has become fashionable to avoid all physical models of quantum events; it is considered poor taste to contaminate the quantum world with classical concepts. This noble idea of the so-called Copenhagen interpretation of quantum theory has resulted in a probabilistic computational model that, not only defies, but denies comprehension.

4.4.2 The Planetary Model

The other outstanding problem of classical physics was to understand the way in which atoms emit radiation only at discrete wavelengths, as formulated for hydrogen by the Balmer formula:

$$\frac{1}{\lambda} = R\left(\frac{1}{2^2} - \frac{1}{n^2}\right) \quad n = 3, 4, 5 \ldots$$

The earliest known interpretation of the Balmer formula was proposed in 1904 by Hantaro Nagaoka, apparently unaware of Planck's work. He assumed that an atom was constructed like the planet Saturn with its rings. Most of the mass remains concentrated in a positively charged nucleus (planet) with charge balance provided by the electrons that circle the nucleus in fixed orbits, likened to the particle rings. He recognized the complication arising from the fact that, unlike Saturn, the subatomic units are electrically charged and that the orbiting (accelerated) electrons would be expected to radiate away their energy. His proposed resolution of the problem is accepted today as the essential difference between classical particles and quantum objects.

Nagaoka's remedy was to superimpose simple harmonic displacements on the circular orbits of an electron, which is equivalent to describing the electron as a standing wave with wavelength $\lambda = 2\pi r/n$, in terms of the orbital radius r and a series of integers, $n = 1, 2, \ldots$. The same solution was rediscovered twenty years later by de Broglie to rationalize Bohr's atomic model. Although Nagaoka's model accounted for the Balmer formula and predates the famous nuclear model of the atom, proposed by Rutherford in 1911, it was ignored by the physics community at the time, and is virtually forgotten by now.

Between Rutherford and Bohr, Nagaoka's model was presented ten years later in the form of a miniature solar system, postulated to be stable when in a stationary state of mechanical and electrostatic equilibrium between electron and nucleus. This equilibrium is identified by minimization of the total energy,

$$E = \frac{p^2}{2m} - \frac{e^2}{r},$$

consisting of kinetic and potential energy parts, as a function of r. The expression is converted into wave formalism by assuming the de Broglie relationship between momentum and wavelength, $p = h/\lambda$, for standing waves at a distance r_n from the nucleus, such that $n\lambda = 2\pi r_n$, and hence

$$E = \frac{n^2h^2}{8\pi^2 m r_n^2} - \frac{e^2}{r_n} \tag{4.17}$$

The condition $dE/dr = 0$ identifies the stationary states with radii

$$r_n = \frac{n^2h^2}{4\pi^2 me^2} \equiv n^2 a_0$$

in terms of the quantum number $n = 1, 2, \ldots$ and a_0, known as the first Bohr radius.

Allowed energies are obtained by substituting r_n back into (4.17):

$$E_n = -\frac{me^4}{2(n\hbar)^2} = -\frac{Rhc}{n^2}$$

where the Rydberg constant, $R = \dfrac{2\pi^2 me^4}{h^3 c}$, accounts for the Balmer formula if the difference between Bohr levels, $\Delta E = Rhc\left(1/n_1^2 - 1/n_2^2\right)$, is assumed to match the frequency $\nu = \Delta E/h$, of radiation, emitted as the electron changes energy levels. This condition ensures that any atom only emits or absorbs radiation at characteristic frequencies, known as a line spectrum, which is widely used in chemical and astronomical analyses, known as spectroscopy. As shown in the next section, the origin of the Bohr formulae derives from the wave motion of the electron.

The Nagaoka and Bohr atomic models are based on the assumption of self-similarity between atoms, planetary rings and solar systems. Although only partially successful these models were sufficiently accurate to pave the way for development of the more detailed wave-mechanical model.

Once a reliable mathematical model of electronic motion in atoms has been established, it would be possible to return the favour and use this model to upgrade the celestial mechanics of the solar system. We shall return to this topic in due course.

4.4.3 Wave Mechanics

The postulate of Nagaoka and de Broglie, and the discovery of electron diffraction suggested that the appearance of integer quantum numbers relates to the periodicity of wave motion, which is also characterized by integers, and that the behaviour of quantum particles should be described by the general wave equation, which in one dimension reads:

$$\frac{\partial^2 u(x,t)}{\partial x^2} - \frac{1}{c^2}\frac{\partial^2 u(x,t)}{\partial t^2} = 0. \tag{4.18}$$

This equation can be solved by separation of the variables, assuming

$$u(x,t) = f(x) \cdot e^{-i\omega_0 t}$$

Hence

$$\begin{aligned}
\frac{\partial u}{\partial x} &= e^{-i\omega_0 t}\frac{\partial f(x)}{\partial x} \\
\frac{\partial^2 u}{\partial x^2} &= e^{-i\omega_0 t}\frac{\partial^2 f(x)}{\partial x^2} \\
\frac{\partial u}{\partial t} &= -i\omega_0 f(x)e^{-i\omega_0 t} = -i\omega_0 u \\
\frac{\partial^2 u}{\partial t^2} &= -\omega_0^2 f(x)e^{-i\omega_0 t} = -i\omega_0\frac{\partial u}{\partial t}
\end{aligned}$$

Substitution of these derivatives into (4.18) leads to two new equations:

(i)
$$\frac{\partial^2 u}{\partial x^2} - \frac{1}{c^2}\frac{\partial^2 u}{\partial t^2} = \frac{\partial^2 f(x)}{\partial x^2} + \left(\frac{\omega_0}{c}\right)^2 f(x) = 0.$$

which is the Helmholtz equation

$$\left(\frac{d^2}{dx^2} + k_0^2\right) f(x) = 0 \quad , \text{ where } \quad k_0 = \omega_0/c\,.$$

(ii)
$$\frac{\partial^2 u}{\partial x^2} + \frac{i\omega_0}{c^2}\frac{\partial u}{\partial t} = 0\,.$$

By introducing the quantum postulate $\hbar\omega_0 = E - V$, the energy in excess of a constant potential, and the velocity of a matter wavefront, in Hamilton-Jacobi formalism (Goldstein, 1980), $c = \sqrt{T/2m}$, these equations (in 3D) transform into the familiar set of Schrödinger equations:

$$\frac{\hbar^2}{2m}\nabla^2\psi + (E - V)\psi = 0 \tag{4.19}$$

$$\frac{i\hbar}{2m}\nabla^2\Psi = \frac{\partial\Psi}{\partial t} \tag{4.20}$$

Equation (4.20) in Ψ, the so-called time-dependent wave equation, is often considered to be a diffusion equation, as it shows only a first time derivative.

In the axiomatic approach to quantum mechanics these equations are obtained by substituting differential operators for the classical variables of momentum and energy

$$p \rightarrow -i\hbar\nabla, \quad E \rightarrow i\hbar\frac{\partial}{\partial t}$$

into the Hamiltonian expression $\qquad H = \dfrac{p^2}{2m} + V\,.$

For systems of chemical interest the amplitude function ψ that occurs as a solution of (4.19) is postulated to give a complete description, provided the potential energy V, is correctly specified. In reality, the only chemically significant problem that has been solved is of an electron associated with an isolated stationary proton, with potential energy $V = e^2/r$, in atomic units. The differential wave equation is separable in spherical polar coordinates. Separate solutions, as functions of radial (r) and angular (θ, φ) coordinates, describe the quantized energy and angular momentum of the electron as:

$$
\begin{aligned}
E(n) &= E_1/n^2 \\
L^2(l) &= l(l+1)\hbar^2 \\
L_z(m) &= m\hbar
\end{aligned}
$$

where $n = 1, 2, 3 \ldots, l = 0, 1, \ldots, (n-1), m = 0, \pm 1, \ldots, \pm l$.

In more careful analyses, taking into account the equivalence of space and time variables, as demanded by special relativity, the wave functions become *spinors* and an additional quantum condition, that restricts the so-called spin angular momentum of the electron to values of $\pm\frac{1}{2}\hbar$, appears.

By assuming that the dynamic variables follow the same rules as for hydrogen in many-electron atoms, there was the expectation that the periodic table of the elements could be reduced to the Schrödinger solution for hydrogen. Apart from a superficial correlation, which is commonly assumed to vindicate this expectation, it has now been shown that the neglect of general-relativistic curvature of space-time prevents such reduction. Once this defect has been rectified the atomic model will be used to investigate commensurability in the self-similar solar system.

Despite the spectacular success of quantum theory in correctly predicting many of those atomic properties that baffled classical physics, the meaning of state functions and the mechanism of so-called quantum jumps between stationary states have remained problematical. The major inhibiting factor has been the reluctance to abandon the classical concept of indivisible point particles as a basis of the wave-like properties of matter. The compromise concepts of wave/particle duality and probability density have stimulated an illogical belief in ghost-like phenomena in order to rationalize quantum behaviour.

Most of the problems are avoided by accepting that the wave equation works because it describes wavelike entities rather than infinitesimal hard spheres. By accepting that an electron has a wave structure, the formalism of ordinary wave theory immediately defines a density function $\rho = \Psi\Psi^*$ and a current density

$$
\boldsymbol{j} = \frac{\hbar}{2mi}\left(\Psi^*\nabla\Psi - \psi\nabla\Psi^*\right), \tag{4.21}
$$

from which follows a continuity equation as in classical hydrodynamics

$$\frac{\partial \rho}{\partial t} + \mathrm{div}\boldsymbol{j} = 0 \tag{4.22}$$

A general expression for a one-electron wave function over all available states

$$\Psi = \sum_k c_k \psi_k e^{2\pi i \nu t} \tag{4.23}$$

may be used to calculate the current density over two states k and l:

$$\boldsymbol{j} = \frac{\hbar e}{mi} \sum_{k,l} c_k c_l \left(\psi_l \nabla \psi_k - \psi_k \nabla \psi_l\right) e^{2\pi i(\nu_k - \nu_l)t} \tag{4.24}$$

If only a single eigenvibration is excited, the current disappears and the distribution of electron density remains constant. Otherwise an electron flows from one state to another in an exchange that involves a photon to keep the energy in balance.

This flow of electricity can hardly be described as a quantum jump. More realistically the vibrations of the two affected states (emitter and acceptor) are seen to interact and generate a beat (wave packet) that moves to the state of lower energy. The virtual photon that links two equilibrium states turn into a real photon that carries the excess energy, either into or away from the system.

Electron Spin

Schrödinger's equation appears incomplete in the sense of lacking an operator for spin, only because its eigenfunction solutions are traditionally considered complex variables. The wave function, interpreted as a column vector, operated on by square matrices, such that

$$\begin{bmatrix} e^{i(\omega t - kx)} & 0 \\ 0 & e^{-i(\omega t - kx)} \end{bmatrix} \begin{bmatrix} \phi_1 \\ \phi_2 \end{bmatrix} = \begin{bmatrix} \phi_1 e^{i(\omega t - kx)} \\ \phi_2 e^{-i(\omega t - kx)} \end{bmatrix}$$

abbreviated to $\begin{bmatrix} \phi_1 e^+ \\ \phi_2 e^- \end{bmatrix}$, represents a spinor that moves in the x-direction. By forming the derivatives:

$$\frac{\partial \phi}{\partial t} = i\omega \begin{bmatrix} \phi_1 e^+ \\ \phi_2 e^- \end{bmatrix}$$

$$\frac{\partial^2 \phi}{\partial x^2} = k^2 \begin{bmatrix} \phi_1 e^+ \\ \phi_2 e^- \end{bmatrix}$$

it follows that (in three dimensions):

$$-i\frac{\partial\phi}{\partial t} = \frac{\omega}{k^2}\nabla^2\phi$$

This is Schrödinger's equation, providing $(\hbar k)^2 = 2m\hbar\omega$, *i.e.*

$$-i\frac{\partial\phi}{\partial t} = \frac{\hbar}{2m}\nabla^2\phi \quad , \quad \text{as in (4.20):} V = 0$$

which shows $\hbar\omega = E = p^2/2m$, $k = 2\pi/\lambda$, $p = h/\lambda$. This result is interpreted to show that a region of the continuum, which rotates in spherical mode, interacts with its environment by generating a wave-like disturbance at half the angular frequency of the core. The angular momentum on the surface of a unit sphere is $L = m\omega$. At $\lambda = 2\pi$, $k = 1$, the spin angular momentum follows as $L = \hbar/2$, with intrinsic magnetic moment $\mu = \hbar e/2mc$.

4.4.4 Bohmian Mechanics

The connection between wave mechanics and hydrodynamics, expressed by equations (4.20) and (4.22), was developed in more detail by Madelung, writing the time dependence of Ψ as an action function, $\Psi = \psi e^{2\pi i\nu t} \rightarrow Re^{iS/\hbar}$, which seperates (4.18) into a coupled pair that resembles the field equations of hydrodynamics:

$$\frac{\partial S}{\partial t} + \frac{(\nabla S)^2}{2m} - \frac{\hbar^2\nabla^2 R}{2mR} + V = 0 \tag{4.25}$$

$$\frac{\partial R^2}{\partial t} + \nabla\cdot\left(\frac{R^2\nabla S}{m}\right) = 0 \tag{4.26}$$

which describe the irrotational flow of a compressible fluid, assuming R^2 to represent the density $\rho(x)$ of a continuous fluid with stream velocity $v = \nabla S/m$. It was shown that both density and flux vary periodically with the same periodicity as $\nu_{ik} = (E_i - E_k)/h$, that results from superposition of states i and k. This means that radiation is not due to quantum jumps, but rather happens by slow transition (current) in a non-stationary state.

An attractive feature of the hydrodynamic model is that it obviates the statistical interpretation of quantum theory, by eliminating the need of a point particle. It is worth noting that the assumption of a point electron derives from the observation that it responds as a unit to an electromagnetic signal, which must therefore propagate instantaneously through the interior of the electron, thought to be at variance with the theory of relativity. However, by now it is known from experiment that non-local (instantaneous) response is possible in quantum systems and the initial reservation against Madelung's proposal and Lorentz's definition of an electron as a flexible sphere should fall away.

On reinterpretation it was pointed out by David Bohm that equation (4.25) differed from the classical Hamilton-Jacobi equation only in the term

$$V_q = -\frac{\hbar^2\nabla^2 R}{2mR}\,. \tag{4.27}$$

The quantity V_q, called *quantum potential* vanishes for classical systems as $h/m \to 0$. A gradual transition from classical to quantum behaviour is inferred to occur for systems of low mass, such as sub-atomic species. All dynamic properties of classical systems should therefore be defined equally well for quantum systems, although the relevant parameters are hidden (Bohm, 1952).

Quantum Potential

As for the classical potential, the gradient of quantum potential energy defines a quantum force. A quantum object therefore has an equation of motion, $m\,\ddot{x} = -\nabla V - \nabla V_q$. For an object in uniform motion (constant potential) the quantum force must vanish, which requires $V_q = 0$ or a constant, $-k$ say. $V_q = 0$ defines a classical particle; alternatively $-(V + V_q) = T$, the kinetic energy of the system. Hence $\hbar^2\nabla^2 R/2mR = -E$, which rearranges into

$$\nabla^2 R + \frac{2mE}{\hbar^2}R = 0$$

Schrödinger's equation for a free particle.

The quantum potential concept is vitally important for understanding the structure of an electron and of quantum systems in general. The fact that the amplitude function (R) appears in both the numerator and denominator of V_q implies that the effect of the wave field does not necessarily decay with distance and that remote features of the environment can affect the behaviour of a quantum object.

The quantum potential for a many-body system:

$$V_q = \sum_{i=1}^{n} -\frac{\hbar^2}{2mR}\left(\frac{\nabla_i^2 R}{m_i}\right)$$

depends on the quantum state of the entire system. The potential energy between a pair of entities, $V_q(x_i, x_j)$ is not uniquely defined by the coordinates, but depends on the wave function of the entire system, Ψ. This condition defines a holistic system in that the whole is more than a sum of the parts. The instantaneous motion of one part depends on the coordinates of all other parts at the same time. That defines a non-local interaction of the type assumed to exist within an indivisible electron, and now inferred to occur in all quantum systems. If the system is distorted locally, the entire system responds instantaneously. As the quantum potential is not a function of distance, the behaviour of a composite system depends non-locally on the configuration of all constituents, no matter how far apart.

The contentious issue of quantum-particle trajectories is put into perspective by the Bohmian model. One interpretation is that the quantum electron has an unspecified diffuse structure, which contracts into a classical point-like object when confined under external influences. The observed trajectory, as in a cloud chamber, may be considered to follow the centre of gravity.

In a two-slit experiment an electron wave passes through both slits to recombine, with interference, but without rupture. The interference pattern disappears on closure of one slit or when the slits are too far apart, compared to the de Broglie wavelength. It now behaves exactly like a classical particle, when forced through a single slit (Holland, 1993).

Stationary States

Writing the wave equation in two equivalent forms:

$$\Psi(x,t) = \Psi_0 e^{-iEt/\hbar}$$

$$\Psi(x,t) = R(x,t)e^{iS(x,t)/\hbar}$$

and noting that $R(x,0) = R_0(x)$; $S(x,0) = S_0(x)$; $\Psi_0 = R_0 e^{iS_0/\hbar}$, it follows that:

$$\begin{aligned} S(x,t) &= S_0(x) - Et \\ R(x,t) &= R_0 \end{aligned} \tag{4.28}$$

The unexpected conclusion is that a real wave function, $\Psi_0 = \psi$, implies $S_0(x) = 0$ and hence the momentum $\nabla S = p = 0$ and $E = V + V_q$. Those states with $m_l = 0$ all have real wave functions, which therefore means that such electrons have zero kinetic energy and are therefore at rest. The classical (electrostatic) and quantum forces on electrons in such stationary states are therefore balanced and so stabilize the position of the electron with respect to the nucleus.

For the hydrogen atom in the ground state, $R(r) = Ne^{-r/a_0}$ and hence,

$$\frac{d^2R}{dr^2} = \frac{N}{a_0^2}e^{-r/a_0},$$

such that, from (4.27), $V_q = \hbar^2/2ma_0^2$. In general

$$V_q = \frac{\hbar^2}{2mr^2}, \tag{4.29}$$

and the quantum force on the electron:

$$F_q = \frac{\partial V_q}{\partial r} = -\frac{\hbar^2}{mr^3}$$

whereas the electrostatic force $F = e^2/r^2$. These forces are in balance when

$$\frac{\hbar^2}{mr^3} = \frac{e^2}{r}; \quad r = \frac{\hbar^2}{me^2} = a_0 ,$$

the Bohr radius. This means that $V = V_q$ at $r = a_0/2$, halfway between proton and electron.

4.4.5 Antimatter

An important feature of relativistic wave equations is that their energy spectra are no longer limited to positive energies. It poses the awkward possibility that, unless all negative energy levels are blocked, any electron would simply cascade down the energy ladder into a bottomless pit.

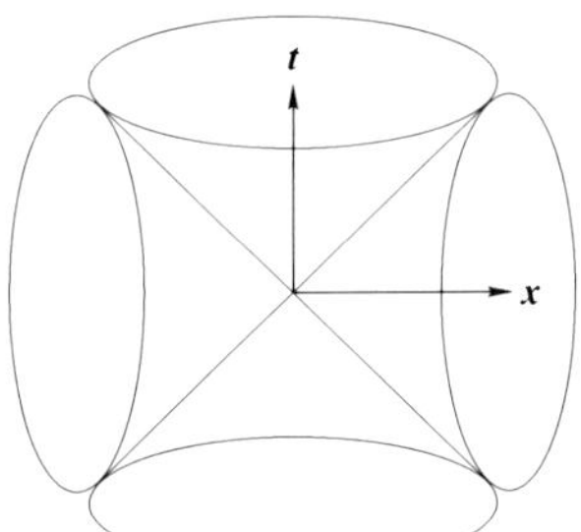

Figure 4.2: *Two-dimensional representation of Minkowski space-time*

An expensive way to prevent the cascade was proposed by Dirac. He suggested that all negative-energy levels are fully saturated with electron pairs, in line with the exclusion principle, postulated before in order to rationalize elemental periodicity. He allowed for the possibility of vacancies, or holes, in the negative part of the spectrum. These would appear equivalent to positively charged electrons. As an electron moves into the vacant site it radiates energy and appears to annihilate the supposed positive electrons. By this reasoning each elementary particle should be mirrored by an anti-particle of opposite charge, but identical mass. The existence of anti-particles with exactly the predicted properties was demonstrated experimentally to interact as shown in equation (1.1). It is interesting to note that in Dirac's (1930) original proposal he identified the negative-energy vacancies as protons.

Perhaps this is the right time to reconsider the universal imbalance between matter and antimatter if we recall a similar dissymmetry in the Minkowski space of special relativity.

The common two-dimensional representation is done in terms of a time axis and one space axis, which is interpreted as three space directions at the same time. The so-called time cone, and by implication the complementary space cone, extends into a further undefined dimension, perpendicular to the x, t plane of the diagram. To get the complete picture it is necessary to superimpose three mutually perpendicular Minkowski bodies of this type, which is only posssible in four dimensions. In this superposition the time axis does not remain fixed and becomes entangled with the space directions as seen in three-dimensional space. The three-dimensional surface of the generalized light cone of Figure 4.2 becomes a surface in four-dimensional space-time that separates all space into two equivalent regions.

It is tempting to identify the two regions with matter and antimatter domains, but until the appropriate topology of space-time has been decided the geometry of the interfacial surface remains conjectural. However, the conclusion that space and time coordinates cannot be treated as independent, as in expanding-universe cosmologies, is firm. The simplification introduced by assuming a universal time, independent of non-Euclidean curvature, is simply too restrictive and unrealistic to serve as a basis for sensible cosmology.

4.5 The Vacuum

The two theories of physics project a face of inscrutability to the uninitiated, which means virtually the entire world population. The science writers who have tried to bridge the gap did so in a woefully uncritical fashion. A notable exception was the philosopher Karl Popper (1965) who made a ruthless analysis of the near-dogmatic nature of quantum theory.

The root of the problem can be traced back over centuries and the interminable debate on the structure of the vacuum. The utter confusion that exists over concepts such as the void, elementary matter, atoms, the æther, and action at a distance, all aspects of the vacuum was discussed in elaborate detail by the philosopher Roger Boscovich (1763). He reached the painful conclusion that, despite the mathematical genius of Newton and Leibniz, the existence of hard point-particles, moving and interacting in a structureless void, was a logical impossibility. Suffice it to remark that Popper's criticism of the Copenhagen interpretation of quantum theory is, in many ways, a modern paraphrase of Boscovich's dilemma. An additional aggravating factor is the textbook belief that the special theory of relativity proves the vacuum to be a void.

The physics community have distanced themselves from the debate by accepting quantum theory as a mathematically useful tool, without agonizing over the physical interpretation. For the chemist who deals with three-dimensionally structured objects, like molecules, this approach creates a dilemma. Modern chemistry is best understood in terms of experimentally measured electron-density distributions, awkward, if not impossible, to visualize in terms of zero-dimensional objects. The alternative wave model, not only makes intuitive sense, but also eliminates poorly defined concepts such as probability densities and quantum jumps.

The wave nature of the electron and the physical implications thereof were discussed recently in some detail (Boeyens, 2010). As in the theory of general relativity it is accepted that an empty universe is featureless and flat, but that curvature of space-time causes wavelike distortion of the vacuum. The equivalent of an infinite plane wave in flat space develops interference effects, like wave packets, in curved space, interpreted as units of mass and energy.

Wave Packets

A wave packet generated by a superposition of waves is shown in Figure 4.3. The tangent curve follows the amplitude of the $1/r$ Coulomb potential, which reflects the actual charge density, except when $r \to 0$. The secondary waves propagate with the group velocity v_g of the system and the primary waves have phase velocity v_ϕ, such that $v_g v_\phi = c^2$, where $c = 1/\sqrt{\epsilon_0\mu_0}$, is the velocity of light in the vacuum. Such a wave packet (Wolff, 1995):

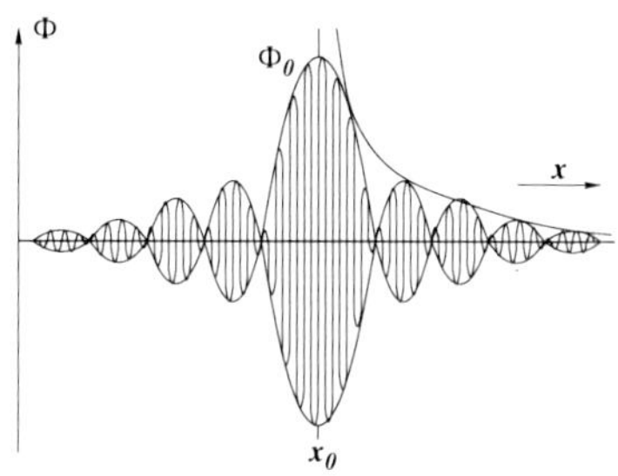

Figure 4.3: *One-dimensional section through a spherical wave packet with converging and diverging components*

$$\Phi = Ae^{i\omega t}\left\{\frac{\cos kr}{kr} \text{ or } \frac{\sin kr}{kr}\right\} \tag{4.30}$$

describes elementary waves, equivalent to the postulated elementary distortions of space, *i.e.* the elementary particles of atomic physics. Interpreted as an electron, the distance between nodal points represents $\lambda_{dB} = h/m_e v_g$, the de Broglie wavelength of a free electron and $\lambda_C = 2\pi/k = h/m_e c$, the Compton wavelength. The amplitude of the standing wave is proportional to the electronic charge.

$\Phi_0 = A$, in eqn.(4.30), represents a wave packet with charge proportional to 0 or $\pm A$. Electrons and protons, despite their difference in mass have charges of $\pm e$. The neutron is neutral. The field intensity $\Phi\Phi^* = A^2(\sin kr/kr)^2 = C/r^2$ defines the force between charges, in line with Coulomb's law, except when $r \to 0$. The breakdown of Coulomb's law, which occurs naturally for charged wave structures is equivalent to the special renormalization postulate of quantum field theory.

4.5.1 Interaction Theory

The most attractive part of the wave-field vacuum is that it provides a natural explanation to the vexing question of action at a distance. The seminal idea stems from the work of Tetrode (1922) who concluded that

> "...emission and absorption are mutually causative events and by each emission it is already predetermined, when, where and how the absorption will follow".

The same idea was advanced independently some years later by Gilbert Lewis (1926), stating

> "...that an atom never emits light except to another atom, and that it is as absurd to think of light emitted by one atom regardless of the existence of a receiving atom as it would be to think of an atom absorbing light without the existence of light to be absorbed".

The central idea in both proposals is that signals between emitter and absorber, going respectively forward and backward in time, could serve to establish two-way contact and facilitate the exchange of radiant energy as a *transmission* rather than emission. Both authors emphasized the fact that on the relativistic light cone, the vanishing world scalar

$$ds^2 = c^2dt^2 - dx^2 - dy^2 - dz^2 = 0\,,$$

implies that emitter and absorber, irrespective of spatial separation remain in virtual contact.

Two atoms in virtual contact, like two atoms in physical contact, have no problem to establish whether their relationship allows the transfer of energy, even though the time of emission and absorption may be separated by millenia.

The reader who feels uneasy about the time-inversion symmetry is reminded that the simplest mathematical distinction between matter and antimatter relates to an inverted time parameter. This is the convention used in the analysis of interactions by Feynman diagrams (Gottfried & Weisskopf, (1984).

The Estonian academician G.I. Naan (1964), on the basis of the Bohr-Lüders (1954) theorem, argued that the universe cannot exist without an element of CPT (Charge conjugation–Parity–Time) inversion symmetry, which implies the co-existence of material and anti-material worlds. Any interaction in the material world must be mirrored in the anti-world, but without direct contact between the two domains. Because of the inversion symmetry all conservation laws are automatically satisfied as invariant, at magnitudes of zero.

By this model each electromagnetic interaction was interpreted (Boeyens, 2005) as having an inverse in the anti-world. The CPT image of the conventional emitter is an anti-world absorber that transmits negative energy backwards in time, as shown in Figure 4.4.

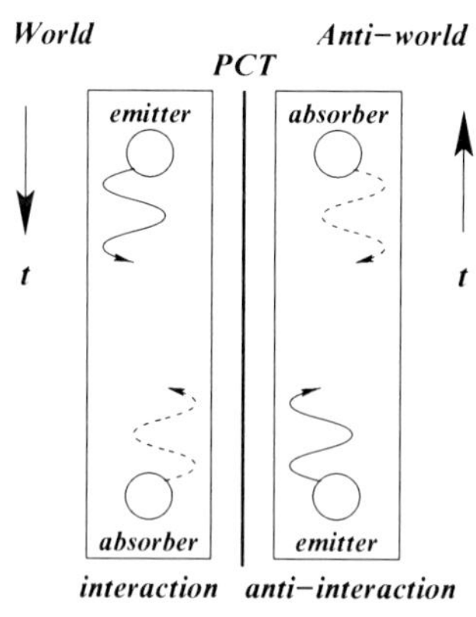

Figure 4.4: *Interactions in the world and anti-world*

This definition of an electromagnetic interaction corresponds exactly with the absorber theory of Wheeler and Feynman (1945). According to this theory an accelerated charge in otherwise charge-free space does not radiate electromagnetic energy unless acted on by a field arising from other particles. These fields are represented by one-half the retarded plus one-half the advanced solutions of Maxwell's wave equation. The law of force is symmetric with respect to past and future. It is assumed that there are sufficiently many particles to absorb completely the radiation emitted from the source. The consequences of this model have been considered in detail by Cramer (1986) and the formalism was shown to be, not only equivalent to standard quantum field theory, but also able to account for non-local interactions of the Einstein-Podolsky-Rosen (EPR) type. In simple words the interaction amounts to a handshake between emitter and absorber by means of a pair of retarded and advanced waves. These are the $+t$ and $-t$ solutions of equation (4.18). The advanced wave that retrogresses in time is emitted by the absorber when the retarded wave from the emitter arrives, and it reaches the emitter at the exact moment when the retarded wave is first emitted. The net effect is the transfer of one quantum of energy from emitter to absorber. The interaction is confined to the period between times of emission of the retarded (t_1) and advanced (t_2) waves and to the space between emitter and absorber. It can be viewed as a standing wave that exists for the period $t_2 - t_1$ between emitter and absorber, and is also known as a photon.

Cramer (1980) models non-local interaction at a distance as the vector sum of advanced and retarded waves, involving two absorbers and a single emitter, as in Figure 4.5, to generate a space-like connection between the two absorbers:

$$S_{12} = R_2 + A_1 \, .$$

The non-local space-like interaction S_{12} is the vector sum of a retarded signal

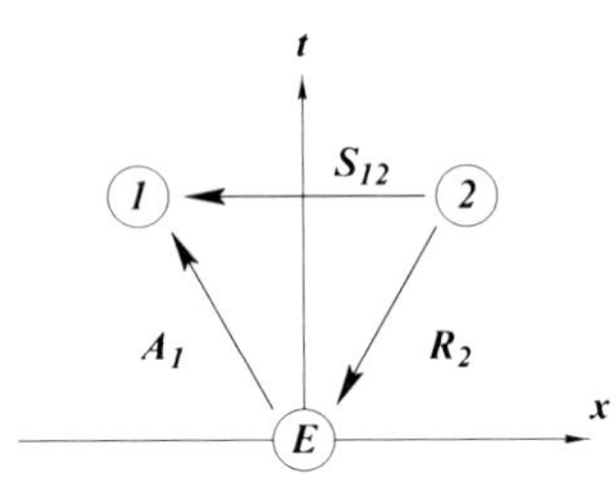

Figure 4.5: *Non-local interaction*

R_2 and the advanced response A_1. An experimental verification consists of measuring the spin of one of a previously correlated pair at position 2, and thereby predict the spin state at position 1. This interaction does not involve a transfer of energy.

Energy transfer only occurs if the emitter and absorber are at different energy levels. Most potential emitters and absorbers in the universe are at the same vacuum level. Waves emitted from such sites interfere amongst themselves to establish what is commonly known as the radiation field, which is in universal equilibrium. Any wave emitted by an electron is therefore balanced by the residual radiation from all other sources in the universe, constituting a resultant spherical wave that converges to the electron. Superposition of the diverging and converging spherical waves generates the wave packet of Figure 4.3. The spherically averaged wave packet represents the electron as shown schematically in Figure 4.6.

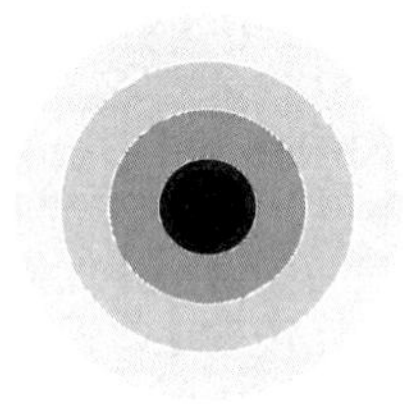

Figure 4.6: *Wave structure of a free electron with de Broglie wavelength* $\lambda_{dB} = h/m_e v_g$

The electron is here identified with a local maximum in the radiation field, immersed in a sea of virtual photons. By comparison, quantum field theory describes elementary particles as excitations of fields whose ground state is the vacuum.

The same ideas were used by Hoyle and Narlikar (1964) to explain Mach's principle in the context of general relativity. It is argued

> "...that particle couplings are propagated essentially along null geodesics – i.e. at *no* distance in four-dimensional sense".

To show that the mass of a particle a is not entirely an intrinsic property but arises from interaction with all other particles b in the universe, they write:

$$m_a(A) = \lambda_a \sum_{b \neq a} m^{(b)}(A) \,. \tag{4.31}$$

This means that at a typical world point A on the world line of a particle a, the mass it acquires is the net sum of contributions from all other particles in the cosmos. Note that, if a were the only particle in the universe, then $m_a = 0$, as in flat space.

If the superposition of point-source solutions is determined by a Green function[5], then by analogy with Wheeler–Feynman absorber theory the total interaction results from a symmetrical combination of retarded and advanced Green functions. On defining these Green functions to be compatible with space-time geometry (Riemann tensor) the interaction is shown to be consistent with Einstein's field equations.

This construction of a mass point is identical with Wolff's wave-packet.

4.6 Astrophysics

Without additional data the theory of gravitation by itself cannot predict any astronomical structures beyond the solar system. A serious lack is the estimation of distances to celestial objects, not part of the solar system.

By the end of the 18th century a sufficient number of distances had been measured to deduce that the solar system was part of a disc-like galaxy. By following the relative motion of stars with respect to nebular clouds it was discovered that, like the solar system, the entire Galaxy rotates about a centre.

At this stage it was still debatable whether the nebulae formed part of the Galaxy or not. As more powerful telescopes came into play some nebulae were discovered to have galactic structures of the same type as the Milky Way and with the introduction of spectroscopic methods into astronomy the apparent size of the known universe was extended by orders of magnitude.

4.6.1 Spectroscopy

The technique whereby starlight is resolved into a spectrum of wavelengths was developed into a major tool for the exploration of the cosmos. The quantum relationship between frequency of radiation and electronic levels dictates that atoms emit or absorb radiation at narrowly defined characteristic frequencies, readily identified in an observed spectrum. By this technique it was possible to demonstrate the presence of known elements in the sun by comparison with known atomic line spectra. However, on using the same technique for the analysis of galactic light it was observed that although the same absorption pattern was observed, the actual frequencies were shifted to lower values, towards the red part of the optical spectrum. Hubble made the

[5]The inverse of a differential operator.

momentous discovery that these so-called redshifts appeared to depend on the distance to the galaxy of interest, where this was known.

Using Hubble's law in the form that relates the shift in wavelength to distance, the linear relationship,

$$\Delta\lambda = HD\,,$$

distance measurement could be extended to any object with a measurable redshift, providing Hubble's constant, H, is known. It is important to note that the only astrophysical measurement in this application is the magnitude of the redshift. Hubble's law and all elaborations thereof are unproven assumptions.

Astrophysical spectroscopy is not limited to visible frequencies but also extend to ultraviolet, X-ray, γ-ray, infrared, microwave and radio frequencies, special applications of which are to be discussed.

4.6.2 Cosmic Rays

Particles, approaching from all directions, with relativistic energies and mass, known as cosmic rays, constitute a potentially important source of astrophysical data. Primary cosmic rays consist almost entirely of fully ionized atomic nuclei in abundancies that differ significantly from the solar abundances of the elements. The flux is isotropic and for low-energy particles it remains constant in time. Judging by their speeds, that approach c, cosmic rays should disappear from the Galaxy in a relatively short time, unless they are constantly regenerated. However, the source of cosmic rays remains conjectural and despite their high energy density they find no mention in cosmological theories. This is in stark contrast with the comparatively trivial isotropic microwave radiation, considered a pillar of standard cosmology.

The major component of cosmic rays is relativistic protons, which, not surprisingly, are accompanied by an equivalent amount of relativistic electrons. When such fast electrons move in a magnetic field, as in a synchrotron, they emit electromagnetic radiation at a frequency proportional to the magnetic field strength. This fact has been used to associate the isotropic radiofrequency radiation, which arrives from the Milky Way, to an interstellar magnetic field that pervades the Galaxy.

Another, linearly polarized, radio source in the Crab nebula, where a supernova explosion was observed in the year 1054, leaving behind a pulsating neutron star, could be of similar origin as the radiation in the Milky Way. The likelihood that the pulsar drives the acceleration of both relativistic electrons and of cosmic rays could imply that most cosmic rays in the Galaxy are also accelerated in pulsars associated with other remnants of supernovae. This possibility raises many unanswered questions of cosmological importance. We shall return to consider the nature of other poorly understood quasi-stellar objects as a function of their enormous gravitational and magnetic fields.

The origin of X-rays, γ-ray bursts and radiowaves that pervade the Galaxy is as mysterious as the cosmic-ray sources. Evidence is mounting that in most cases these phenomena are linked to violent explosions, thought to include neutron-star quakes, quark nuggets, supernovae, matter-antimatter annihilation, clumping of quasars and colliding stars. Of all space radiations radio phenomena have been studied in most detail.

4.6.3 Radio Galaxies

Radio sources are difficult to locate precisely in the sky because of their long wavelength and the low flux density in some cases. Parallax measurements are complicated by the same factors and the only reliable distances are obtained when optical counterparts to radio sources can be identified. Because of the large difference in resolution unique matching of radio and optical sources often is, either complicated, if not impossible.

The most important result of radio astronomy, to date, has been the identification of quasi-stellar objects, to be discussed next. In addition countless radio galaxies and blank radio fields have been identified. The blank fields are good radio sources without visible counterparts. It is not clear whether the blank fields contain radio galaxies or quasi-stellar objects.

A common feature of many radio sources is that they occur in pairs, symmetrically disposed, on opposite sides of an optical galaxy. The first and best example of such a pair is known as Cygnus A. This configuration suggests that an explosion in the optical galaxy drives the activity by generating fast particles that move out along jets, causing synchrotron emission in the radio-emitting regions.

A staggering amount of energy is required to drive such a process even if current estimates are wrong by orders of magnitude. Uncertain factors include the estimates of distance, electron flux, magnetic field strength and the role of cosmic rays. The most efficient process to generate such energies would be

(i) by the annihilation of equal masses of matter and antimatter. Current estimates require the annihilation of the rest-mass of a million solar masses. Other, less efficient processes, also listed by Sciama (1971), include:

"(ii) Catastrophic gravitational collapse followed by expansion (and perhaps including significant nuclear reactions.)

(iii) Rapid collisions in an assembly of closely packed stars.

(iv) A chain reaction of supernova explosions.

(v) Hydromagneto-gravitational instability (a large-scale version of a process proposed to account for solar flares.)

(vi) Instabilities associated with rapidly rotating highly magnetised bodies (in analogy with the rotating neutron star models of pulsars).

This fascinating problem is unlikely to yield to a stroke of genius".

What a surprising statement from an otherwise unbiased author.

4.6.4 Quasars

The search for radio galaxies led to the unexpected discovery of what became known as quasi-stellar objects, QSO's, or *quasars*. The first quasars to be discovered were all strong radio sources, but the majority known today are radio silent. More characteristically, most of them are X-ray sources. However, the most characteristic property of quasars is the unusually large redshifts of their optical emission spectra. Analyses of the physical properties of quasars and speculation about their nature are tainted by the conviction that these redshifts are fully accounted for by Hubble's law.

Apart from redshift, the two most conspicuous properties of quasars are their small angular diameters and high optical luminosity. Should the Hubble distance be real the absolute optical luminosity of a quasar, although a thousand times smaller in diameter, exceeds the output of even the brightest galaxy by orders of magnitude.

To complicate the picture, several quasars are observed to be variable in both radio and optical emission. The period of variation limits the physical size of the object to values far below the size implied by the angular diameter and the assumed Hubble distance.

A plausible rationalization of the discrepancy is that the quasar expands at a relativistic rate. However, an increasing number of recent observations of quasars that vary in intensity over periods of less than an hour suggest that, either the quasar is smaller than a few light-minutes in size, or else the disturbance, which causes the pulsation, travels many times faster than light. Hubble's law applied to quasar redshifts has another awkward consequence. The enormous distances from earth, implied by the uniformly large redshifts, also imply that all quasars are far removed in time; came into existence billions of years ago, and never again. This idea of an evolving universe served as a powerful argument against proposals of a steady-state universe. Record-breaking redshifts (Peterson, 1989) have now moved some quasars so far back in time as to be virtually contemporaneous with the big bang itself. What is inferred to be a very young quasar at its time, has a spectrum of the same type, indicating elements in the same abundance, as in older quasars. The rapid evolution of such an object is hard to rationalize. Big-bang theory does not allow the appearance of heavy elements before the disappearance of at least one generation of stars, with lifetimes in the billions of years.

We shall return to further anomalies associated with quasars and their spectra. At this stage it is worth noting that Hubble's law applied to quasars, whatever they are, is an unsolved problem of astrophysics.

4.7 References

Bergmann, P.G. (1976): *Introduction to the theory of relativity*, Dover edition, NY.

Boeyens, J.C.A. (2005): *New Theories for Chemistry*, Elsevier, Amsterdam.

Boeyens, J.C.A. (2010): *Emergent Properties in Bohmian Chemistry*, in M.V. Putz (ed.), *Quantum Frontiers of Atoms and Molecules*, NOVA, New York.

Bohm, D. (1952): *A Suggested Interpretation of the Quantum Theory in Terms of "Hidden variables"*, Phys. Rev. 1952 (85) 166–179, 180–193.

Boscovich, R.J. (1763): *A Theory of Natural Philosophy*, English translation of First Venetian Edition of, *Theoria Pilosophiæ Naturalis* (1763), by J.M. Child (1921), MIT Press, Cambridge, Mass. (1966).

Cramer, J.G. (1980): *Generalized absorber theory and the Einstein–Podolsky–Rosen paradox*, Phys. Rev.D, 1980 (22) 362–376.

Cramer, J.G. (1986): *The transactional interpretation of quantum mechanics*, Revs Mod. Phys., 1986 (58) 647–687.

Dirac, P.A.M. (1930): *A Theory of Electrons and Protons*, Proc. Roy. Soc., 1930 (126) 360–365.

Einstein, A. (1950): *The Meaning of Relativity*, 4th ed., 1950, Methuen, London.

Einstein, A. & W. Mayer (1931): *Einheitliche Theorie von Gravitation und Elektrizität*, Sitz. Ber. Preuss. Akad., 1931, 541–557.

Einstein, A. & N. Rosen (1935): *The Particle Problem in the General Theory of Relativity*, Phys. Rev., 1935 (48) 73–77.

Goldstein, H. (1980): *Classical Mechanics*, 2nd ed., 1980, Addison-Wesley, Reading Mass., p.486.

Gottfried, K. & V.F. Weisskopf (1984): *Concepts of Particle Physics*, Clarendon, Oxford.

Holland, P.R. (1993): *The Quantum Theory of Motion*, University Press, Cambridge.

Hoyle, F. & J.V. Narlikar (1964): *A new theory of gravitation*, Proc. Roy. Soc. 1964 (A282) 191–207.

Lewis, G.N. (1926): *The Nature of Light*, Proc. Nat. Acad. Sci. U.S., 1926 (12) 22–29.

Lüders, G. (1954): *On the equivalence of invariance under time reversal and under particle-antiparticle conjugation for relativistic field theories*, Dan. Mat. Fys. Medd., 1954 (28) 1–17.

Naan, G.I. (1964): *The Symmetric Universe*, Publication of Tartu Astronomic Observatory, Volume XXXIV, #6, 1964, Tartu, 13 pages, (Translated into English by Olga Vasilyeva.)

Peterson, I. (1989): *Quasar Illuminates the Most Distant Past*, Science News, 1989 (136) 340.

Pauli, W. (1981): *Theory of Relativity*, translated from the German original by G. Field, Dover edition, NY, 1981.

Popper, K.P. (1930): *Quantum Theory and the Schism in Physics*, Reprinted by Routledge, London (1995).

Schwarz, M. (1972): *Principles of Electrodynamics*, 1972, McGraw-Hill; Dover edition, 1987.

Sciama, D.W. (1971): *Modern Cosmology*, University Press, Cambridge.

Tetrode, H. (1922): *Über den Wirkungszusammenhang der Welt*, Z. Phys., 1922 (10) 317–328.

Wheeler, J.A. & R.P. Feynman, *Interaction with the Absorber as the Mechanism of Radiation*, Revs Mod. Phys., 1945 (17) 157–181.

Wollf, M. (1996): *Beyond the Point Particle – A Wave Structure for the Electron*, Galilean Electrodynamics, 1995 (6) 83–91.

Chapter 5

Chemical Evidence

5.1 Chemistry and Cosmology

The only chemists known to have contributed seriously to cosmology, are Svante Arrhenius (1908) with the eminently sensible suggestion that new stars and planets arise from the debris of previous cycles, and Harold Urey (1952) who advanced chemical arguments to support the idea of aggregation in the spiral arms of a central core that rotates in a nebular cloud. By way of contrast it is noted that medieval cosmology was intimately entangled with alchemy, astrology, religion, creation and natural philosophy (Priesner & Figala, 1998). During the Renaissance a link to Christian theology was added.

The fundamental assumption was that *Chaos* represented *apeiron*, the boundless, infinite and indeterminate original state of the Universe, before the beginning of space and time. From Chaos emerged the first god Chronos, who produced the World Egg of Alchemy. As the god Zeus hatched from the cosmic egg, the shell transformed into Heaven, the skin into Earth, leaving behind *Materia prima*, the primeval matter that was turned into the four elements through the action of *forma*, which gave form to shapeless matter. Philosophically, creation brought order, or *Cosmos*, into Chaos.

Another interpretation pictures transformation of the egg into four metals, which by the operation *opus magnum* produced the philosopher's stone. In the other direction it produced the four elements and their associated Platonic solids, as shown in the scheme overleaf.

The cube is constructed by the fusion of four right-angled triangles and the building block of the other solids is the isosceles triangle consisting of six right-angled triangles of a second kind:

J.C.A. Boeyens, *Chemical Cosmology*, DOI 10.1007/978-90-481-3828-9_5,

$$
\textit{Lapis Philosophorum} \leftarrow \left\{
\begin{array}{ccccccc}
\mathrm{Cu} & \leftarrow & \mathrm{Shell} & \rightarrow & \mathrm{Earth} & \rightarrow & \mathrm{Cube} \\
\mathrm{Fe} & \leftarrow & \mathrm{Skin} & \rightarrow & \mathrm{Air} & \rightarrow & \mathrm{Octahedron} \\
\mathrm{Pb} & \leftarrow & \mathrm{White} & \rightarrow & \mathrm{Water} & \rightarrow & \mathrm{Icosahedron} \\
\mathrm{Sn} & \leftarrow & \mathrm{Yolk} & \rightarrow & \mathrm{Fire} & \rightarrow & \mathrm{Tetrahedron}
\end{array}
\right.
$$

This explains why earth (*caput mortuum*) differs from the interconvertible triad of other elements. This restriction was relaxed in another interpretation whereby each element is convertible into an immediate neighbour by changing one of its two qualities:

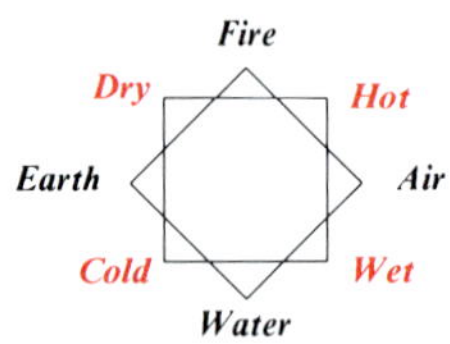

Conversion of an element into its antagonist is impossible.

The four elements are not the same as the substances with the same name that occur on Earth. The primary elements can be combined in any proportion to produce the known variety of homogeneous substances. Whereas the four elements are in constant interaction with each other and hence interconvertible, heavenly bodies do not consist of earthly elements, are immutable, circle the Earth on perfect orbits and consist of *quintessence*, the fifth heavenly element. The alchemists adopted the four-element theory, which offered the possibility of transmutation of metals. As even the elements can be interconverted the same must be true for metals, which are homogeneous bodies and, although not elements, are made up of the four elements. The Arabian alchemist Ibn Sina (Avicenna), who questioned the transmutation of metals, explained that metals are formed by the combination of sulphur (air + water) and mercury (fire + earth), which derive from the four elements. In order to make the theory applicable, not only to metals, but to all substances, Paracelsus[1] added salt as the third element of alchemy. Again, his *tria prima* of salt (body), sulphur (soul), and mercury (spirit) are not

[1]Theophrastus Bombastus von Hohenheim (1493–1541).

the chemical substances of the same name, but abstractions of the respective qualities: incombustibility and non-volatility; inflammibility; fusibility and volatility.

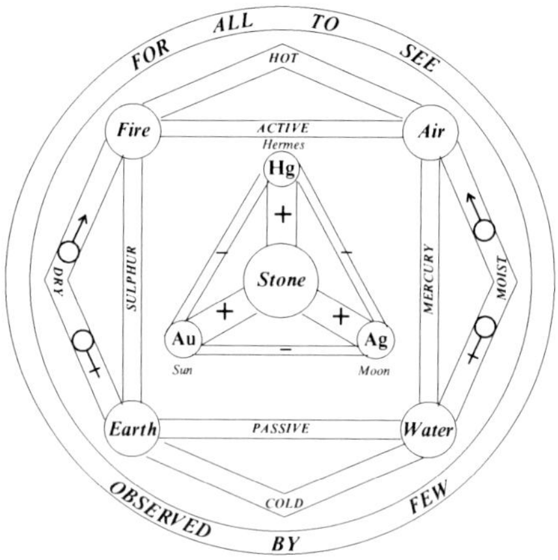

Figure 5.1: *Adapted from the original in Leiden, reproduced in Priesner & Figala (1998)*

Paracelsus also gave new meaning to alchemy by interpreting the book of Genesis as a creation event that represents the separation of up and down, good and bad, male and female, providing evidence of divine alchemy. The *opus magnum*, not only produces gold, but also leads to the purification, refinement and salvation of people as well as minerals and metals. The philosopher's stone becomes Christ, the saviour. Like human destiny is controlled astrologically by the progress of planets through the constellations, the substances and processes of alchemy are linked to the stars.

In the same way that the macrocosm is constructed around the central Earth, the microcosm of people and alchemy is built around the central *Lapis Philosophorum*. The role of the philosopher's stone in the regulation of *opus magnum* changed with time, being interpreted progressively as yeast, medicine, redeemer, accelerator and finally as catalyst.

The cosmological importance of alchemy is demonstrated by the cosmic diagram of the Philosopers' Stone, (Figure 5.1).

The pattern is there for all to see, but only appreciated by a few initiates[2].

[2]Omnibus sed paucis luceo.

The nature of the philosopher's stone depends on the positive interaction (+) of Sun, Moon and Hermes Trismegistos, the mythical founder of alchemy, whose name appears on the Rosetta stone and who represents the syncretic link between the Greek god Hermes and the Egyptian moon god Thot. Active masculine qualities are separated from passive feminine qualities. The relationship between Avicenna's two-element model and the four elements of Aristotle is shown. The total pattern resembles the incommensurable nesting of Platonic solids within the celestial sphere.

This incommensurable relationship between polygons and the limiting circle reflects the unbridgeable gap between gods and humans, interpreted to demonstrate that heavenly bodies move in perfect circles and differ from earthly objects in composition. Alchemy, astrology, the Aristotelean model of the cosmos and the four-element theory assume a hidden mystical link between heaven and earth, only revealed through pure belief in its magic.

This tradition was finally broken by Kepler who developed the contention of Pythagoras and Plato that a proportional relationship exists between polyhedron and circle. In time, this approach led to modern astronomy and, in the hands of Robert Boyle (1661) and Antoine Lavoisier (1789), to chemistry. However, the victory of science over magic has by no means been final[3]. Modern big-bang cosmology, like the chain of assumptions in computational quantum chemistry (Rouvray, 2009), are rapidly reverting to the occult approach. An alternative model, emerging in chemistry, is based on the golden ratio, which both Kepler and Leonardo da Vinci referred to as the *Divine Proportion.*

5.2 Periodicity

Chemists of the 19th century stumbled onto one of the fundamental properties of the world – an all-pervasive periodicity. Mathematically, periodicity

[3]There is sufficient evidence to show that Boyle, like his contemporary, Newton, never gave up the idea of transmuting base metals into gold. Modern-day alchemists, following Paracelsus and relying on cold fusion, are still experimenting with the reaction between salt, sulphur and mercury to make gold as in the following scheme (with proton count in parentheses):

$$\begin{aligned}\mathrm{NaCl}(28) + \mathrm{S}(16) &\rightarrow \mathrm{NaS}(27) + \mathrm{Cl}(17)\uparrow \\ \mathrm{NaS}(27) + \mathrm{Hg}(80) &\rightarrow \mathrm{NaCl}(28) + \mathrm{Au}(79)\end{aligned}$$

develops in any closed function. Periodic physical systems include wave motion, crystal structure, planetary orbits, life cycles, weather patterns, and finally, practically all systems of interest. It may reasonably be assumed that observed periodicity is always the result of some closed system. In the final analysis we are looking at a closed universe.

Chemical periodicity was first observed to occur as a function that regulates the properties of the chemical elements. At the time the periodic laws were assumed to apply equally to families of molecular fragments, called radicals, as well as to the elements. By the middle of the 19th century Jean-Baptiste-André Dumas argued that groups of elements can act in unison as a single element and, using atomic (and molecular) weights on the hydrogen (H=1) scale, demonstrated related periodicities of hydrocarbons and elements (Stewart, 1922), as in Table 5.1. The true nature of elemental

Table 5.1: *An early notion of chemical periodicity*

n	Hydro-carbon	MW $(a+nb)$	Element	AW $(a+b+nc)$	n
0	H_2	2	Nitrogen	14	
1	CH_4	16=2+14	Phosphorus	14+17=31	0
2	C_2H_6	$30 = 2 + 2 \times 14$	Arsenic	14+17+44=75	1
3	C_3H_8	$44 = 2 + 3 \times 14$	Antimony	14+17+88=119	2
4	C_4H_{10}	$58 = 2 + 4 \times 14$	Bismuth	14+17+176=207	4

periodicity only became apparent after the discovery of atomic numbers. By 1914 Johannes Rydberg characterized the noble gases in terms of a simple formula based on atomic number, interpreted as the number of extranuclear electrons in an atom.

The Rydberg sequence, shown in Table 5.2, was later explained, in part, by the wave-mechanical solution of the electronic energy levels for the hydrogen atom. The calculated sequence, in terms of the quantum numbers $n = 1, 2, 3\ldots$ and $l < n$, defines the sub-levels $s(l = 0)$, $p(l = 1)$, $d(l = 2)$ and $f(l = 3)$ with multiplicities of $2(2l + 1)$ and predicts the noble-gas electron configurations correctly, assuming (ns^2np^6), also shown in Table 5.2.

However, the inferred electronic configuration for the elements of periodic groups 1 to 10 and for the lanthanides is not in line with this sequence. It is therefore of interest to note that the observed periodic pattern occurs in the arrangement of Ford circles, as defined by the Farey sequence of rational fractions.

Table 5.2: *The Rydberg sequence*

Element	Formula	n	Atom	Configuration	Z
Helium	$2(1^2) = 2$	1	He	$1s^2$	2
Neon	$2(1^2 + 2^2) = 10$	2	Ne	$[\text{He}]2s^2 2p^6$	10
Argon	$2(1^2 + 2^2 + 2^2) = 18$	3	Ar	$[\text{Ne}]3s^2 3p^6$	18
Krypton	$2(1^2 + 2^2 + 2^2 + 3^2) = 36$	4	Kr	$[\text{Ar}]3d^{10} 4s^2 4p^6$	36
Xenon	$2(1^2 + 2^2 + 2^2 + 3^2 + 3^2) = 54$	5	Xe	$[\text{Kr}]4d^{10} 5s^2 6s^6$	54
Niton	$2(1^2 + 2^2 + 2^2 + 3^2 + 3^2 + 4^2) = 86$	6	Rn	$[\text{Xe}]4f^{14} 5d^{10} 6s^2 6p^6$	86

A pair of unimodular rational fractions h_i/k_i ($i = 1, 2$) in which the determinant $\begin{vmatrix} h_1 & h_2 \\ k_1 & k_2 \end{vmatrix} = 1$, define two kissing Ford circles with radii and central y-coordinates of $1/2k_i^2$ and x-coordinates h_i/k_i. All adjacent fractions in a Farey sequence,

$$\mathcal{F}_n = \left\{ \frac{0}{1}\ \frac{1}{n}\ \frac{1}{n-1} \cdots \frac{1}{2} \cdots \frac{n-2}{n}\ \frac{n-1}{n}\ \frac{1}{1} \right\}$$

are unimodular and generate an infinite set of Ford circles. The Farey se-

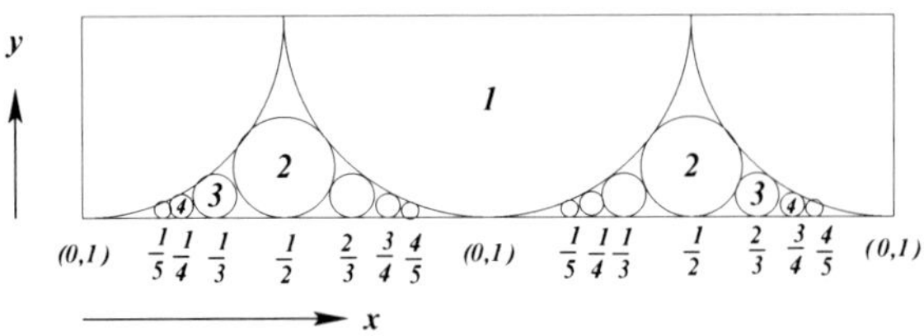

Figure 5.2: *Ford-circle mapping of the periodic table*

quence defined on an affine lattice (Figure 3.4) is represented by a symmetrical periodic arrangement of Ford circles. The numbered set of circles in Figure 5.2 is a mapping of the periodic table of the elements. Reciprocals of the decreasing radii ($1/y_j$, $j = 1, 4$) of these circles represent the number of elements per periodic group, exactly as in the Rydberg sequence.

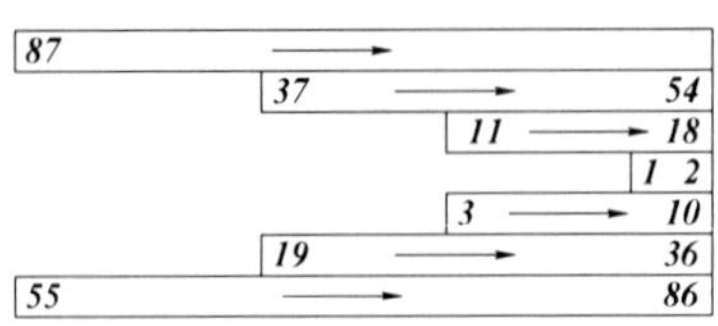

There is an obvious convergence of Ford circles of diminishing size around the central circle at $x = 0, 1$. Self-similar convergence occurs around each of the smaller circles. Of particular importance is the convergence around the circle at $x = 3/5$, shown in Figure 5.3. On one side it follows the unimodular fractions F_n/F_{n+1}, defined by the Fibonacci series:

$$F_n = \{0, 1, 1, 2, 3, 5, 8, 13\ldots\}$$

i.e. $$\left\{\frac{1}{1}, \frac{1}{2}, \frac{2}{3}, \frac{3}{5}, \frac{5}{8}, \cdots \to 0.61803\ldots\right\}$$

which converges to the golden mean, τ. On the other side it converges like $L_n/\left(L_n + F_{n+1}\right)$ with the Lucas sequence

$$L_n = \{2, 1, 3, 4, 7, 11, 18, \ldots\}$$

i.e. $$\left\{\frac{2}{3}, \frac{1}{2}, \frac{3}{5}, \frac{4}{7}, \frac{7}{12}, \frac{11}{19}, \cdots \to 0.5802\ldots\right\}$$

It will be shown that, on interpreting Farey fractions as representing the ratio

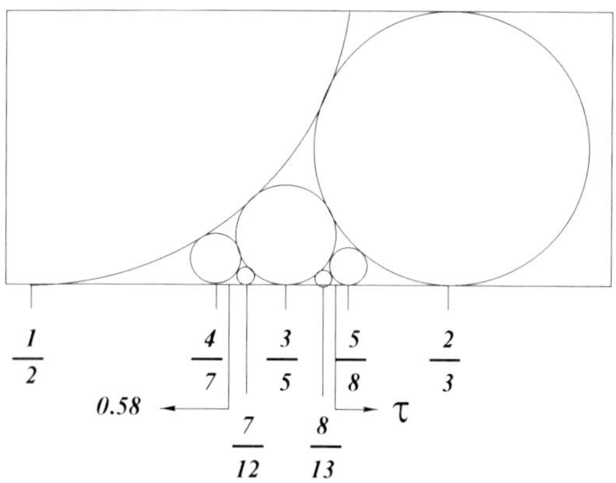

Figure 5.3: *Convergence to τ and to 0.5802...*

of protons:neutrons (Z/N), the observed and wave-mechanically calculated periodic tables of the elements are projected out at these two limiting ratios of $0.61803\ldots$ and $0.5802\ldots$ respectively.

The first indication that the periodicity of atomic matter depends on the proton:neutron ratio was discovered by William Harkins, a decade before the discovery of the neutron. He found that the ratio $(A - Z)/A$ never exceeds 0.62 for any stable nuclide. More precisely, it can be shown that the neutron imbalance, formulated as either Z/N or $(N-Z)/Z$, converges to the same value. From $Z/N = (N-Z)/N$, *i.e.* $Z^2 + NZ - N^2 = 0$, this value follows as $Z = N\tau$, *i.e.* $Z/N \rightarrow 0.61803\ldots$, the golden ratio.

The distribution of stable nuclides in a plot of Z/N *vs* A shows (Boeyens & Levendis, 2008) that the periodicity, observed as a function of atomic number, is a subset of a more general periodic relationship that groups the 264 stable nuclides into 11 sets of 24. When plotted as a function of Z, rather than A, the 11 hem lines that separate the groups of 24 are no longer parallel to the Z/N-axis and their points of intersection with lines parallel to Z, depend on Z/N. By extending the hemlines from $Z/N = 0$ to 1.04 five special ratios are recognized with intersection patterns shown in Table 5.3. The

Table 5.3: *Periodic patterns in a plot of Z/N vs Z*

	Points of Intersection										
Z/N	1	2	3	4	5	6	7	8	9	10	11
1.04	14	24	32	38	56	62	78	88	96	102	102
	$4f$	$3d$	$1s$	$5f^6$	$4d$	$3p$	$6f$	$5d$	$3s$	$7f^6$	
1.00	14	24	32	38	54	60	76	84	94	100	100
	$4f$	$3d$	$1s$	$3p$	$5f$	$4p$	$5p$	$6p$	$5d$	$4p$	
0.62	10	18	28	36	38	46	48	54	62	70	80
	$2p$	$3p$	$3d$	$4p$	$5s$	$4d$	$5s$	$5p$	$4f^6$	$4f$	$6s$
0.58	10	18	28	36	36	46	46	52	60	68	78
	$2p$	$3p$	$3d$	$4p$		$4d$		$4f^6$	$4f$	$5p$	$5d$
0	4	10	24	32	10	24	4	10	16	24	48

construction is shown in Figure 5.4(a). The points along $Z/N = \tau \simeq 0.62$ are immediately recognized as the atomic numbers that signal the saturation of prominent electronic energy levels according to the observed periodic table of the elements. The electronic levels for the H atom, as calculated by the Schrödinger equation, assuming an exclusion principle, do not correspond exactly with the observed periodic levels, being:

$1s^2$	$2s^2$	$2p^6$	$3s^2$	$3p^6$	$3d^{10}$	$4s^2$	$4p^6$	$4d^{10}$	$4f^{14}$	$5s^2$	$5p^6$	$5d^{10}$
2	4	**10**	12	**18**	**28**	30	**36**	**46**	**60**	62	**68**	**78**

This predicted sequence of energy levels is reproduced exactly by the points of intersection along $Z/N = 0.58$. As for $Z/N = 0.62$ one of the periods is closed at the configuration $4f^6$. The points at $Z = 38$ and 48 do not occur at $Z/N = 0.58$ and these are precisely the points at which the observed

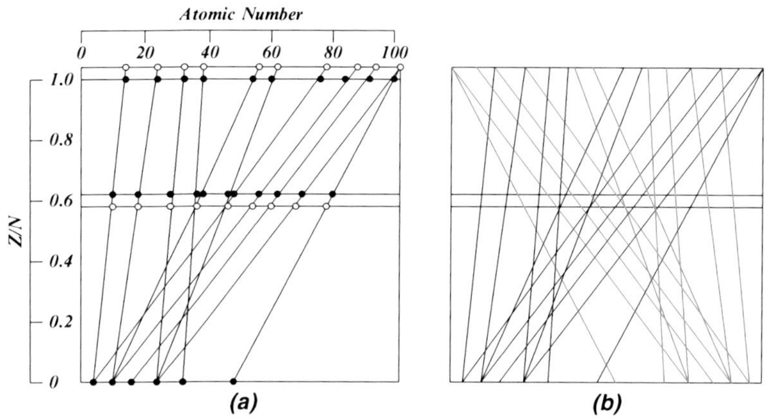

Figure 5.4: *The 264 non-radioactive nuclides, which occur as 11 groups of 24, are plotted in a field defined by the axes $x = Z$ and $y = Z/N$. The straight hemlines that separate the groups of 24 are those shown in plot (a). Reflection of this pattern in the symmetry line at $Z = 51$ generates diagram (b)*

periodicity differs from the Schrödinger spectrum. The periodic table of the elements predicted by this graphical analysis is shown in Figure 5.5.

The atomic numbers along $Z/N = 1.04$ correspond to an inversion of the H energy levels, as in the sequence:

4*f*	3*d*	2*p*	1*s*	5*f*	4*d*	3*p*	2*s*	6*f*	5*d*	4*p*	3*s*
14	**24**	30	**32**	46	**56**	**62**	64	**78**	94	**96**	102

Only numbers 38 and 102 from table 5.3 are not accounted for, but, as before, they correspond to $5f^6$ and $7f^6$ respectively.

The differences between $Z/N = 1$ and 1.04 are of a minor nature, like those between 0.62 and 0.58. In fact, the arrangement at 1.0 can be interpreted as inverse to 0.62. The points at $Z/N = 0$ have been interpreted in terms of nucleon energy levels.

The points along 1.04 appear to be arranged symmetrically around $Z = 51$. In fact, if full symmetry is established by the addition of a few points, an inverse relationship can be defined in either direction: $0 \rightarrow 102$, or backwards.

Periodic Table of the Elements

	1	2	3	4	5	6	7	8	
1s	1 H	2 He							
2s	3 Li	4 Be	5 B	6 C	7 N	8 O	9 F	10 Ne	*2p*
3s	11 Na	12 Mg	13 Al	14 Si	15 P	16 S	17 Cl	18 Ar	*3p*
4s	19 K	20 Ca							
	21 Sc	22 Ti	23 V	24 Cr	25 Mn	26 Fe	27 Co	28 Ni	*3d*
4s	29 Cu	30 Zn	31 Ga	32 Ge	33 As	34 Se	35 Br	36 Kr	*4p*
5s	37 Rb	38 Sr							
	39 Y	40 Zr	41 Nb	42 Mo	43 Tc	44 Ru	45 Rh	46 Pd	*4d*
5s	47 Ag	48 Cd	49 In	50 Sn	51 Sb	52 Te	53 I	54 Xe	*5p*
6s	55 Cs	56 Ba	57 La	58 Ce	59 Pr	60 Nd	61 Pm	62 Sm	
	63 Eu	64 Gd	65 Tb	66 Dy	67 Ho	68 Er	69 Tm	70 Yb	*4f*
	71 Lu	72 Hf	73 Ta	74 W	75 Re	76 Os	77 Ir	78 Pt	*5d*
6s	79 Au	80 Hg	81 Tl	82 Pb	83 Bi	84 Po	85 At	86 Rn	*6p*
7s	87 Fr	88 Ra	89 Ac	90 Th	91 Pa	92 U	93 Np	94 Pu	
	95 Am	96 Cm	97 Bk	98 Cf	99 Es	100 Fm	101 Md	102 No	*5f*

Figure 5.5: *Compact form of the Periodic Table*

0	6	14	24	32	38	40	46	56	62	64	70	78	88	96	102
$\rightarrow$	$4f^6$	$4f$	$3d$	$1s$	$5f^6$		$5f$	$4d$	$3p$	$2s$	$6f^6$	$6f$	$5d$	$3s$	
	$3s$	$5d$	$6f$	$6f^6$	$2s$	$3p$	$4d$	$5f$		$5f^6$	$1s$	$3d$	$4f$	$4f^6$	$\leftarrow$
102	96	88	78	70	64	62	56	46	40	38	32	24	14	6	0

It is tempting to interpret the two-way periodicity with respect to matter and antimatter and to identify atomic numbers 0 and 102 to define a closed system. The induced periodicity is given by

$$z = \sin 2\pi \left(\frac{i-3}{32} \right) , \; i = 0, 102.$$

This function implies that the natural elements are restricted to atomic numbers $0 \rightarrow 101$, with the exception of numbers 43 and 61. The total of 100 natural elements include the neutron at $Z = 0$. Nodal points of the periodic function are regularly spaced at intervals of 16 between atomic numbers 3 and 99. It has been known for more than a century that the periodic table of the elements starts with such a non-periodic section. An equivalent pattern is obtained by reflecting the points on the zero line in the point $Z = 51$.

On looking for a rationale behind the hundred natural elements one arrives at an almost mystical interpretation of material periodicity. Of relevance are the fixed number of elements (100), the 11×24 periodicity of the stable nuclides and the 6×16 periodicity at $Z/N = 0, 1$.

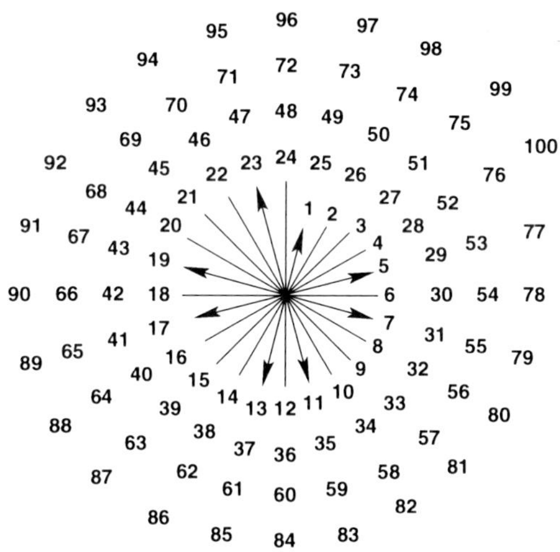

Figure 5.6: *Natural numbers arranged on a spiral of period 24. Arrows identify the eight radial prime-number directions*

In Figure 5.6 we arrange the natural numbers along a spiral with a pitch of 24. All prime numbers, except for 2 and 3, occur on eight radial lines as $p = 6n \pm 1$. By mapping the natural elements to these radial lines the periodicity of $16 = 2 \times 8$ is accounted for at the same time as the nuclide periodicity of 24. This arrangement is known as Plichta's prime-number cross. It has the remarkable property that the sum of all numbers over any complete cycle is given by

$$\sigma_{j+1} = \sum_{n=24j}^{24(j+1)} n = (2j+1)a\,,\ a = 300\,,\ j = 0, 1, 2\ldots$$

The coefficients $(2j+1)$, $j = 0 \rightarrow 3$, match the numbers of s, p, d, f electron pairs for a total of a atoms. The constant $a = 300$ is interpreted to count the maximum number of stable isotopes of the $a/3 = 100$ natural elements. It is important to note that in forming the sums σ_{j+1}, the numbers $24j$ and $24(j+1)$ are both counted in each sum, which defines a closed atomic cycle, as inferred before.

The unexplained features of only 81, rather than 100, stable elements observed in Nature, and the relationship between the periodic functions that occur at $Z/N = 1.04, 1.0, 0.62$ and 0.58 are explained later on.

5.2.1 Cosmic Implications

Figure 5.4 signifies more than elemental or nuclide periodicity. It summarizes the appearance of ponderable matter in all modifications throughout the universe. Following the extended hemlines from top left at $Z/N = 1.04 \rightarrow$ bottom left at $0 \rightarrow$ top right at $Z/N = 1.04 \rightarrow$ bottom right at 0, and back to top left, the involuted closed path, which is traced out, is mapped to the non-orientable surface of a Möbius band in Figure 5.7. The two sides of the double cover are interpreted to represent both matter and antimatter.

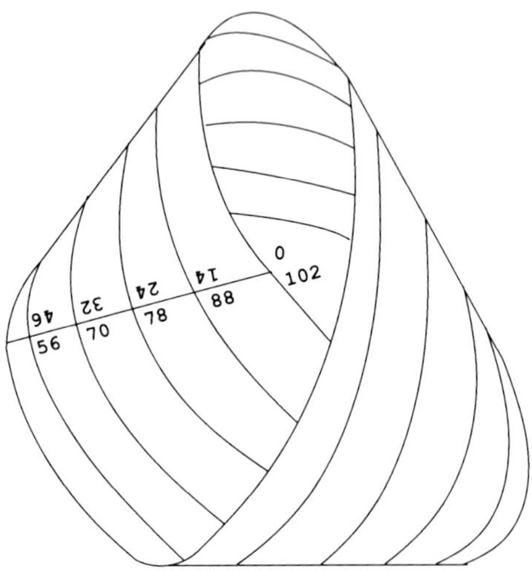

Figure 5.7: *The Periodic Function mapped to a Möbius band*

Intersection of the hemlines with the line at $Z/N = 0$ was shown (Boeyens and Levendis (2008) to generate the magic numbers of nuclear physics. If therefore, the ratio $Z/N = 0$ is associated with nuclear matter (neutrons), it is necessary to interpret this ratio to also coincide with a domain of high pressure and curvature, which implies that Z/N at $1.04 \equiv Z/N$ at 0.

As the total field is inferred closed in both the Z, as well as the Z/N directions, the Möbius model is incomplete and should be expanded into a projective plane, which cannot be embedded in 3-dimensional space. Like the physical universe, the cosmic distribution of matter should then also be specified in four-dimensional space-time. The reconstruction of Figures 5.4 and 5.7 can therefore, at best, be seen as a three-dimensional caricature of the actual four-dimensional distribution in the curved Minkowski space of general relativity.

In our pseudo-Euclidean tangent space it is customary to distinguish between time-like, space-like and light-like events, well aware that this is another caricature of curved space-time. For convenience, space-like events are usually ignored as physically unreal. There is no justification for this assumption in four-dimensionally curved space-time.

The separation of time-like and space-like events creates the impression of two types of response to increased curvature of space-time. If only space coordinates are curved it results in the inversion of the time coordinate and the conversion of matter into antimatter. This situation will be encountered in the Schwarzschild solution of the gravitational field equations, which serves as a model of a black hole, and assumed here to account for an inversion at $Z/N = 1.04$. It resembles the limitless time-like accumulation of matter, resulting in a space-time singularity.

The alternative of increased space-like curvature results in relentless compression that compacts all matter into infinitesimal space with parity inversion at $Z/N = 0$. Our contention is that these alternatives are inseparable in curved space-time where inversion of total CPT symmetry occurs on compression.

Analysis of the periodicity of atomic matter therefore guides us to a projective model of a closed universe in the double cover of four-dimensional projective space-time. Transport across the interface, or along the involution, results in the inversion of CPT symmetry.

5.2.2 Chemical Redshifts

Response of the electronic configuration of atoms to extreme pressure may turn out to be of crucial importance in cosmology, which is currently based almost exclusively on a single observation — the spectroscopic variability of galactic light.

Electromagnetic radiation provides the only contact that we have with objects outside the solar system. To interpret the incoming signals it is necessary to make a spectroscopic analysis of the radiation in terms of its wavelength, or frequency. Interpreted in terms of the Bohr frequency condition

$$\Delta E = h\nu = R\left(\frac{1}{n_1^2} - \frac{1}{n_2^2}\right) , \tag{5.1}$$

an atom in an excited state, on relaxation, emits radiation of characteristic frequency, which is spectroscopically observed as a line spectrum. Alternatively, an atom in a cool environment absorbs the same frequency from a polychromatic light beam, giving rise to dark (Fraunhofer) lines in the observed spectrum. Both types of spectrum are observed from astronomical sources. The Fraunhofer lines in sunlight gave the first indications of the elemental composition of the solar corona. The same absorption lines occur in galactic starlight. Although the characteristic patterns are easily recognized, they appear shifted to lower frequencies, said to be redshifted.

Emission lines are observed in the light from so-called active galaxies, also known as Seyfert galaxies. Emission spectra are also emitted by quasars or quasi-stellar objects (QSO). Like the absorption lines of normal galaxies these emission lines are also redshifted.

A feasible cause of redshifts is a relative change in the energy levels of an atom. As noticed before such a change could be effected under conditions of high pressure, if it affects the quantum levels n_1 and n_2 of (5.1) differently – which it does.

The shift in electronic energy levels of an atom depends not only on pressure (space-time curvature), but also on the angular-momentum quantum number l. Spectroscopic changes induced by high pressure must clearly be a complicated function. However, the response of a given absorption frequency should be a regular function of environmental pressure, although not necessarily linear. Calculations (Boeyens, 1994) have shown that increased pressure pushes all electronic levels towards the ionization limit, causing a general decrease in level spacing, ΔE. This implies a redshift that increases with pressure. These chemical redshifts, which depend on the local gravitational field at the source, should not be confused with relativistic gravitational redshifts.

By this argument there is a fundamental difference between the emission redshifts of quasars and the absorption redshifts of galaxies. In the case of quasars the chemical redshift results from the high gravitational field inside the quasar. In the case of the more diffuse galaxies the absorption that leads to measured redshifts occurs in the nebular intragalactic gas clouds, at lower gravitational fields. This difference immediately explains the many examples of physically connected pairs, of quasar and galaxy, with wildly different redshifts (Arp, 1982).

More recently a controversy flared up over anomalous Fraunhofer lines of metal atoms such as Fe, Cr and Zn, associated with quasar light on passing through an interstellar cloud (Webb et al., 2001). The puzzling feature here is that the absorption maxima are not uniformly displaced in the same direction or by the same amount. The mystery disappears when we recall that each atom responds differently to compression (Boeyens, 1994).

The gravitational field that exists in the nebula causes a characteristic shift for each metal, which cannot be explained by assuming a constant redshift for all metals. No other satisfactory explanation of the observed effect has been reported. Where others look at a time-dependent fine-structure constant or variable c or $\hbar$ to account for the modified spectra, we ascribe the observation to a simple response to space-time curvature.

Any astronomically measured frequency shift consists of several components, including the chemical shift, described here. Other contributions include relativistic gravitational redshifting, a distance-dependent redshift caused by the topological curvature of space-time, and a Doppler shift where the source is in relative motion.

All cosmological models ignore the chemical shift and consequently overemphasize the importance of the other components. In most analyses no distinction is made between the topological and Doppler shifts, treated together as a so-called cosmological redshift. The result of this is a vastly inflated astronomical distance scale.

5.3 Self-Similarity

One of the most important features of chemical periodicities is the prominent role of the golden ratio and its relationship with the curvature of space. As a special case we mention the self-similarity that exists between the Ford circles of Figures 5.2 and 5.3. Whereas the latter are barely visible on the scale of Figure 5.2 exactly the same pattern is seen to emerge in the two cases. This is reminiscent of the self-similarity that has been noticed between biological growth structures and galactic spirals. We note the operation of self-similar symmetry between periodic atomic structures, nautilus shells, the morphological variability of biological species (Thackeray, 2007) and spiral galaxies, from which we infer a general curvature of space, characterized by the golden mean. The logical assumption is that all systems which develop in the curved surface of space must acquire self-similar shapes that reflect the topology of space.

An immediate test of the hypothesis is provided by viewing the solar system as intermediate between biological and galactic spirals and looking for self-similar growth patterns at all three levels.

5.3.1 The Solar System

The rings of Saturn and Kepler's model of planetary motion inspired the early atomic theories of Nagaoka, Rutherford and Bohr. The pioneering models have now been superseded by more fundamental wave-mechanical descriptors, which characterize extranuclear atomic charge distributions through a set of integer quantum numbers, characteristic of standing waves. The vivid similarity between atomic and celestial structures, which faded in the process, has now been highlighted again (Boeyens, 2009) by the observation that the motion of planets, moons and rings in the solar system is also controlled by quantum-like integers.

The first indication of a regular numerical pattern, which summarizes the mean planetary orbital radii in astronomical units (1au$=1.496\times10^8$km) by the formula

$$r = a + bc^n \quad (a = 0.4,\ b = 0.3,\ c = 2,\ n = -\infty, 0, 1, 2, \ldots),$$

was reported independently by Titius and Bode in the 18th century. There was no known planet on the orbit predicted by the formula for $n = 3$. The subsequent discovery of the minor planet Ceres and the entire asteroid belt on this orbit, between Mars and Jupiter, inexplicably failed to convince the sceptics, who still dismiss the regularity as coincidence. A simpler, more accurate version of the Bode–Titius (BT) law, as rational fractions, has now been reported. The results are tabulated below.

	Sol	Merc	Venus	Earth	Mars	Ceres	Jup.	Sat.	Uran.	Nep.
Obs	0	.39	.72	1.0	1.52	2.9	5.2	9.5	19.2	30.1
B-T		0.4	0.7	1.0	1.6	2.8	5.2	10.0	19.6	38.8
Frac	$\frac{0}{1}$	$\frac{1}{80}$	$\frac{1}{40}$	$\frac{1}{30}$	$\frac{1}{20}$	$\frac{1}{10}$	$\frac{1}{6}$	$\frac{1}{3}$	$\frac{2}{3}$	$\frac{1}{1}$
Calc	0	.38	.75	1.0	1.5	3.0	5.0	10.0	20.0	30.0
N						τ	1	2	4	6
n		8	4	3	2	1				

The fractional radii are obtained by defining the mean orbital radius of Neptune at unit distance from the Sun.

The fractions that describe the orbits of the four inner (rocky) planets are an order of magnitude ($\frac{1}{10}$) smaller than those of the outer gas giants. On taking the mean orbital radius of Jupiter as unity, the mean orbital radius of each outer planet assumes an integer value, N. On this scale the mean radius for the asteroid belt corresponds to the golden mean, $\tau = 0.61803\ldots$. The mean radii of the inner planets on this scale are given as τ/n.

The many moons of Saturn are readily shown to orbit the planet at commensurable distances, exactly analogous to those of the planets orbiting the sun. Assuming the moon Hyperion to reside at a relative orbital distance of 5/6, which means $N = 5$, the other major moons, likewise assume integer N-values.

By the same reasoning the orbits of the major moons that orbit all other planets are seen to obey the same rules, as shown in Table 5.4.

Table 5.4: *The major satellites in the solar system*

n		Sol	Uranus	Saturn	Jupiter	Mars	Neptune	N
0	$\frac{1}{1}$	Neptune	Oberon					6
1	$\frac{5}{6}$		Titania	Hyperion	Callisto	Deimos		5
2	$\frac{2}{3}$	Uranus		Titan				4
3	$\frac{4}{7}$							
4	$\frac{1}{2}$		Umbriel		Ganymede		Triton	3
5	$\frac{3}{8}$							
6	$\frac{1}{3}$	Saturn	Ariel	Rhea	Europa	Phobos		2
7	$\frac{1}{4}$		Miranda					
8	$\frac{1}{5}$			Dione				
9	$\frac{1}{6}$	Jupiter	Puck	Tethys	Io		Proteus	1

A myriad of minor moons and rings have the same disposition with respect to the parent planet that the asteroid belt and the inner planets have with respect to the sun. Closer examination of the orbital relationship within these groups, and also of the accumulation of the asteroids with respect to their mean orbit, demonstrate the same commensuration pattern as the planets around the sun, only on a different scale.

The obvious inference is that the distribution of matter around a central attractor occurs according to a fixed pattern, independent of scale. It is readily demonstrated that this pattern is conditioned by the golden ratio and a regular divergence angle on a golden spiral. The golden spiral in Figure 5.8 is shown segmented into steps of 36°, starting from a position marked 0 on the edge of a golden rectangle.

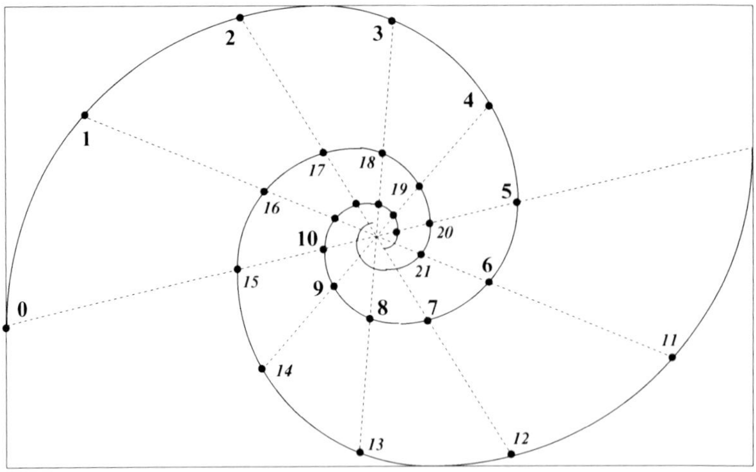

Figure 5.8: *Double golden spiral with self-similar positions on different scales, with divergent angle of 36°*

These steps are represented by the index n in Table 5.4. Each value of n represents an allowed orbital distance for a satellite from its parent attractor. The planets have indices of Neptune(0), Uranus(2), Saturn(6), Jupiter(9), Asteroids(12), Mars(15), Earth(18), Venus(21) and Mercury(24). Because of the self-similar symmetry of the golden spiral this progression can be continued indefinitely on a continuously increasing scale.

In Figure 5.9 the inner and outer planets are placed along separate spiral arms on different, but self-similar scales. The outer planets are on a diverging spiral, starting from τ, whereas the inner planets are on the complementary converging spiral.

The self-similarity between the orbits of planets and of their satellites suggests that each planetary system is like a miniature solar system and each solar system is like a miniature spiral galaxy. The intimate relationship between spirals and elliptic orbits is visualized by rotating successive ellipses on a concentric set by a constant angle to generate two virtual spirals, as in

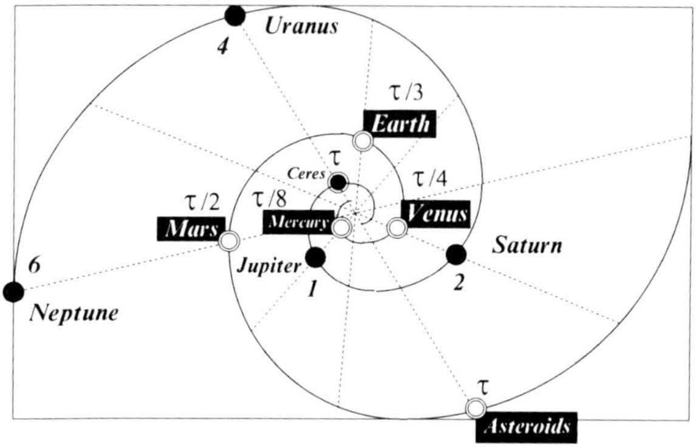

Figure 5.9: *Planetary orbits defined as products of N-numbers and the mean Jovian orbit. Orbits of the inner planets are shown on a larger, self-similar, scale*

Figure 5.10.

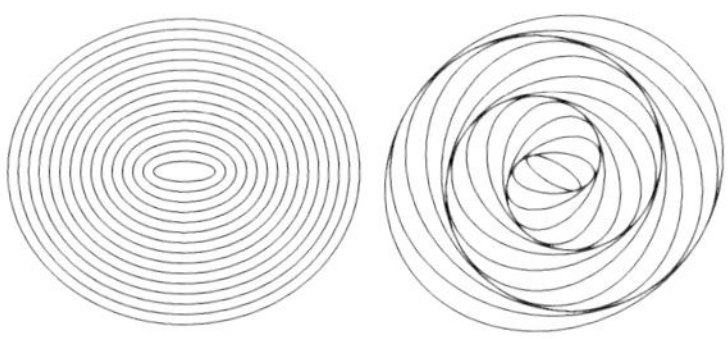

Figure 5.10: *Simulation of a double spiral by rotating adjacent ellipses in a concentric set through a constant angle*

This observation inspired the notion that the divergence angle derived here may be a consequence of optimized accretion of material in a cloud which rotates about a central attractor. Such an optimization problem, governed by second-order dynamics, can be solved by logarithmic spirals. It is inferred that our quantum-like numbers define eigenvalues for an optimization of planet or satellite formation under central-field gravitation as control parameter. As in the Bohr model of the atom, allowed orbits are selected by angular-momentum eigenvectors.

We define dimensionless astronomical units by setting Kepler's constant $k = 1/4\pi^2$. In terms of mean radius, r and orbital period, T, the orbital velocity is

$$v = 2\pi r/T\ ,$$

to give $v^2r = 4\pi^2r^3/T^2 = 4\pi^2k = 1$. The angular momentum, $L = mvr$, and hence, numerically,

$$\left(\frac{L}{m}\right)^2 = (vr)^2 = r\,.$$

Now assume the square of the angular momentum vector, perpendicular to the ecliptic, to be quantized like quantum-mechanical electronic orbital angular momentum, $L_z = m_l\hbar$; thus

$$\left(\frac{L}{m}\right)^2 = l'H = r\,,$$

where $l' = \tau/l$ for inner planets, and $l' = l$ for outer planets as in Table 5.5, which compares observed orbital radii of the planets with calculated radii, assuming the analogue of Planck's constant, $H = 4.9$au.

Table 5.5: *Relative planetary mass, orbits and angular momenta*

Planet	l	Mass	$r = l'H$ au	r(obs) au	L
Mercury	8	0.0553	0.38	0.39	0.03
Venus	4	0.8150	0.76	0.72	0.71
Earth	3	1.000	1.00	1.00	1.00
Mars	2	0.1074	1.51	1.52	0.13
Minors	1		3.0	2.9	
Jupiter	1	317.8	4.9	5.2	705
Saturn	2	95.16	9.8	9.5	298
Uranus	4	14.50	19.6	19.2	64
Neptune	6	17.20	29.4	30.3	93
Pluto	8	0.01	39.1	39.4	0.06

A similar trend is found in each of the planetary systems. The major moons with $N \geq 1$ carry most of the angular momentum, with one or two dominating. Like the asteroid belt and the inner planets around the sun, the major planets have a large number of insignificant moons and rings with $N \leq \tau$. As in the solar system with the Kuiper belt beyond the orbit of Neptune, lesser moons on remote irregular orbits are associated with the four major planets.

Self-similar ordered regions with well-defined satellite orbits are periodically separated by regions of disorder. Should the analogy with microscopic systems hold, periodic variation of this type must also be evident, for instance, in the structure of atoms.

5.3.2 Universal Symmetry

An atom is commonly thought to comprise a nucleus and extranuclear charge cloud; these regions known to be regulated by the same periodic law. The electromagnetic interaction that keeps the whole in balance is seldom recognized as an integral component of an atom. More precisely formulated however, this field consists of virtual photons (standing waves), constantly exchanged between the nucleus and the electronic charge, readily identified as a region of disorder that separates the self-similar hadronic and leptonic regions.

We conclude that the organization of atomic matter is self-similar with the arrangement of planetary systems and conditioned by the same design principles recognized in biological growth and in galaxies. We propose that any of these systems can be described correctly by a golden spiral. This means that infinitesimally small and astronomically large systems only differ in scale. This brings us back to a closed universe, where, as in projective space, there is only one point at infinity: large and small become the same. The closure is responsible for the appearance of periodicity.

This conclusion has far-reaching cosmological implications. It means that the topology and geometry as perceived in the solar system cannot be extrapolated indefinitely towards either astronomical or sub-atomic distances. It is well known that both inverse-square laws of gravitational and electromagnetic interactions break down into singularities at zero separation. Space-time at infinitesimal distances does not behave as anticipated, being warped in such a way as to merge with the infinitely large. The best available model to rationalize this strange behaviour is projective geometry.

The proposed self-similarity of atoms, planets, galaxies, *etc.* is most clearly expressed in the structure of the solar system and readily visualized as the gravitational effect of a large mass, such as the sun, on its environment in space-time, as shown in Figure 5.11(a). Like a pebble dropped into a pond it creates a periodic wave pattern. Each of the troughs represents a local potential minimum for the accumulation of matter and the distance at which the major planets would grow on a rotating spiral in a dust cloud. The relative masses of Jupiter (300), Saturn (100), Uranus (20), Neptune (20) and Pluto (0), clearly follow the expected wave pattern with quantum numbers 1,2,4,6,8. As the major planet grows it creates a local wave pattern, similar to (a), but on a smaller scale, in its immediate environment. Not only does it create local minima for the development of satellite moons, but also perturbs the original wave pattern which now develops secondary minima in the region between Jupiter and the Sun, (b).

A broad minimum develops on each side of Jupiter with effective quantum

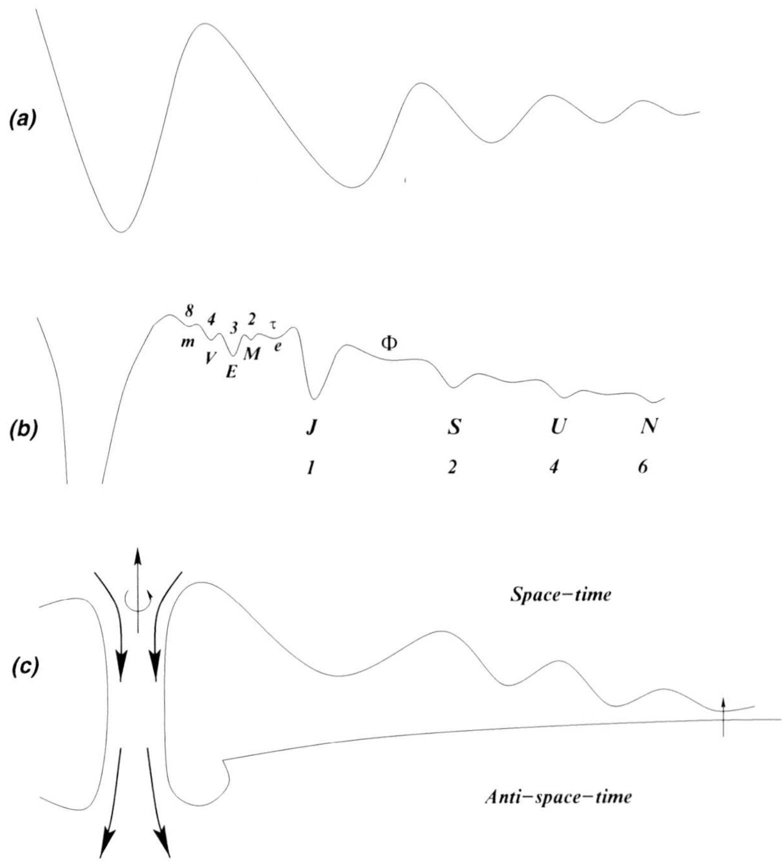

Figure 5.11: *Schematic drawing of distortions in the gravitational field due to (a) an isolated mass; (b) a solar system; (c) a spiral galaxy centred on a black hole*

numbers of τ and $\Phi = 1/\tau$. The asteroid belt occurs in the first of these and we predict a minor asteroid belt at Φ.

The main planets between τ and Sol follow the wave pattern generated by the perturbation of the solar pattern by the presence of Jupiter. The central inner planet, Earth is the most massive of these. Each planetary system is surrounded by a wave pattern similar to (a), with moons on well-defined quantized orbits and with rings and shepherding satellites in the inner sector.

The inner planets only have satellites in the deepest secondary minima.

Continued refinement of this proposed wave structure in space-time must result in a fractal function, or self-similarity, wherein the curve, at any scale, contains a smaller copy of itself.

Figure 5.12: *Globular cluster M3. Courtesy: National Optical Astronomy Observatory/Association of Universities for Research in Astronomy/National Science Foundation*

In Figure 5.11(c) we examine the implications of a massive black hole at the centre of a spiral galaxy. In this case the vacuum interface is ruptured completely, with the implication that the matter which is sucked into the black hole emerges as antimatter. It requires little imagination to recognize in the globular cluster $M3$, shown in Figure 5.12, the result of such a penetration – an anti-black hole releasing copious amounts of matter into intergalactic space to form all of the familiar stellar objects under the influence of gravity. A substantial fraction of the excess atomic hydrogen could end up as molecular hydrogen, not observed as invisible dark matter.

5.3.3 Quantized Redshifts

A spiral galaxy, such as M51, shown in Plate 5.1 has a structure, which is self-similar to the generating spiral of the solar system. Major stars in the spiral arms are expected to occupy quantized orbits like the planets in the solar system. This conjecture is confirmed by the observation of two distinct redshifts in the light from the different arms of M51 and other spiral galaxies. More importantly the jump in redshifts tends to have a constant value for all of these galaxies (Tifft & Cocke, 1987).

Plate 5.1:
Whirlpool galaxy M51 (Courtesy Hubble Heritage Team, ESA, NASA)
Golden spiral superimposed.

Scrutiny of larger samples show that redshift breaks also occur at intervals of 1/2, 1/3 or 1/6 of the originally observed value.

On extending the search to pairs or groups of associated galaxies the same phenomenon was observed. The redshift differences bunched up near multiples of the originally observed jump, j. It was correctly inferred that "if redshift is solely a Doppler effect, then the differences between the measured values for members of pairs should show no jumps".

Final confirmation came from a radio survey of binary galaxies which established redshift differences between pairs grouped around j, $2j$ and $3j$ at high confidence levels.

When measurements were extended to widely separated galaxies, taking the motion of the solar system into account, the quantization of redshifts was established as a general phenomenon, including shifts at simple fractions j/n. The important conclusion is that observed galactic redshifts cannot have a Doppler origin, which means that galaxies in clusters are not moving apart as suggested by the Doppler interpretation. The need of dark matter to stabilize a cluster gravitationally therefore falls away.

Even without knowing the cause of galactic redshifts their quantum nature rules out the Doppler interpretation and confirms that galaxies have structures self-similar to that of the solar system. The major source of galactic light is stars on characteristically quantized orbits around the galactic nucleus and whether redshifting occurs because of chemical, gravitational or optical effects, it will be an intrinsic property of this orbital structure.

Tifft's results have been confirmed independently by several workers and interpreted in terms of concentric shells of galaxies expanding away from Earth, at the centre of the cosmos. This interpretation may resonate with anthropic arguments and creation science, but is less convincing than the explanation based on a quantized gravitational field within each galaxy.

5.4 Nucleogenesis

The relevance of nuclear synthesis and abundance to cosmology was analyzed by Boeyens and Levendis (2008) by comparison of two rival models of nucleogenesis, as outlined below.

5.4.1 The Alternative Models

A central, but elusive aim of nuclear research for almost a century has been the understanding of nuclear synthesis, particularly in the hands of astrophysicists. The dearth of experimental data to drive such enquiry has often resulted in the bending of synthesis theory to follow current thinking in cosmogony, rather than the other way around. This situation has not changed materially since the publication of two comprehensive reviews of the problem in 1950, after the decline of Lemaître's, but before the advent of Gamow's big-bang theories. This period provided an opportunity to consider rival theories without prejudice.

Two back-to-back reviews (Ter Haar, 1950; Alpher & Hermann, 1950) compared the strengths and weaknesses of the models and reached opposite conclusions. Ter Haar concluded that "the equilibrium theory offers a better solution how to account for the observed abundances of the chemical elements than the $\alpha - \beta - \gamma$ theory"[4].

Not surprisingly, Alpher and Herman argued for "a number of indications that element formation probably took place in an early prestellar state of the universe, in which it is difficult to conceive of the existence for a sufficiently long time of the physical conditions required for an equilibrium among nuclei". They added: "Difficulties of the kind mentioned and in particular the freezing-in problem have led to the development of non-equilibrium theories

[4]The $\alpha - \beta - \gamma$ notation refers to the seminal big-bang paper under the names Alpher, Bethe, and Gamow (1948), published on 1 April.

of element formation, unattractive as they may be in requiring the discussion of specific nuclear reactions".

Alpher's preference is clearly conditioned by cosmology rather than science, but in the event it led to the total elimination of the equilibrium model from subsequent enquiry. A complicating factor in the argument is the continuously changing evidence provided by astronomical observation. In both theories, the success is measured in terms of predicted nuclear abundances and invariably these predictions depend on the nature and characteristics of known types of star, assumed as the seat of nucleogenesis. As more powerful telescopes identify new types of heavenly body, new possibilities of nuclear synthesis open up and the model has to be reworked. This process continues for the $\alpha - \beta - \gamma$ model only. The equilibrium model was abandoned before the discovery of quasars and black holes, that obviously provide more attractive environments for nuclear synthesis. The only mechanism for the dispersal of freshly synthesized material is still assumed to be supernovae and this assumption could also stand reassessment.

The abundance criterion

The main premise of both models is that a successful theory should predict the correct cosmic abundances of all nuclides. This idea, that abundances hold the clue to nucleogenesis probably stems from a proposal by Harkins (1931) in 1915 that the abundance of an element depends on two factors:

(i) the structure of the nuclei of its atoms; and

(ii) the relative abundance of the materials used in the formation of the element in question.

The Harkins theory of nuclear structure was formulated before the discovery of the neutron, but predicted the existence of this particle. An atom was described in terms of a mass number, P, that specifies the number of protons in a nucleus, an atomic number Z, that specifies the number of extranuclear electrons and the number of nuclear electrons, N. Another fundamental quantity was defined as the isotopic number

$$n = P - 2Z = 2N - P = N - Z$$

which today would be called the neutron excess of a nucleus.

The essence of Harkins theory is that all atomic nuclei are built up from α-particles and clusters H_x of hydrogen atoms, with $x = 0, 1, 2, 3$ to yield mass numbers of $4n$, $4n - 2$, $4n \pm 1$, together with charge compensating electrons. The formula was shown to hold for the most abundant isotopes of the elements of atomic number less than 30.

In his early work Harkins tested his ideas against the prediction, made on the basis of the He-H nuclear model, that there should be a marked difference in the abundance of elements with even and odd atomic numbers respectively. This prediction was tested against elemental abundances measured in meteorites, which showed that the even-numbered nuclei were more abundant by a factor of 70.

Harkins discovered a classification of the stable nuclides in terms of the ratio n/Z and showed that this ratio never exceeds the value 0.62 in atomic species. The same classification was rediscovered independently many years later and the maximum was shown, more precisely, to be the golden ratio $\tau = 0.6180...$

Cosmic abundance

Although the models of Harkins are now all but forgotten, the central idea that any theory of nuclear synthesis should account for the observed abundancies of nuclides, remains. Analytical techniques have been refined to the point where *cosmic abundances* can be stated with confidence for all naturally occurring nuclides. Cosmic abundances are derived from a variety of measurements on systems under vastly different conditions, *e.g.* meteorites, cosmic rays, stellar spectra, interstellar clouds and nebulae, supernova remnants, the earth's crust, the surface of the sun, and others.

Abundance stipulates the fraction of a given sample that represents the species of interest. To define cosmic abundance it is therefore necessary to find an agreed reference standard accessible in all situations listed above.

In modern work the mass density relative to 10^6 silicon atoms is used as the standard and the nuclear fraction for species j is defined as

$$X_j = \frac{N_j A_j}{\rho N_A}$$

Finally, the ratio of the number of nuclei of species j to the total number of nucleons in the system

$$Y_j = \frac{X_j}{A_j} = \frac{N_j}{\rho N_A}$$

which is a mole fraction.

Abundances on the Si6 scale are defined as

$$\log y_i = \log f_{\mathrm{Si}} + \log Y_i \,.$$

For $y_{\mathrm{Si}} = 10^6$ and assuming $Y_{\mathrm{Si}} = 2.529 \times 10^{-5}$ as the mole fraction of silicon, $\log f_{\mathrm{Si}} = 10.5971$.

The Anders–Grevesse table of cosmic abundances is based on this scale.

Nuclear Synthesis

A modern view of nucleogenesis (Arnett, 1996) discounts the equilibrium model as a viable proposition, having been discarded on the strength of an earlier conclusion (Evans, (1955):

> "The general trend of these isobaric abundances clearly approximates an exponential decrease with increasing A, until $A \sim 100$, above which the relative abundance is roughly independent of A. There is no overwhelming distinction between even-A and odd-A".

Also that, in "theories which invoke a great event rather than a continuum of creation" equilibrium conditions of synthesis would lead to an exponential decrease in abundance, such that for heavy nuclei the predicted abundance " is $\sim 10^{50}$ smaller than the observed abundances".

On re-examination of the abundances there is little evidence to support these conclusions. There is a clear distinction between even-A and odd-A nuclei, already noted very early on, as well as between $A = 4n$ and $A = 4n+2$ nuclei. These differences are most prominent at $A < 150$ and disappear altogether for the heavy nuclei. This is the exact behaviour to be expected for equilibrium systems that develop from different starting materials, but use a common building block.

It is instructive to examine the model that was used to simulate nuclear synthesis in a state of statistical equilibrium which was subsequently frozen down. The probability of the system being in a given quantum state with N neutrons and Z protons with chemical potentials μ and λ respectively is given by the grand canonical ensemble

$$e^{[\mu N + \lambda Z - E(N,Z)]/kT}$$

with partition function

$$Q = \frac{(2\pi AkT)^{3/2} g}{h^3}$$

that treats the system as an ideal gas of protons and neutrons. Qualitatively, the concentration of nuclei in the ground state is given by

$$c_A = Qe^{\mu A/kT} = bA^{3/2}e^{-cA}$$

which represents the abundance as a function of mass number.

Comparison with modern abundance data shows that an exponential function is not such a disastrously wrong predictor of abundances as to discount all equilibrium models, particularly if nuclear synthesis is driven by neither a great event nor continuous creation. A clear resemblance between abundances in the $4n$, $4n+2$ and $4n \pm 1$ series points to an equilibrium model.

Detailed comparison of the models of nucleogenesis is complicated by the two assumptions that cosmic abundance is the same as solar system abundance and that all nuclear synthesis after the big bang has happened in the familiar types of star that eventually disintegrate as supernovae. The second assumption is the most problematical. It provides the theoretical basis of the non-equilibrium or big-bang model of nuclear synthesis. It assumes that the big bang produced all matter in the form of light elements, mainly hydrogen and helium. Initially it was assumed that all nuclides originated from the big bang, but the difficulty to bridge the gap due to the non-existence of nuclides of mass number 5 and 8 was interpreted to prevent any significant build-up to heavier elements by proton and neutron capture in the limited few minutes of reaction time.

The modern big-bang model postulates that all matter emerged from that event as hydrogen and helium, which at a later stage fused together in active stars to produce heavier nuclei. The reaction products would remain at the centre of the star so that the composition of the star gradually becomes richer in the heavy reaction products. An instability eventually sets in as the star runs out of fuel and a consequent gravitational collapse may initiate a catastrophic nuclear explosion that releases the heavy elements synthesised at the core, during a supernova. It is necessary to assume that material, enriched in heavy elements and ejected by many stars, accumulates in space, to become available for further condensation into new stars.

The role of the big bang in the non-equilibrium scenario is to account for the high abundance of H and He, relative to the heavier elements. The stellar synthesis of all other elements is assumed to proceed according to a proposal, originally made by Hoyle to refute big-bang synthesis. It is therefore of interest to note that an alternative explanation of helium abundance exists. Only a small percentage of stars are massive enough to become supernovae. Many other aging stars simply blow off their outer layers while heavy elements remain trapped in their core remnants. The outer layers consist mainly of helium, produced by hydrogen-burning chain reactions, with the net result:

$$4\mathrm{p} \rightarrow {}^{4}\mathrm{He} + 2e^{+} + 2\nu$$

Given sufficient time, helium should therefore accumulate in interstellar space. It appears that the helium abundance problem is largely self-generated by the theory that restricts the age of the universe to 10^{10} years.

The process of element building in stars is supposed to proceed by a variety of nuclear reaction sequences, such as

$$3\,{}^4\text{He} \rightarrow {}^{12}\text{C} + \gamma$$

The hydrogen-burning cycle may be interpreted as catalytic conversion of H into He, with ^{12}C as catalyst:

$$\begin{array}{rcl} {}^{12}\text{C} + p & \rightarrow & {}^{13}\text{N} + \gamma \\ {}^{13}\text{N} & \rightarrow & {}^{13}\text{C} + e^+ + \nu \\ {}^{13}\text{C} + p & \rightarrow & {}^{14}\text{N} + \gamma \\ {}^{14}\text{N} + p & \rightarrow & {}^{15}\text{O} + \gamma \\ {}^{15}\text{O} & \rightarrow & {}^{15}\text{N} + e^+ + \nu \\ {}^{15}\text{N} + p & \rightarrow & {}^{12}\text{C} +^4 \text{He} \end{array}$$

Towards the end of the hydrogen-burning cycle the temperature has increased sufficiently for helium burning to kick in. The main products, called *ashes*, of helium burning are ^{12}C, ^{16}O and ^{22}Ne. After helium, nuclear fuels that ignite in consecutive stages of increasing temperature, with their ashes, are:

$^{12}\text{C} \rightarrow {}^{20}\text{Ne},\ {}^{24}\text{Mg},\ {}^{16}\text{O},\ {}^{23}\text{Na},\ {}^{25,26}\text{Mg}$

$^{20}\text{Ne} \rightarrow {}^{16}\text{O},\ {}^{24}\text{Mg},\ {}^{28}\text{Si},\ldots$

$^{16}\text{O} \rightarrow {}^{28}\text{Si},\ {}^{32}\text{S},\ldots$

$^{28}\text{Si} \rightarrow {}^{56}\text{Ni},\ A \approx 56$ nuclei.

$^{56}\text{Ni} \rightarrow {}^{56}_{26}\text{Fe}{+}2p$

^{56}Fe at the end of the chain is considered the most abundant and most stable isotope produced by fusion reactions. To account for the build-up of elements heavier than iron, it is necessary to resort to the idea of neutron-capture reactions. The neutrons may be supplied by (α, n) reactions, such as

$^{13}\text{C}{+}\,{}^4\text{He} \rightarrow {}^{16}\text{O}{+}\text{n}$

$^{17}\text{O}{+}\,{}^4\text{He} \rightarrow {}^{20}\text{Ne}{+}\text{n}$

$$^{21}\text{Ne}+{}^{4}\text{He} \longrightarrow {}^{24}\text{Mg}+\text{n}$$

In the notation X(p,γ)Y that represents the nuclear reaction X+p→Y+γ, the more important types of reaction, responsible for nuclear synthesis in stars, include

$$(\text{p},\gamma),\ (\text{p},\text{n}),\ (\text{p},\alpha),\ (\text{n},\gamma),\ (\text{n},\alpha),\ (\alpha,\gamma)$$

and their inverses, *e.g.*

$$^{16}\text{O}\ (\alpha,\gamma)\ {}^{20}\text{Ne}$$

$$^{20}\text{Ne}\ (\alpha,\gamma)\ {}^{24}\text{Mg}$$

$$^{18}\text{O}\ (\text{n},\gamma)\ {}^{19}\text{F}$$

$$^{22}\text{Ne}\ (\alpha,\gamma)\ {}^{26}\text{Mg}$$

$$^{22}\text{Ne}\ (\alpha,\text{n})\ {}^{25}\text{Mg}$$

Also represented are heavy-ion reactions with emission of proton, α, neutron or deuteron, *e.g.*

$$^{12}\text{C}+{}^{16}\text{O} \rightarrow {}^{26}\text{Al}+{}^{2}\text{H}$$

There are myriads of possible nuclear reactions that might and probably do occur in the interior of stars and which could and probably do contribute to the formation of heavier, from lighter nuclei. The predominant type of reaction, responsible for the formation of elements beyond the iron group, is generally assumed to be neutron capture followed by β-decay to stable nuclides. One aspect in common to all possible reactions, and especially to neutron capture, is the non-discrimination between nuclei of odd and even mass number that feature together in one complicated reaction network. There is nothing to suggest that a web of reactions, entangled in this fashion, should produce separate series of products with comparable abundance distributions.

The final word on nucleogenesis has obviously not been spoken. There are just too many loose ends and unwarranted assumptions to provide a consistent picture. Too many alternative mechanisms are ignored without mention or comment. The role of black holes, quasars, Seyfert galaxies and white holes, all of which could participate in a chain of nuclear synthesis, is not understood and therefore ignored. Even cosmic ray abundances, matched on the scale of solar abundances show up some important discrepancies. Both H and He have low abundances in cosmic rays, whereas the elements Li, Be and B are 5 orders of magnitude more abundant. The relatively low abundance of the elements between Ca and Fe does not show up like that in cosmic radiation. The relative abundance of odd-Z elements is uniformly higher in cosmic rays. As a result there is much less variation in cosmic-ray abundances than in solar and meteoric abundances. The general trend is a decrease in abundance from He to Fe with regular, though minor variation between even and odd-Z elements.

5.4.2 The New ΛCDM Model

The common version of big-bang theory requires heavy elements to be synthesized in massive stars and subsequently released, during supernova explosions, for incorporation into next-generation stars. Radiation sources that antedate the appearance of second-generation stars should hence be metal-free. According to this criterion many astronomers would agree that the time required for the appearance of metal-enriched galactic sources would be no less than 5×10^9 years. The observation of a γ-ray burst (Salvaterra et al. 2009) with a redshift of $z = 8.1$ implies that it originated when the universe was only about 4% of its current age, *i.e.* about 560 000 years after the big bang.

The surprising aspect of this observation is that spectroscopically measured metal content of the parent galaxy corresponds to that of modern stars. By way of explanation the team invoke a simulation (Springel, et al. 2005) which is interpreted to show that:

> "...stars and galaxies, already enriched in metals, are in place only $\sim$ 600000 yr after the Big Bang".

However, in their massive "Millenium Simulation" the closest that Springel's team comes to such a conclusion is

> "...that quasars can plausibly form sufficiently early in a ΛCDM universe to be compatible with observation, that their progenitors were already massive by $z \approx 16$..."

The nature of this "observation" is not mentioned.

The simulation assumes galaxies to originate from cold dark matter, augmented with a dark-energy field (in the form of a cosmological constant, Λ – the value of which is not disclosed), together with cosmic inflation, driven by weak density fluctuations in the early rapidly expanding universe. To quote:

> "...critical aspects of galaxy-formation physics, [...] such as the ejection and mixing of heavy elements, [are considered] uncertain and beyond the reach of direct simulation; these must be treated

> by phenomenological models whose form and parameters are adjusted by trial-and-error as part of the overall data-modelling process".

To my mind this rationalization of the observation, and many others like it, is an admission that the big-bang nucleogenesis model is refuted by the observation of GRB 090423, and is now replaced by the assumption that heavy elements occurred during the inflation epoch by mechanisms that differ from those operating in ordinary stars.

The currently favoured new mechanism consists of a simulation based on the suggestion, described by Carr et al. (1984), that the cosmos began with a generation of very massive stars (VMS) rather than the debris of the big bang. These huge stars had masses in excess of $100M_{\odot}$ and because of the much higher pressures and temperatures at their cores, they would have burnt up their fuel much faster than smaller stars. Thus, they would have burned themselves out long ago, probably surviving as black holes. Corliss (1994) comments:

> "...that, like the Big Bang itself, the generation of massive stars [...] came from nowhere, like something that was pulled out of a magicians hat!"

Maddox (1984) considers the proposal as an alternative to the big bang. However, in the new ΛCDM approach it becomes part and parcel of big-bang cosmology and the basis for the simulation of nucleogenesis. It is assumed (Oh et al., 2001) that the initial population of metal-free VMS's caused early Fe enrichment and produced the ionizing fluxes that reionized the primordial molecular hydrogen, which acted as the coolant, needed to promote formation of VMS's in the first place. The authors make the breathtaking statement:

> "At present, the best evidence for VMS's [...] may be from their distinctive nucleosynthetic products..."

To summarize: In order to support the big bang, the observation of heavy metals in the very early universe needs to be explained. These elements can only be produced in stars. Suitably parametrized simulations identify a certain class of VMS's, if they exist, as a possible source of metal enrichment to match observation. Therefore VMS's existed in the early universe! The only remaining problem is that the temperature of the hot big bang is too high for clumping to occur within the primordial dark matter without a suitable coolant. Voilá: Primordial molecular hydrogen could do the job, provided it is destroyed again before it disrupts post-inflation events.

Additional assumptions are needed to account for the explosion of VMS's and reionization of the hydrogen gas. The explosion is ascribed to electron/positron pair instability, which is very different from core-collapse supernovae. It is said (Heger & Woosley, 2002):

> "Provided that such stars existed and retained their high mass until death, the outcome is uniquely calculable".

The mystery deepens as we recall that cold dark matter is non-baryonic. It is explained that:

> "[C]oherent oscillations in the primordial plasma [...] leave an imprint in the linear power spectrum of the dark matter",

called 'baryon wiggles'. These wiggles supposedly represent a small percentage of baryonic matter in the CDM and which now represents the visible matter in the universe. The CDM, of which the origin is still undisclosed, remains as the dark-matter halo associated with each galaxy. The conversion of the wiggles into heavy elements does not merit further elucidation.

Given the large number of scientists, affiliated with the most prominent astrophysical institutions of the world, and associated with this exercise, it may safely be concluded that the so-called Λ*CDM cosmology* has unceremoniously replaced the traditional big-bang theory as the new standard cosmology. We reflect that this new cosmology depends on a chain of unsubstantiated assumptions, which include nonbaryonic CDM, primordial molecular hydrogen, short-lived VMS's that disappear into black holes after pair-instability supernovae, and reionization of the hydrogen gas before the universe first became transparent. The evidence is therefore doomed to remain hidden forever.

The chain is simply too long for comfort, and not only inconsistent with classical big-bang theory, but also with inflation. The logical conclusion is that standard cosmology no longer has an explanation to account for nucleogenesis, apart from what exists in supercomputers, totally divorced from reality.

5.4.3 The Periodic Model

Two significant aspects of the symmetry observed in the analysis of periodicity are the inverted electronic energy levels and the approach of $Z/N \rightarrow 1$ for all nuclides. The inversion is explained by the computational observation that electronic sub-levels respond differently to compression of an atom.

The most revealing result was obtained by Goldman and Joslin (1992) who reported the sublevel sequence

$$1s < 2p < 3d < 2s < 4f < 3p < 3s < 4d < 5f < 4p$$

on simulated compression of the H atom to a radius of $0.1a_0$. At this compression the $2p$ and $3d$ levels already lie below $2s$ while $4f$ lies lower than $3p$ and $3s$. With increased compression the inverted sequence observed in Figure 5.4 at $Z/N = 1.04$ must inevitably ensue. We conclude that the symmetrical periodicity can only occur under conditions of extreme pressure, which can only be imagined in the interior of massive stellar objects.

This conclusion immediately leads to an explanation of the Z/N ratio that approaches unity. The most common nuclide with unit ratio is the α-particle, ${}^4\mathrm{He}^{2+}$. The fusion of α-particles in an equilibrium process under extreme pressure would produce atomic nuclides with uniform $Z/N \simeq 1$, which is a testable hypothesis.

The simplest case of nucleogenesis by the equilibrium fusion of α-particles must, not only produce nuclides with $Z/N = 1$, but also with mass numbers $A = 4m$. Should these nuclides be released into interstellar space many of them would be unstable in a low-pressure environment and decay by positron emission or electron capture. Because of the inherent instability of nuclei with an odd number of both protons and neutrons such a decay must consist of at least two steps, such that

$${}^{4m}_{2m}X \xrightarrow{-2\beta^+} {}^{4m}_{2m-2}X\ ,$$

without change in mass number. However, scrutiny of known nuclides shows that mass numbers of $4m$, $4m-2$ and $4m \pm 1$ are all observed. This could happen if the build-up of heavy nuclei starts from the elementary combinations of neutrons and anti-neutrons: n^*n^*, n^*, n^*n, n, to produce the $4m-2$, $4m-1$, $4m$ and $4m+1$ series on α-particle addition. By following each of the four synthetic routes, with due allowance for radioactive decay, over 51 steps, the result is two series of even A, consisting of 81 nuclides each, and two odd series of 51 each. There is a one-to-one correspondence between the calculated and 264 observed nuclides. The only two elements not produced in any of the four sequences are Z=43 and 61. It is equally plausible (Boeyens, 2008) to start from 300 stable isotopes of 100 natural elements and produce the 264 isotopes of 81 stable elements by following the radioactive decay route.

The reciprocal relationship between matter and the curvature of space implies that the occurrence of different periodic functions at $Z/N = 1, \tau$ and 0.58 arises from a variation of the electronic configuration of atoms with the local curvature of space. The one-electron Schrödinger solution for the H atom takes account of only the coulombic interaction with a single proton and therefore represents the atom in empty space, which is flat space by definition. By contrast the different arrangement at $Z/N = 1$ refers to highly curved space and the intermediate situation at $Z/N = \tau$ implies the moderately curved space that we live in.

5.4.4 Nuclear Abundance

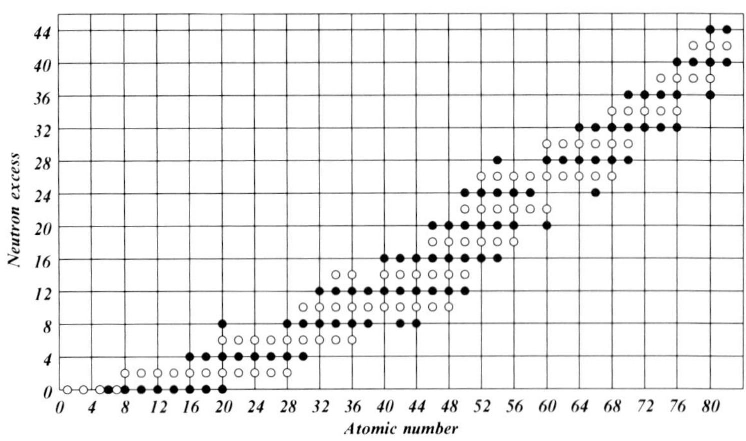

Figure 5.13: *Filled and open circles represent the stable nuclides with $A = 4m$ and $4m + 2$ respectively. Nuclides with odd A are not shown and they occur within the same field*

The mechanism of nucleogenesis must be reflected in the relative cosmic abundance of atomic nuclei, even though abundances are notoriously difficult to estimate. It is therefore instructive to note that each of the four modular groups $A(\text{mod}4)$ of stable nuclides separates into parallel sequences, according to neutron excess, $N_e = N - Z = A - 2Z$. For the nuclides with $A = 4m$, $N_e = 4k$, $k = 0, 11$. The distribution of the nuclides with $A = 4m$ and $4m + 2$ is shown in Figure 5.13. Each of the sequences is flanked on both sides by unstable radioactive nuclides. It is inferred that the relative nuclear stability must tend to a maximum in the middle of each sequence. This trend should be mirrored by the measured binding energies per nucleon and nuclear abundance, when compared for nuclides with common N_e.

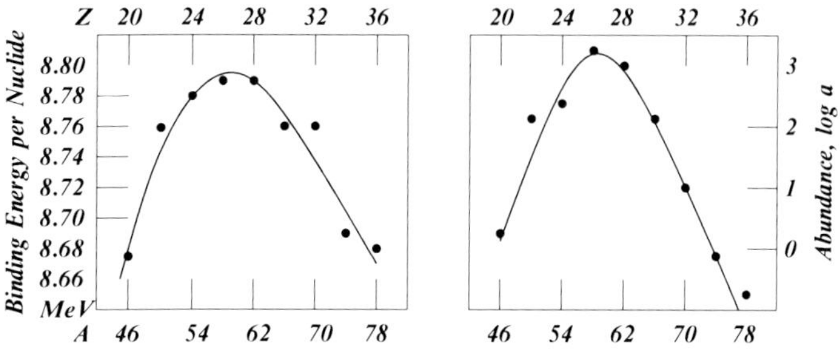

Figure 5.14: *Nuclear binding energies and solar abundance of nuclides in the* A=*4m+2 series with* $N_e = 6$

Now we are ready to test the hypothesis of nuclear synthesis by α-addition in an equilibrium process, which requires that the relative abundance of each nuclide within a common group must be directly related to its relative stability. The amazing reality is that, despite the uncertainty associated with the measurement of solar abundances, the correlation between nuclear stability and estimated abundance is totally convincing. The binding energy per nucleon and the reported abundance for the nuclides in the $A = 4m + 2$ series, with $N_e = 6$, *i.e.* for the relevant isotopes of Ca to Kr, are shown in Figure 5.14 as a typical example. The correlation is unmistakable. There is no evidence of any discontinuity at Fe, as required by non-equilibrium models of nucleogenesis.

Each of the sequences defined by $N_e = 0 \rightarrow 44$, covering all stable nuclides, shows a similar correlation between nuclear stability and abundance, although, not in all cases, as convincing as in Figure 5.14.

The final conclusion is clear: Uniform correlation between nuclear stability and abundance cannot result from nucleogenesis in a large number of unrelated processes under a variety of reaction conditions, as required by the big-bang mechanism. The suggested alternative of nuclear synthesis by an equilibrium process of systematic α-addition points at a completely different cosmological model and to the direction which this enquiry must follow, while remaining consistent with physical theory.

5.5 References

Alpher, R.A., H. Bethe & G. Gamow (1948): *The origin of chemical elements*, Phys. Rev., 1948 (73) 80–82.

Alpher, R.A. & R.C. Hermann, (1950): *Theory of the Origin and Relative Abundance Distribution of the Elements*, Revs. Mod. Phys., 1950 (22) 153–212.

Arnett, D. (1996): *Supernovae and Nucleosynthesis*, University Press, Princeton.

Arp, H.C. (1982):*Quasars, Redshifts and Controversies*, Interstellar, Berkeley.

Arrhenius, S.A. (1908): *Worlds in the Making*, Harper, London.

Boeyens, J.C.A. (1994): *Ionization radii of compressed atoms*, J. Chem. Soc. Faraday Trans., 1994 (90) 3377–3381.

Boeyens, J.C.A. (2008):*Chemistry from First Principles*, Springer.com.

Boeyens, J.C.A. (2009):*Commensurability in the solar system*, Physics Essays, 2009 (22) 493–499.

Boeyens, J.C.A. & D.C. Levendis, *Number Theory and the Periodicity of Matter*, 2008, Springer.com.

Boyle, R. (1661): *Sceptical Chymist.*

Carr, B.J., J.R. Bond & W.D. Arnett (1984): *Cosmological consequences of Population III stars*, Astrophys. J., 1984 (277) 445–469.

Corliss, W.R. (1994): *Science Frontiers: Some anomalies and curiosities of Nature*, Sourcebook Project, Glen Arm, USA.

Evans, R.D. (1955): *The Atomic Nucleus*, McGraw–Hill, N.Y.

Goldman, S. & C. Joslin, *Spectroscopic Properties of an Isotropically Compressed Hydrogen Atom*, J. Phys. Chem. 1992 (96) 6021 – 6027.

Harkins, W.D. (1931): *The periodic system of atomic nuclei and the principle of regularity and continuity of series*, Phys. Rev., 1931 (38) 1270–1281.

Heger, A. & S.E. Woosley (2002): *The nucleosynthetic signature of population III*, Astrophys. J., 2002 (567) 532–543.

Lavoisier, A.L. (1789): *Traité élémentaire de chimie.*

Maddox, J. (1984): *Alternatives to Big Bang*, Nature, 1984 (308) 491.

Oh, S.P., K.M. Nollett, P. Madau & G.J. Wasserburg (2001): *Did very massive stars preenrich and reionize the universe?*, Astrophys. J., 2001 (562) L1–L4.

Priesner, C. & K. Figala (1998): *Alchemie*, Beck, München.

Rouvray, D. (2009): *Is quantum chemistry the new alchemy?* Chemistry World, 2009, p. 62.

Salvaterra, R. & 44 co-authors. (2009): *GRB 090423 at a redshift of* ~ 8.1, Nature 2009 (461) 1258–1260.

Springel, V. & 16 co-authors. (2005): *Simulations of the formation, evolution and clustering of galaxies and quasars*, Nature 2005 (435) 629–636.

Stewart, A.W. (1922): *Some Physico-Chemical Themes*, Longmans, London.

Thackeray, J.F. (2007): *Approximation of a biological species constant?*, S.A. Jour. Sci. 2007 (103) 489.

Ter Haar, D (1950): *Cosmological Problems and Stellar Energy*, Revs. Mod. Phys., 1950 (22) 119–152.

Tifft, W.G. & W.J. Cocke (1987): *Quantized Galaxy Redshifts*, Sky and Telescope, 1987 (73) 19–21.

Urey, H.C. (1952): *The Planets: Their Origin and Development*, Yale Univ. Press, New Haven.

Webb, J.K., M.T. Murphy, V.V. Flambaum, V.A. Dzuba, J.D. Barrow, C.W. Churchill, J.X. Prochaska & A.M. Wolfe, *Further Evidence for Cosmological Evolution of the Fine Structure Constant*, Phys. Rev. Lett. 2001 (87) 09103, astro-ph/0012539.

Chapter 6

Standard Cosmology

There is a perception that standard cosmology is firmly based on the general theory of relativity, which is said to supersede the dictates of special relativity. However, this is a fabrication that only serves to corroborate the big-bang fantasy. Ironically, many opponents of the big-bang model fail to recognize this fundamental fallacy at the root of the expanding-universe avowal.

General relativity is the theory that gave physical content to Riemann's formulation of curved mathematical space and identifies the four-dimensional metric tensor with the gravitational field. The four dimensions of general relativity are the same as in the Minkowski space of special relativity. The velocity of light remains a constant in free space and the inability to specify simultaneous events remains in force.

Expanding universe cosmologies, in contrast, are based on a special metric that assumes a universal time, independent of the curved spatial manifold, as a solution of Einstein's gravitational field equations. In reality this is a second-generation generalization of Einstein's static-universe cosmology, once assumed to be the only possible solution of the field equations on a cosmic scale. However, as emphasized by Fuller and Wheeler (1962):

> "[t]he Einstein field equations, being differential equations, are purely local in character. They tell nothing about the topology of space with which one is dealing..".

The only solution of the field equations without ruinous approximations was obtained by Schwarzschild. It serves as a model for isolated objects and is too localized for cosmology. A concise critical summary of the cosmological models was recently published by Mamone Capria (2005) and our more superficial treatise that follows will concentrate only on those aspects of immediate relevance.

J.C.A. Boeyens, *Chemical Cosmology*, DOI 10.1007/978-90-481-3828-9_6,

6.1 Solution of the Field Equations

The gravitational field is described in general relativity by the set of equations (4.11). The right hand side depends on the description of matter in the system of interest and the corresponding solution consists of finding that form of the fundamental tensor $g_{\mu\nu}$ that satisfies (4.11). The first successful solution of cosmological interest, obtained by Schwarzschild, is text-book material, described in detail by Adler *et al.* (1965). The time-independent spherically symmetric line element is of particular importance as a model of the basic one-body problem of classical astronomy. This element, of the form:

$$ds^2 = \left(1 - \frac{2m}{r}\right)\left(dx^0\right)^2 - \frac{dr^2}{1 - 2m/r} - r^2\left(d\theta^2 + \sin^2\theta d\varphi^2\right) \tag{6.1}$$

in spherical coordinates, satisfies the field equations $R_{\mu\nu} = 0$ for empty space and converges to the pseudo-Euclidean Minkowski metric at infinity. The integration constant

$$m = \frac{\kappa M}{c^2}$$

in the case of a central point mass M and κ the gravitational constant.

The solution to (6.1) is of direct importance for modelling planetary motion on a spherical orbit. It is noted that the coefficient of dr^2 becomes infinite on the spherical shell $r = 2m$ and, the solution therefore goes singular at both $r = 0$ and $r_s = 2m$, known as the Schwarzschild radius. For any known macroscopic body the radius $r_s = 2\kappa M/c^2$ falls well inside the body where free-space equations are not valid and (6.1) does not yield an appropriate solution.

The Schwarzschild solution in nonempty space estimates the stress or energy-momentum tensor $T_{\alpha\beta}$ in terms of an incompressible perfect fluid medium with the same symmetry as before and serves as a simple model of a star. To get the complete picture the interior solution for a sphere of perfect fluid of radius r_0 is joined continuously with the free-space solution that applies at $r > r_0 > 2m$. As before $m = \kappa M/c^2$, where M is the mass of the fluid sphere.

Since the radius of the fluid sphere must be greater than $2m$, the mass M is restricted to

$$M < \frac{c^2 r_0}{2\kappa}\,.$$

This is interpreted to mean that an arbitrary amount of fluid cannot be concentrated into a sphere of fixed radius without producing catastrophic effects on the metric.

6.1.1 The Black Hole

These catastrophic effects are the subject of relativistic theories of black holes. In Newtonian theory a black hole appears as the mass of a star increases to the point where its escape velocity exceeds the speed of light. In relativistic theory the singularity at the Schwarzschild radius complicates this description. The problem is avoided by demonstrating that the singularity at $r = 2m$ disappears with a suitable choice of coordinates, as in Figure 6.1.

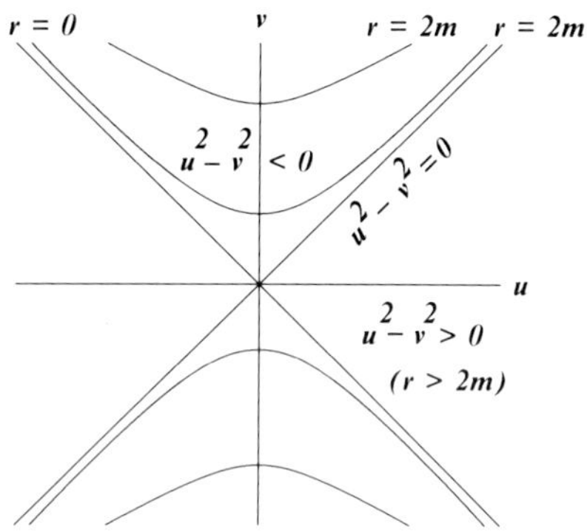

Figure 6.1: *Coordinates to remove the singularity at* $r = 2m$

A procedure to map space-time events to a Euclidean plane is described by Synge (1950). The Euclidean plane has orthogonal cartesian axes: $-\infty \leq u, v \leq \infty$, and a new variable $(u^2 - v^2)$ is introduced to replace the radial parameter, r, in construction of the Schwarzscild line element. By using this coordinate system Kruskal (1960) defined the line element

$$ds^2 = f^2(du^2 - dv^2) + r^2 d\omega^2$$

where f depends on r alone. It is demonstrated that the singularity at $r = 2m$ is avoided and only remains at $r = 0$.

The important conclusion is that a region without singularity opens up between the two hyperbolic curves, $r = 0$. It acts as a non-Euclidean bridge, of the type proposed by Einstein and Rosen, between two otherwise Euclidean spaces, as shown in Figure 6.2. The null geodesics ($r = \pm 2m$) meet at the origin, which means an inversion of time flow, $+t \to -t$.

The physical interpretation is that matter, which seems to disappear into a black hole, has an escape route through a non-Euclidean throat that connects one part of the universe to a related region of opposite chirality, which implies conversion of matter into anti-matter.

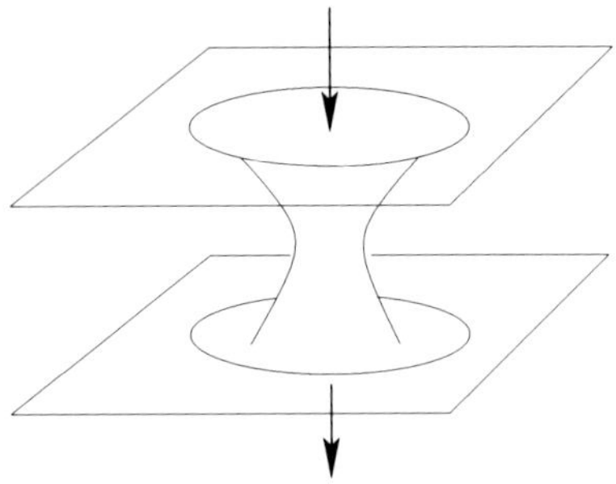

Figure 6.2: *Einstein–Rosen bridge through a black hole*

As a test particle crosses the coordinate singularity at $r = 2m$ the radial coordinate becomes time-like, which means that it resembles a stationary particle in Minkowski space. The consequence of this is that the residence time within the inversion zone (or bridge) is extended almost indefinitely.

6.2 Einstein's Universe

In order to avoid the problem of boundary conditions at infinity Einstein, in his original cosmological model, based on general relativity, assumed space to be closed, in which case there is no infinity. As a working model he selected what is known as a topological 3-*sphere*, which is the 3-surface of the four-dimensional hypersphere of radius R_0, defined by the set of points in Euclidean space that satisfies the equation

$$x_1^2 + x_2^2 + x_3^2 + x_4^2 = R_0^2 \tag{6.2}$$

These are not coordinates in four-dimensional space, as (Einstein, 1950) the

> "... fourth space dimension has naturally no significance except as a mathematical artifice."

This space, like a circle, is unbounded without being infinite. A sphere of radius R is known to have constant curvature of $K = 1/R^2$.

Einstein added a privileged time coordinate such that the line element of space-time is:

$$ds^2 = -R_0^2 d\sigma^2 + c^2 dt^2 \,,$$

where $d\sigma^2$ is the metric of the unit 3-sphere S^3,

$$d\sigma^2 = \sum_{a,b=1}^{3} \left(\delta_{ab} + \frac{x^a x^b}{R_0^2 - (x_1^2 + x_2^2 + x_3^2)} \right) dx^a dx^b \,.$$

The coordinate system in which this metric is expressed is obtained by parallel projection along the x_4-axis in $\mathbb{R}^4$. Topologically, this space-time could be seen as a 4-dimensional cylinder in Minkowskian $\mathbb{R}^5$.

Constraining the coordinates to the surface of the hypersphere means that

$$x_1 dx_1 + x_2 dx_2 + x_3 dx_3 + x_4 dx_4 = 0 \,,$$

which allows the elimination of x_4 from the line element

$$d\sigma^2 = dx_1^2 + dx_2^2 + dx_3^2 + dx_4^2$$

to give

$$d\sigma^2 = \sum_{i=1}^{3} (dx_i)^2 + \frac{d\left[\sum_{i=1}^{3} (x_i)^2\right]^2}{4x_4^2} \,.$$

In terms of the usual spherical coordinates

$$x_1 = r\cos\varphi\sin\theta \qquad x_2 = r\sin\varphi\sin\theta \qquad x_3 = r\cos\theta$$

with $R_0^2 = r^2 + x_4$, it becomes:

$$\begin{aligned} d\sigma^2 &= dr^2 + r^2\left(d\theta^2 + \sin^2\theta d\varphi^2\right) + \frac{r^2 dr^2}{R^2 - r^2} \\ &= \frac{d\rho^2}{1 - r^2/R^2} + r^2\left(d\theta^2 + \sin^2\theta d\varphi^2\right) \end{aligned}$$

The curvature of the spherical 3-space is $K = (1/R^2) = 1$ and by scaling the line element to the surface of radius S, the metric is written as:

$$ds^2 = c^2 dt^2 - S^2 \left(\frac{d\rho^2}{1 - r^2} + r^2\left(d\theta^2 + \sin^2\theta d\varphi^2\right) \right) . \tag{6.3}$$

For a more detailed analysis of Einstein's assumptions and methodology the reader is referred to the recent review (Capria, 2005). In particular to the statement that:

> "... the all-important assumption of the existence of a privileged time coordinate (which will be called *cosmic time*), [ ] was clearly a robust injection of Newtonianism into 'general relativity'."

A former colleague and biographer, Banesh Hoffmann (1972) is even more forthright:

> "It undoubtedly comes as a surprise to see Einstein himself reintroducing absolute rest and absolute simultaneity in this way. In solving his immediate cosmological problem, he apparently subverted his whole previous structure.
>
> As for the absolute cosmic time and absolute cosmic rest, they were a price he dared to pay in order to treat the universe as a whole. And people who later extended his work paid a comparable price".

Many of those coming later apparently did so without realizing what price they paid. We read (Pecker, 2001):

> "The fact that there is a *unique cosmic time* is inherent in the very concept of cosmology."

In order to test the feasibility of the proposed metric, the line element (6.3) is used to calculate the Einstein tensor

$$G_{\mu\nu} = R_{\mu\nu} - \tfrac{1}{2} g_{\mu\nu} R = -\kappa T_{\mu\nu} \,.$$

Details of the calculation are given by Narlikar (2002), leading to the equalities:

$$\frac{1}{R_0^2} g_{ab} = 0, \qquad -\frac{3}{R_0^2} = -\kappa c^2 \rho$$

where the matter density ρ is considered constant in both space and time.

These equalities cannot lead to a sensible solution. From the first equality it follows that R_0 should be infinite, which implies that the 3-space degenerates into ordinary Euclidean 3-space; but the second equality suggests $\rho = 0$, meaning that the universe is empty. To remedy this absurdity Einstein modified the field equations, as in (4.11), by introducing the so-called *cosmological constant* Λ, to read:

$$R_{\mu\nu} + \Lambda g_{\mu\nu} = -\kappa \left(T_{\mu\nu} - \tfrac{1}{2} T g_{\mu\nu} \right) \tag{6.4}$$

This form leads to the new equalities:

$$\frac{1}{R_0^2} g_{ab} = \Lambda g_{ab}, \qquad -\frac{3}{R_0^2} + \Lambda = -\kappa c^2 \rho \,,$$

from which follows: $\Lambda = 1/R_0^2 = \kappa c^2\rho/2$, and since the volume of the 3-sphere, $V = 2\pi^2 R_0^3$, the total mass of the Einstein universe is calculated as

$$M = 2\pi^2 R_0^3 \rho = \frac{4\pi^2}{\kappa c^2 \sqrt{\Lambda}}.$$

The radius of the universe

$$R_0 = \sqrt{1/\Lambda} = \frac{1}{c}\left(\frac{2}{\kappa\rho}\right)^{\frac{1}{2}}.$$

6.3 De Sitter's Solution

The gravitational field equations in empty space,

$$R_{\mu\nu} = 0$$

are satisfied in Minkowski space-time and also in Schwarzschild space-time near infinity, as it should in the absence of any matter. By this argument the free-space form of the modified equations,

$$R_{\mu\nu} = -\Lambda g_{\mu\nu}$$

creates a problem if an acceptable solution exists, because that would define inertia in space-time without matter.

Such a solution was found by Willem de Sitter on assuming a four-dimensional spherical *space-time* with constant curvature, rather than a spherical 3-*space*. He started with five-dimensional Minkowski space to define the hypersurface

$$-x_0^2 - x_1^2 - x_2^2 - x_3^2 + c^2 x_4^2 = -R_0^2$$

As before, parallel projection along x_0 leads to the metric:

$$ds^2 = \left(\eta_{\mu\nu} - \frac{\eta_{\mu\rho}\eta_{\nu\sigma}x^\rho x^\sigma}{R_0^2 + \eta_{\rho\sigma}x^\rho x^\sigma}\right) dx^\mu dx^\nu, \quad \mu, \nu = 1 \to 4,$$

where $\eta_{\mu\nu}$ is the matrix of the Minkowski metric in $\mathbb{R}^4$. By computations very similar to those performed for Einstein's metric, one finds:

$$R_{\mu\nu} = -\frac{3}{R_0^2} g_{\mu\nu}$$

such that the modified equation (6.4) reduces to:

$$R_{\mu\nu} + \Lambda g_{\mu\nu} = \left(-\frac{3}{R_0^2} + \Lambda\right) g_{\mu\nu} = -\kappa\left(T_{\mu\nu} - \tfrac{1}{2}T g_{\mu\nu}\right).$$

It follows that if $\Lambda = 3/R_0^2$ equation (6.4) is satisfied with a vanishing stress-energy tensor. In simple terms – there exists a gravitational field, not induced by matter, as a property of empty space.

Furthermore, noting that for $x_4 = t$, the hypersurface

$$x_0^2 + x_1^2 + x_2^2 + x_3^2 = R_0^2 + c^2t^2$$

has a radius $R(t) = \sqrt{R_0^2 + c^2t^2}$, with R_0 =constant. This equation can be explained in two possible ways. The common, more popular, interpretation views $R(t) = \sqrt{R_0 + c^2t^2}$ as a variable radius of the four-dimensional hypersurface. The variable part of the radius, $r(t) = ct$, changes at a rate $\dot{r} = c$. Hence $\dot{r}/r = 1/t$. This is interpreted to mean that the hypersurface expands from an initial value of R_0 at time $t = 0$ to a value[1] R at time T, at a rate given by the relation:

$$\frac{\dot{R}}{R} = \frac{1}{T}, \tag{6.5}$$

in integrated form $R = R_0e^{t/T}$. It will be shown how this interpretation was incorporated in the expanding-universe model that relates galactic redshifts to a Doppler effect.

The second interpretation relates $R(t)$ to space-time curvature,

$$K = \frac{1}{R^2(t)} = \frac{1}{R_0^2 + c^2t^2},$$

that varies as a function of time. For:

$$\left.\begin{array}{ll} t = 0, & K_0 = 1/R_0^2 \\ t \neq 0, & K < K_0 \\ t = \pm\infty, & K \to 0 \end{array}\right\}$$

Any time-like displacement implies a gradual decrease in curvature that proceeds throuh a minimum of zero before returning to the maximum K_0. This variation describes the topology of a Möbius strip, or more generally, of projective space. The integrated curvature over the double cover averages to zero, in accord with de Sitter's empty universe, which implies zero curvature. This result is in line with Naan's proposed universe, consisting of equal parts matter and antimatter, which also integrates to zero. In de Sitter's (1917) own words:

[1]Note that $R_0 \neq 0$ as in later versions of standard cosmology.

> "The four-dimensional world, which we have for the sake of symmetry represented as spherical, is in reality hyperbolic, and consists of two sheets, which are only connected with each other at infinity. The formulae embrace both sheets, but only one of them represents the actual universe."

It is tempting to add that the second sheet represents Naan's anti-world, which lies on the opposite side of the projective double cover.

Introduction of the cosmological term therefore results in an empty universe with a time-dependent metric – exactly the opposite of Einstein's intention to produce a matter dominated static universe.

6.4 Friedmann's Generalization

The contradictions inherent in the schemes of Einstein and de Sitter were never resolved, despite what Capria (2005) calls a "circuitous return [to] basic Newtonian concepts", eventually to be superseded by a more general formulation of the problem.

By recalculating the line element (6.3), starting from a pseudo-Euclidean hypersurface with constant negative curvature

$$x_1^2 + x_2^2 + x_3^2 - x_4^2 = -S^2$$

the result differs from (6.3) only in the sign of K.

By assuming the Gaussian line element

$$ds^2 = dx_0^2 - e^{G(x^0,r)}d\sigma^2$$

Robertson and Walker independently obtained the expression

$$ds^2 = c^2dt^2 - S^2(t)\left(\frac{dr^2}{1-Kr^2} + r^2\left(d\theta^2 + \sin^2\theta d\varphi^2\right)\right) \tag{6.6}$$

which contains both forms of (6.3) as special cases and remains valid for any constant curvature. The simplest form corresponds to $\dot{S}(t) = 0$ and $K = 1$ – the Einstein model. The metrics with $K = -1$ or 0 represent negative or zero curvature of hyperbolic and Euclidean space, respectively.

The general solution of the field equations (6.4), by using the Robertson–Walker metric, was obtained by Aleksandr Friedmann on substituting (6.6) into (6.4). Details of the procedure are outlined by Narlikar (2002). By considering the matter distribution, as galaxies in space, to be like dust, the stress-energy tensor $T_{\mu\nu}$ has only one non-zero component: $T_{44} = \rho c^4$, and after several manipulations the final result is Friedmann's equation:

$$\dot{S}^2 = \frac{\Lambda c^2 S^2}{3} + \frac{A}{S} - c^2 K$$

where $A = \kappa c^4 \rho S^3/3$.

For a static universe with $K = 1$, $\dot{S} = 0$, $S(t) = R_0 = \sqrt{1/\Lambda}$, *i.e.*

$$\frac{c^2}{3} - c^2 = -\frac{A}{R_0}$$

giving
$$R_0 = \frac{3A}{2c^2} = \frac{\kappa c^4 \rho R_0^3}{2c^2},$$

the critical radius for equilibrium as

$$R_0 = \frac{1}{c}\left(\frac{2}{\kappa\rho}\right)^{\frac{1}{2}} = \sqrt{\frac{1}{\Lambda_c}}, \quad \text{as before.}$$

The whole variety of cosmologies consistent with Friedmann's equation will not be discussed here. Instead we focus on a few popular features at the roots of standard cosmology. The most important consensus is that the Friedmann solutions demonstrate that a static universe is inherently unstable and not compatible with the astronomical observation of spectroscopic redshifts. Noting that the cosmological constant was originally introduced to generate an acceptable static solution it is now generally accepted that Λ has no meaningful role in cosmology. However, the theorem of Vermeil (1917) shows that the mean curvature of an n-dimensional Riemannian manifold is only represented by the rank 2 symmetric tensor that contains the fundamental tensor and its first and second derivatives. Recast into familiar tensor notation (Weyl, 1922), it reads:

$$E_{\mu\nu} = c_1 R_{\mu\nu} + c_2 S g_{\mu\nu} + c_3 g_{\mu\nu}\,.$$

If in addition $E_{\mu\nu}$ is required to have zero divergence (Capria, 2005), then

$$E_{\mu\nu} = c_1\left(R_{\mu\nu} - \tfrac{1}{2} S g_{\mu\nu} + \Lambda g_{\mu\nu}\right)$$

which is the same as (6.4), and therefore the mathematically correct form of the gravitational field equations.

In all cases $\Lambda = 0$ predicts a universe of radius $R(t)$ that expands from an initial value of zero at time $t = 0$. For the special case, $K = 1$, the function $R(t)$ describes a cycloid, which means that initial expansion is followed by contraction, such that $R(t)$ returns to zero as a periodic function of time. For Euclidean and hyperbolic spaces, $K \leq 0$, $R(t)$ increases monotonically from zero to infinity. The situation $K = 0$, the so-called Einstein-de Sitter space-time, is favoured for a variety of reasons. In this case the three-dimensional world is Euclidean, at variance with the principles of general relativity. It is in fact not even possible to distinguish what value of K would best fit astronomical observation. The one feature that all values of K have in common is the discontinuous birth of the universe, which is also the most controversial. It coincides with a singularity of infinite mass density, which appears because of an oversimplification of, not only the mathematical model, but also the physical model of galactic matter. The dust model that assumes zero pressure can certainly not be valid in what is supposed to be a high-density early universe.

The velocity of light in a radial Robertson–Walker direction can be calculated by putting $ds = 0$, as in Minkowski space, together with angular coordinates $d\theta$ and $d\varphi$ equal to zero, such that

$$c^2 dt^2 = R^2(t)\frac{dr^2}{1 - Kr^2}$$

i.e.

$$\frac{c^2\left(1 - Kr^2\right)}{R^2(t)} = \left(\frac{dr}{dt}\right)^2 \neq c^2 .$$

This is not a requirement of general relativity, as often stated, but of the expanding Friedmann model, which is not covariant.

6.5 Mach's Principle

The biggest surprise about de Sitter's solution was that, although a legitimate solution of Einstein's gravitational field equations, it violates Mach's principle, which Einstein considered to be the pillar that supports the general theory of relativity.

The principle is summarized by the statement that the geometry of space is determined by the matter distribution, or alternatively, that the inertia of one body is due to the presence of all other bodies in the universe. The principle is recognized in Einstein's equation by the relationship between $T_{\mu\nu}$, which characterizes the matter-energy distribution and $G_{\mu\nu}$ that specifies the geometry of space.

In relativistic terminology the principle is best expressed as the balance between matter density and the curvature of space. The assumption of uniform matter density implies that space is non-Euclidean with constant curvature. Euclidean space implies zero density. Returning to Mach's terminology, only curved space can explain inertia. When de Sitter obtains a non-vanishing curvature tensor for an empty universe it therefore means that the equilibrium condition, implicit in Einstein's equations, is not satisfied. The only way to restore equilibrium is to expand the space indefinitely – approaching zero curvature, exactly in line with de Sitter's result.

The predicted de Sitter expansion of empty space should not be interpreted as a property of the universe. The Einstein–de Sitter model describes another non-equilibrium situation in the form of an expanding material universe with a universal time coordinate, orthogonal to a flat 3-dimensional subspace. The tendency here would be to restore equilibrium by spreading the excess density in order to match the pseudo-Euclidean geometry.

The most likely configuration must surely be a universe in equilibrium, but not necessarily static equilibrium. Furthermore, the assumption of universally uniform distribution of matter is not necessarily correct, as it ignores the symmetry between matter and antimatter. The single most important mathematical difference between matter and antimatter is in their temporal response. Whatever the topology required to accommodate equal amounts of matter and antimatter, it cannot contain a privileged cosmic time coordinate.

6.6 The Expanding Universe

Friedmann's solution of the gravitational field equations remained dormant until the work of Hubble and other astronomers became firmly interpreted as positive evidence that the universe is in a state of uniform expansion. Probably its most attractive feature was the inference that the expansion must have started at a singular point in space-time, interpreted to provide scientific evidence of the ideological notion of a creation event in the beginning. The earliest known record of speculation about explosive birth of the universe dates back to Edgar Alan Poe's essay Eureka of 1848. It must have been known to the Abbé Lemaître who had sufficient mastery of mathematics to dress it up as science. As theoretical cosmology it experienced several cycles of growth and decline, supported at times by secondary evidence as a model of *inter alia*, cosmic radiation, nucleogenesis, the anthropic principle, unified theory in particle physics, galactic redshifts, cosmic evolution and finally the black-body microwave background radiation.

To keep the record straight, what is generally known as *standard cosmology* refers to the model proposed by Einstein and de Sitter (1932). It consists of a Friedmann solution with $\Lambda = 0$, $K = 0$ and zero pressure, without reference to Friedmann or Lemaître. The space-time is topologically like $\mathbb{R}^4$ and the radius of the universe has the form:

$$R(t) = R(t_0)(t/t_0)^{\frac{2}{3}},$$

expanding from a singularity at time zero, the primeval atom of Lemaître.

The more recent, penultimate[2] model, known as the big-bang theory, is supported by a number of suitably selected and refined observations, briefly discussed in the following.

6.6.1 Galactic Redshifts and Hubble's Law

One of the oldest persistent paradoxes of cosmology is succinctly summarized by the challenge to explain "why the night sky is dark". The problem, as precisely formulated by Olbers, originates from the reasonable assumption that luminous matter is evenly distributed through the universe. Each volume element dV should radiate light of intensity I_0dV in all directions. By known optics the radiation density should decrease with distance from an emitter like the inverse square of the distance. An observer at any point in the universe is surrounded by a spherical shell of volume $4\pi r^2dr$ at a distance r, from which a total radiation equal to $I_0(4\pi r^2)dr$ is emitted. Of this, the observer only receives an amount proportional to $I_0(4\pi r^2)dr/r^2 = 4\pi I_0dr$. Integrated over all values of r it now turns out that the energy beamed at the observer is infinite, even for small I_0.

The traditional resolution of the paradox is known as *tired light* theory. Although light from distant sources travels mainly through empty space, during a journey over billions of light years a photon must encounter numerous dust particles, gas molecules, ionized material, atoms, plasma and other cosmic debris, despite the low density of such matter. Each encounter affects the energy of the photon and effectively stretches its wavelength towards the red end of the spectrum and less energy is contributed to the radiation integral.

Not only does tired light resolve Olbers' paradox but it also agrees with spectroscopic analysis of galactic light, which shows redshifts in the light from distant sources. The first qualitative formulation of the effect was recorded by Edwin Hubble who reported a "roughly linear" relationship between observed redshift and the distance of the source. The following interpretation of the effect was offered in concluding a detailed comparison of recessional and non-recessional models of redshift (Hubble & Tolman, 1935):

[2]Discussion of the current ΛCDM version continues in section 6.7.9.

"It also seemed desirable to express an open-minded position as to the true cause of the nebular red-shift, and to point out the indications that spatial curvature may have to play a part in the explanation of existing nebular data".

The linear law, $cz = Hr$, relating the redshift z to distance r through c and the constant H, is known as Hubble's law and H is Hubble's constant. This relationship is of the same form as (6.5), which describes the rate of universal expansion according to de Sitter, rearranged to read:

$$\dot{R} = \frac{R}{T} \equiv RH\,,$$

with Hubble's constant defined as $H = 1/T$.

The two related expressions were soon combined to explain the relative shift in wavelength for galactic light,

$$z = \frac{\Delta\lambda}{\lambda}, \quad cz = DH = V\,,$$

as a Doppler shift arising from a rate of recession, V, between galaxies at a distance D apart. The merit of Hubble's constant is that, once H has been reliably established, accurately measurable redshifts can be used to determine the Hubble distance to galaxies, too remote to measure by conventional methods.

Hubble's observations established that the Milky Way is one of a family of 20 galaxies known as the local group. The largest of these galaxies, the Andromeda galaxy, has a spectrum which is not red-, but blue-shifted. To interpret the observed blue shift, not only of Andromeda, but also of several other galaxies in the local group, it was interpreted as a Doppler shift. It is common to hear the sound, emitted by an approaching fast vehicle, change pitch as the vehicle passes and disappears into the distance. The pitch rises while the vehicle approaches and falls when it recedes. This phenomenon, is known as the Doppler effect, after Christian Doppler who first identified the coloured light of double stars as a velocity-related effect of the same kind. There is an apparent change in the wavelength of light emitted by a source that moves relative to the observer. Waves emitted in the direction of motion appears compressed or blue shifted, whereas those behind the moving source are lengthened or red shifted.

Applied to the blue-shifted spectrum of Andromeda it was inferred that the two galaxies are approaching each other at a velocity of about 100 km s^{-1}. When this eminently reasonable interpretation was uncritically extended to measured red shifts it was concluded from Hubble's law that all galaxies

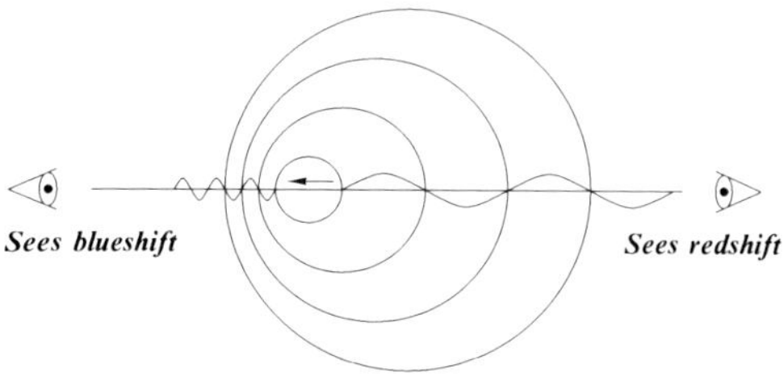

moved apart at a rate that increases as their mutual separation. The expanding universe was born.

Friedmann's expanding-universe solutions of Einstein's equations were assumed to confirm this interpretation.

6.7 The Big Bang

There is an interesting parallel between the development of quantum theory and big-bang cosmology. Both theories developed from a single seminal idea – the atomic nature of radiation assumed by Planck in 1902 developed into quantum theory; the assumed singularity at the beginning of universal expansion was interpreted by Lemaître in 1931 as a creation event, which grew into the big-bang model. From their inception to maturation took about thirty years in each case. During this period several workers, interested in different aspects of the new ideas, established a variety of interpretations to define the underlying physics.

There came a time when each model expanded too rapidly for scientists to keep track of all points of view and for the practitioners to reach a consensus interpretation. In the case of quantum theory the arguments were settled, to the satisfaction of none, on the authority of a few, and ordained as the Copenhagen interpretation. There is no written record to the details of this interpretation, which to this day, remains open to the formulation that best suits individual tastes. One of the authors who claimed responsibility for the interpretation (Heisenberg, 1955) states the purpose of the paper to

> '... subject the various counter-proposals against the Copenhagen interpretation to a short criticism; the details of the "orthodox" quantum theory will be supposed known to the reader.'

Many readers probably had a problem with this.

The situation with standard cosmology is no different. Without real consensus many points of view are defended to reflect the dictates of the

Copenhagen equivalent, known as the Big Bang. In a typical survey (Chown, 1993) three experts were interviewed to explain the basic concepts. In response to the question of how space can expand, one expert replies:

> "Good question. The answer is: space does not expand. Cosmologists sometimes talk about expanding space – but they should know better."

The answer obviously explains nothing and like the rest of the survey leaves us to evaluate an amorphous concept defined by unanswered questions. We may try to examine the way in which some of the obvious questions are dealt with in the cosmological literature:

1. Why did the big bang happen?

The only clue provided by cosmologists, who believe that the universe expands, is the statement that hypothetical inversion of time flow would retrace the history of the world, not as expanding, but as shrinking. The inevitable destiny must then be a state of infinite density as all matter disappears into a singular point. In the same way that this reverse process cannot be followed beyond the singular point, it is argued that the history of the expanding universe could only have started at this point.

One way of looking at a mathematical singularity is as a state of zero information. In the words of undisputed authorities (Barrow & Tipler, 1988)[3]:

> "At this singularity, space and time came into existence; literally nothing existed before the singularity, so, if the Universe originated at such a singularity, we would truly have a creation *ex nihilo.*"

The meaning of this is never explained. If there is nothing to trigger some action, nothing can happen. Before time came into being, by definition, nothing can change. Begging the question the authors explain that:

> "... SAP [Strong Anthropic Principle] requires a universe branch which does not contain intelligent life to be non-existent;"

[3]In the Forward to this book, J.A. Wheeler states that "... the authors provide one of the best short reviews of cosmology ever published."

If I understand this correctly it means that the big bang happened because of a quantum-mechanical probability that intelligent life can evolve, given the correct boundary conditions. To arrive at this conclusion the Universe is described by a wave function $\Psi\big(R(t)\big)$, with a boundary condition imposed at the universal radius $R(\tau)$, close to, but larger than $R(0) = 0$. There is an unexplained quantum jump between $t = 0$ and $t = \tau$. It means that question (1) remains unanswered.

2. When did it happen?

This is easy. Given Hubble's law and constant the time since the big bang is calculated as follows:

From $$z = \Delta\lambda/\lambda = H{\cdot}r/c$$

and $$v = dr/dt = H{\cdot}r = z/c$$

follows: $$T_0 = r/v = 1/H\ ,$$

the Hubble age of the Universe. Hubble's original estimate of $H = 530$ $\text{kms}^{-1}\text{Mpc}^{-1}$ corresponds to an age of $T_0 = 2$ Gyr (2×10^9), which is less than the known geological age of the planet. The most recent estimate, based on a larger database, of $H = 75$ in the same units, predicts an age of about 14 Gyr. Recently observed redshifts, close to 5, imply the existence of objects with an age exceeding 14 Gyr, giving rise to an interminable debate between supporters and critics of the big-bang model.

In the meantime there is another debate among cosmologists to decide between the three possible expansion models with $K = 0$ or ± 1. The importance of this debate is underlined by Figure 6.3 which shows the relationship between Hubble age and actual age, T_a for an expansion model.

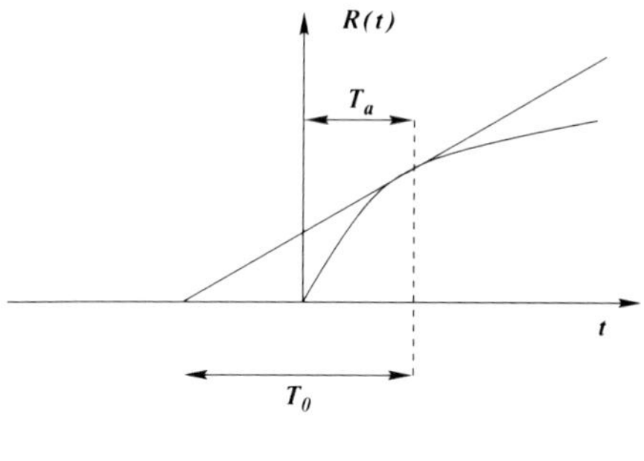

Figure 6.3: *Relationship between actual and Hubble ages of the Universe*

The Hubble age is determined by the tangent to the expansion curve and is always larger than the actual age. For the Einstein–de Sitter model, $K = 0$, $T_a = \frac{2}{3}T_0$. For an oscillating model, $K > 0$, $T_a < \frac{2}{3}T_0$, and for the open ever-expanding model, $K < 0$, $T_a \sim T_0$.

The standard Einstein–de Sitter model predicts an age of the Universe less than 10 billion years. Many ingeneous arguments, based on relativistic effects, mass density, deceleration parameter, cosmological constant and flatness, are forwarded to compare the merits of the competing cosmological models. It is interesting to note that on this basis there is a general preference for the Einstein–de Sitter model, until an age conflict enters the argument, when only the Hubble age is considered to be of relevance, without further explanation.

The question is no longer answered as easily as anticipated.

3. What came before the big bang?

Stephen Hawking (1988), widely recognized as the ultimate authority on big-bang cosmology and the anthropic principle gave a detailed answer to this question in his all-time best seller, that probably needs no further introduction. In his own words (p. 50) he claims credit for the idea

> "...which at last proved that there must have been a big bang singularity provided only that the theory of general relativity is correct ..."

He continues in the same paragraph:

> "...having changed my mind, I am now trying to convince other physicists that there was in fact no singularity at the beginning of the universe..."

One thing about which his mind had not changed at the time is the anthropic principle. He is careful not to support (p. 125) nor reject (p. 126) the strong anthropic principle which claims that the entire universe exists only for an intelligent being to observe and understand it. Judging by the assessment of his own achievements, it is not too difficult to imagine that he had himself in mind as the *anthropos* of the principle. One gets the impression that the book deals, not so much with time, as with Hawking. However, early on (p. 12) it is stated that

> "...the major theme of this book, is the search for a new theory that will incorporate [both quantum mechanics and the general theory of relativity] – a quantum theory of gravity."

He reviews the history of relativity, concluding with the claim (p. 34):

> "...I showed that Einstein's general theory of relativity implied that the universe must have a beginning and, possibly, an end."

He comes close to answering question (3) in stating, (p. 46):

> "...all our theories of science [...] break down at the big-bang singularity, where the curvature of space-time is infinite. This means that even if there were events before the big bang, one could not use them to determine what would happen afterward, because predictability would break down at the big bang. [...]we know only what has happened since the big bang, [...]events before the big bang can have no consequences [...]. We should therefore cut them out of the model and say that time had a beginning at the big bang."

It would be helpful to have an explanation of how events can happen before the beginning of time.

Having contradicted himself on a beginning or no beginning – singularity or no singularity Hawking finds a speculative solution – quantum gravity in imaginary Euclidean space-time (p. 134). He goes on to say (p. 135) that applied

> "...to Einstein's view of gravity, the analogue of the history of a particle is now a complete curved space-time that represents the history of the whole universe [...] these curved space-times must be taken to be Euclidean. [...] if we knew how the Euclidean curved space-times [...] behaved at early times, we would know the quantum state of the universe.

Curved space, by definition is non-Euclidean. Maybe what is defined here as imaginary Euclidean space is the same as Minkowski space. Still, curved Minkowski space is not Euclidean. The dénouement (p. 148) lies in the "histories" that

> "...satisfy the no boundary condition: they are finite in extent but have no boundaries, edges or singularities. In that case, the beginning of time would be a regular, smooth point of space-time and the universe would have begun its expansion in a very smooth ordered state."

To summarize: time began twice – first at the big bang and again later at some smooth point in space-time. The universe, on the other hand, has no beginning and no end, until ... (p. 135):

> "Space-time [is] like the surface of the earth, only with two more dimensions,"

and (p. 150)

> "...the North and South poles correspond to the beginning and end of the universe..."

The obvious intention is to use elliptic geometry (Section 1.2) as a model, which, however, identifies antipodal points and hence a common beginning and end of the proposed universe.

Not knowing exactly what Hawking has in mind it is difficult to comment on his "proposal" (p. 136). Is he, or are his interpreters, trying to hide behind half-baked ideas to protect the image of the "greatest scientist alive"? Maybe he is as confused as all of us, trying to figure out how to make scientific sense out of indefensible dogma. Once the popular media, in awe, report his profound insight into the quantum universe in imaginary curved Euclidean space-time, there is no longer room for doubt or comprehension. If the ultimate proposal is indeed a closed universe without boundaries, beginning or end, in what way, one wonders, does it differ from de Sitter's proposal of 1917?

4. Did space and time exist before the big bang?

The emphatic response of the expert (Steven Weinberg) that space does not expand provides an equally emphatic answer to question 4. If space does not expand (or shrink) it stands to reason that its present extent is the same as it was at the time of the big bang, and *vice versa*. Whatever came into being at the big bang did so at a singular point and by Weinberg's argument, can therefore not be interpreted as spread over all space. The infinite space-time curvature identified by Hawking with the big bang must then clearly be interpreted as only a local distortion around the singular point. The logical conclusion is that space-time antedates the big bang, which Hawking denies.

This conclusion means that either Hawking or Weinberg misinterprets the big bang, unless there was no such thing and they are both mistaken. Or maybe they confuse the mathematical concept of space with the physical vacuum.

5. How does the universe expand?

Martin Rees, a second expert in Chown's survey shows how to answer this question:

> "Think of the Universe in a Newtonian way – that is simply, in terms of galaxies exploding away from each other."

Weinberg elaborates further:

> "If you sit on a galaxy and wait for your ruler to expand, you'll have a long wait – it's not going to happen. Even our galaxy does not expand. You shouldn't think of galaxies as being pulled apart by some kind of expanding space. Rather, the galaxies are simply rushing apart in the way that any cloud of particles will rush apart if they are set in motion away from each other, "

but adds:

> "Every bit of the Universe is rushing away from every other bit."

Does it mean that no bit, smaller than a galaxy, occurs in the universe? And how does the big bang initiate this differential motion? Chown (1993) explains: The matter inside individual galaxies does not take part in the general expansion because it is held together by gravity.

There is no point in trying to make scientific sense of such caricatures of an expanding universe. Most authorities seem to agree that a big-bang singularity occured at the point, $R(t) = 0$, where the explosive birth of the universe imparted radial acceleration to all synthetic products. It explains how lesser cosmologists could draw the wrong conclusion that space-time was created here and carried everything along as it expanded – eventually resulting in all bits racing away from each other. But, supposedly Weinberg knows best.

By contrast it may be helpful to re-examine the consensus model in which the universe is considered made up of galaxies which interact like gas molecules in a situation of zero pressure. By simple ideal-gas kinetic theory the kinetic energy, given by the product pV, and the temperature, $T = pV/R$ must both be zero in such a sample. There is no further argument about this – the distance between a pair of such galaxies must remain constant, unless the measuring scale is shrinking, which means that space expands.

But maybe the pressure is not zero and galaxies are in relative motion. Is it legitimate to ask when and how the galaxies acquired their kinetic energy? The first galaxies were supposedly formed at a considerable time after the big bang and according to Weinberg space could not expand in this pre-galactic epoch. By what mechanism then was matter dispersed, if not by expanding space?

Anticipating the inflation postulate in answer, anti-gravity now enters the argument. Perhaps it is premature to make this assumption until gravity itself is properly understood. The more reasonable assumption of an isotropic universe poses the next question.

6. Where did the big bang happen?

There are no special points in an isotropic space. An isotropic universe therefore denies the possibility of a unique singular point. Either the big-bang singularity occured at all points of space or not at all. Having agreed that space does not expand, it means that a big bang in an isotropic universe occured everywhere. How can such a universe expand any further and what happened to the infinite mass, which was created instantaneously at every point throughout the universe?

This argument is not a malicious distortion of big-bang theory. Without spelling out its unpleasant consequences, it is exactly the argument used in response to one of Chown's (1993) questions:

> "The matter the Earth is made of was in the big bang and mixed in with that matter was the fireball radiation. So why is the radiation from the Big Bang still arriving at the Earth today? Shouldn't it have passed us long ago?"

He could have continued to ask, how did the Earth manage to get here ahead of the radiation which travels at the speed of light? Weinberg answers:

> "Well, radiation that was emitted from our immediate neighbourhood in the big bang did pass us. [...] radiation that was emitted 10 billion light years away passed us 10 billion years after the big bang. Radiation that was emitted 15 billion light years away is just passing us now."

As an aside Weinberg throws out the line:

> "It's slightly complicated because the universe is expanding. [...] The complication occurs because the Universe was smaller in the past."

Slight complication indeed. We are back to the problem of what is expanding. The two opposing points of view are symptomatic of the total disagreement between the standard models of particle physics (Weinberg) and big-bang cosmology, concerning the nature of space (or the vacuum). We shall meet this discord again in the predicted values of the cosmological constant which differ by more than 50 orders of magnitude.

Big-bang cosmology is based exclusively on Friedmann solutions, using the Robertson–Walker metric (6.6). The scale factor $S(t)$ determines the measure between neighbouring points of space-time, as well as the radius of the hypersurface, interpreted as the Universe in Hubble's theorem. To demonstrate this, let $\boldsymbol{v}$ be the velocity of a particle at a radius vector $\boldsymbol{r}$ from a co-moving observer, assuming that the velocity, a function of time, depends on $\boldsymbol{r}$: $\boldsymbol{v}= f(t)\boldsymbol{r}$.

Write $$\boldsymbol{r} = R(t)r_0\,,\ \text{where } r_0 = \boldsymbol{r}(t_0)$$

i.e. $$\boldsymbol{v} = \frac{d\boldsymbol{r}}{dt} = \frac{dR}{dt}r_0 = f(t)\boldsymbol{r} = f(t)R(t)r_0$$

Hence $$\frac{1}{R}\frac{dR}{dt} = f(t)\,,$$

which is Hubble's law and R is the radius of the hypersphere.

In this cosmology expansion of the universe therefore is the same as the expansion of space-time. At a given time two co-moving particles move apart at a velocity which is a linear function of their distance apart. The assumed isotropy is built into Hubble's law. This result has two important implications.

(i) The big-bang scenario is consistent with cosmic isotropy, only if Hubble's law is interpreted in terms of a universe, confined to Euclidean three-dimensional space, which has been expanding from a singular point at time zero. All radiation and matter in the expanded universe originated in close proximity of this point.

Weinberg's interpretation that only galaxies move apart through infinite static space (Chown, 1993) is not supported by this scenario.

(ii) The universe expands with the speed of light. The redshift of incoming radiation is a direct measure of the age and rate of recession of the source.

Assume that the big bang happened at a time T in the past at space coordinates (0,0,0). This means that primeval radiation has travelled a distance $d = cT$ by now. An absorber, which interacts with the same radiation on Earth today, must have moved the same distance at a velocity $v = d/T$ from the big-bang coordinates where it came into being. The only transport mechanism of moving the absorber to its present coordinates, at velocity v, is the universal expansion.

Notice that the only condition that relates the absorber to the source is the radial distance, d, from the origin of coordinates. From the perspective of the absorber the radiation could have been emitted at any point on the spherical surface of radius d, which surrounds the Earth and which is equivalent to saying that all relevant emitters have by now been displaced into this surface. The radiation therefore approaches the Earth isotropically.

Although no record of this exposition could be found in the literature, it probably agrees with the view of those poorly informed cosmologists who "talk about expanding space". The nasty consequence is that the rate at which the universe is thought to expand is much slower than c, but in cosmology it is always possible to argue around unpleasant contradictions. In this instance the idea of *inflation* comes to the rescue, while the *anthropic principle* remains on standby.

7. Is Big bang a relativistic theory?

One of the misleading statements, used to introduce most accounts of big-bang theory, is that the theory is based on general relativity. It is, more precisely, based on the assumption that the universe expands. The only connection between expanding universe and general relativity occurs in de Sitter's 1917 solution (Section 6.3) in terms of a variable hypersphere, which is characteristic of curved four-dimensional space-time. To define this variability as an expanding universe it is necessary to separate time from space coordinates, an operation which is not allowed in relativity theory. The expansion of such a Newtonian universe is deduced from galactic redshifts, but as stated by Parish (1981):

> "...the only proven fact about cosmological redshifts is that they are observable".

It is instructive to note that Milne (1948) derives Hubble's law from two completely different Newtonian models of the universe. In the first model the world is represented by an expanding Euclidean universe of finite volume, having emerged a finite time ago. Alternatively, a stationary world, infinite both in time and space with Lobachevskian geometry, predicts exactly the same observed redshifts. The two alternatives are analogues of those in de Sitter's model.

The demonstration that big-bang theory is not underpinned by relativity does not necessarily invalidate the concept. However, it is the claim, which is made to flaunt respectability, that is disputed here.

8. Are the laws of physics valid in big bang?

Most laws of physics are conservation laws – conservation of mass-energy, momentum, angular momentum and electric charge, to give the best-known examples. Each conservation law can be shown to depend on a symmetry of Nature (Boeyens, 2005). For instance, conservation of energy is a consequence of temporal homogeneity, conservation of momentum is due to the homogeneity of space, angular momentum of the isotropy of space, and the conservation of electric charge is dictated by global gauge invariance.

The most comprehensive symmetry that incorporates all of the foregoing is best known by the acronym CPT (charge conjugation – parity – time). Its most stringent demand is an exact balance between the matter and antimatter of the universe.

A moment's reflection shows that the big bang violates all of the individual conservation laws and tries to outfox the CPT theorem by sleight of hand, dressed up as Grand Unified Theories (GUT's). They find an interval, between $t = 10^{-43}$ and 10^{-35} seconds after the big bang, when a state of non-equilibrium could have wiped out all antimatter. Rather convincing.

Max Born (1962) addressed the problem head-on:

> "Whether there was a creation from nothing is not a scientific question, but a question of belief and beyond experience, as the old philosophers and theologians like Thomas Aquinas knew".

9. Can a unique event be studied scientifically?

Although science cannot advance without original ideas, original ideas by themselves can never develop into science. They need experimental support.

The big bang happened in a flash, never to return for a second show. Cosmologists claim that the event can be reconstructed on smaller scale in the accelerators of particle physics, which claim that data beyond its reach are supplied by the cosmological reconstruction of the event. Between them the two pursuits have failed to answer even the most elementary questions that we have highlighted. The closest they get to experimental support of their ideas is the microwave background, described by some as "the smoking gun". Like the other circumstantial 'evidence', inferred from nucleogenesis, inflation, anti-matter and dark matter, it is inconclusive and open to alternative interpretation. However, the bulk of international funding for fundamental research is directed at an elite group of elementary-particle physicists, who have established a close symbiosis with big-bang cosmologists to whom they provide theoretical underpinning for their ideas. Further discussion demands closer scrutiny of this symbiosis, which, together with other scientific aspects of big-bang theory, is explored next.

6.7.1 Particle Physics

The gauge principle (Section 4.3.1) that was used to link the electromagnetic field to the local structure of space-time can be extended to also describe two other fundamental forces, known as the weak and strong interactions. Without going into details[4] it is sufficient for our purpose to note that the non-Abelian transformation groups $SU(2)$ and $SU(3)$ have been shown to generate the weak and strong interactions respectively. The weak interaction which operates between leptons, such as electrons and neutrinos, is mediated by three gauge bosons. The strong interaction between baryons, in the form of quarks, relies on the exchange of eight gluons.

A development of major significance was the reduction of the known forces of Nature, from four to three, by demonstrating that electromagnetic and weak interactions are generated together by requiring the Lagrangian function that describes all interactions that involve leptons to be invariant under the combined $SU(2) \times U(1)$ group of gauge transformations. Spontaneous breaking of the symmetry causes the appearance of the electromagnetic field, mediated by photons, and the weak interaction, mediated by three gauge bosons.

By the same argument the gluon structure of quantum chromodynamics (QCD) is generated by requiring the quark Lagrangian to be invariant under the $SU(3)$ group of transformations. To achieve unification of the three interactions the grand symmetry group $SU(5)$ is assumed. Gauge invariance of $SU(5)$ generates 24 gauge bosons (12 of which have already been accounted for). The remaining 12 mediate interconversion between leptons and quarks. However, this is a broken symmetry, except at extremely high energies, impossible to reach in particle accelerators. It has been estimated, that even in a hot big bang, the temperature at which the $SU(5)$ symmetry is broken, is reached after 10^{-35} seconds.

By adding experimental data from nuclear physics, wherever possible, particle physicists have managed to calculate a complete timetable of likely events as they unfolded after the big bang. Each epoch is identified by a characteristic size and temperature as the expanding universe cools down. The purpose of the exercise is to correlate the theoretical scheme with experimental evidence that supports the cosmological model. This includes predicted light-nuclei abundances, the nature and temperature of the microwave background radiation and the observed galactic and quasar redshifts.

To avoid any confusion it should be mentioned that GUT's do not include

[4]For a detailed discussion see: Coughlan & Dodd (1991).

gravity at temperatures below $T = 10^{32}$K that limits the epoch of quantum gravity to 10^{-43}s.

6.7.2 Big-bang Nucleogenesis

Predicted nuclear abundance is one of the crucial arguments in deciding the relative merits of alternative cosmologies. In an early version of big-bang cosmology it was claimed as the only means of producing all possible nuclides during the early stages of expansion by continued addition of protons and α-particles, starting with the initial reactions:

$$\begin{aligned} n + p &\rightarrow {}^2\mathrm{H} + \gamma \\ {}^2\mathrm{H} + p &\rightarrow {}^3\mathrm{H} + \gamma \\ {}^2\mathrm{H} +{}^2\mathrm{H} &\rightarrow {}^3\mathrm{He} + p \\ {}^3\mathrm{H} + p &\rightarrow {}^4\mathrm{He} + \gamma \\ {}^4\mathrm{He} +{}^2\mathrm{H} &\rightarrow {}^6\mathrm{Li} + \gamma \\ {}^3\mathrm{H} +{}^4\mathrm{He} &\rightarrow {}^7\mathrm{Li} + \gamma \end{aligned}$$

Because there are no stable nuclides with $A = 5$ or 8, the claim was later toned down, for lack of time, to the big-bang production of only the six light nuclei as above. The synthesis of heavier nuclides is postponed to happen in stars at a more leisurely rate. We note that in a black hole, where the radial coordinate turns into a time coordinate there is more than sufficient time to establish the proposed equilibrium.

The fact that the light nuclides, especially ^{1}H and ^{4}He, are more abundant by orders of magnitude, compared to all heavier nuclides, is interpreted to distinguish two fundamentally different modes of synthesis. The big bang is postulated responsible for primordial synthesis of the light nuclei and stellar nucleosynthesis (Burbidge et al., 1957) to produce all the heavy nuclides.

The stellar mode is based on sound principles and, no doubt, contributes to the overall spread of observed abundances. However, as indicated before, it fails to account for the periodic trends, which favour a comprehensive equilibrium process, that includes the light nuclides.

Elaborate calculations (Krane, 1988) of abundances, based on primordial nucleogenesis, considered one of the three conclusive proofs of the big-bang scenario, are not as telling as claimed and have a healthy injection of speculation. Jay Narlikar (1992) comments on the situation as follows:

> "...the primordial nucleosynthesis of light nuclei assumes the existence of electrons, protons, neutrons, neutrinos and photons, at, say, $t = 0.01$s. The next step is to understand how these

> particles were made. So we push our investigations through the epoch of grand unified theories (GUTs) at say $t \sim 10^{-36}$s. The process, even if it succeeds would again require us to understand the state of the universe at the GUTs epoch in terms of something that existed earlier. This may lead to the Planck epoch of $t \sim 10^{-43}$s when quantum gravity was important. And so on ...
>
> A closed interval terminating at $t = 0$ with the conditions specified here and with no further backward steps allowed is [...] entirely alien to the [...] spirit of enquiry."

The extravagent claim that only big-bang synthesis can account for the relative abundances of H and He is based on a chain of assumptions, long enough to predict any desired result. The logic of the $(\alpha - \beta - \gamma)$ scheme depends on the assumed relationship between black-body temperature and the radius of the expanding universe

$$T_{rad} \propto \frac{1}{R(t)} \propto \frac{1}{\sqrt{t}}$$

at t seconds after the bang, and the entropy density

$$S = \tfrac{4}{3} a T_{rad}^3 \propto n \, ,$$

in terms of Stefan's constant a and the concentration of material particles. Substitution of all relevant constants gives

$$T(K) = \frac{1.5 \times 10^{10}}{\sqrt{t(s)}}$$

and shows that the entropy per particle $s = S/n$ is independent of T_{rad}.

To test the postulates it is assumed that all matter first emerged in the form of neutrons, which in time decay into protons, electrons and neutrinos, with a half-life of 700 seconds. Before depletion of the neutrons it is necessary for the synthesis chain to commence, resulting mainly in the formation of ^{4}He.

The first reaction is rate determining and depends on temperature and matter density. It becomes a race against the expansion of $R(t)$ and depends critically on the value of s at $t = 700$s. Being independent of T this value of s predicts the black-body temperature today, at a particle density of 10^{-7}cm^{-3}, *i.e.*

$$T_{min} \sim 25\text{K}.$$

For smaller values of s the big bang would have produced only He and nothing else.

The limiting factor in the α–β–γ theory appears to be the time of 700 seconds required to turn half the available neutrons into protons for deuterium production. As a possible remedy it was suggested that electron–positron pair production at[5] $T > 10^{10}$K ($t < 2$s) could speed up the formation of protons by the fast process

$$n + e^{+} \rightarrow p + \bar{\nu}\,.$$

For details see Sciama (1971).

Such considerations brought the He/H ratio closer to the experimentally most likely value of 1/11, without lowering the black-body temperature. The cooling rate of the big-bang universe has defeated all efforts to demonstrate the primeval production of heavy elements, which is now accepted to occur in massive stars. The problem is that these same processes also produce large quantities of He, which, added to the big-bang production, far exceeds the theoretical budget (Lerner, 1989).

The equilibrium model does not have this problem. If the process occurs on passage through a black hole all matter is first reduced to an elementary state (analysis) before conversion into its conjugate state on time reversal

$$m \rightarrow m^{*} + \gamma.$$

Enough energy is generated to propel the product into conjugate space, following the inverse synthetic route. Matter (antimatter) is neither destroyed nor created, but circulated.

6.7.3 Microwave Background

Microwave background radiation has replaced nucleogenesis as the indisputable evidence of a big bang, although the calculated value does not match the experimental. The surprising thing about the diffuse black-body heat bath that pervades all space is that it has been known and ignored for decades. It was described by Arthur Eddington (1930) long before there was a big bang. He writes (p. 371):

"Accordingly the total radiation of the stars has an energy-density

$$2000 \times 3 \cdot 83.10^{-16} = 7 \cdot 67.10^{-13}\text{ergs/cm}^{3}.$$

[5]This, by the way, coincides exactly with the epoch of baryogenesis during which the assumed annihilation of all antimatter occured. One gets the suspicion that there are too many balls to keep in the air.

> By the formula $E = aT^4$ the effective temperature corresponding to this density is
>
> $$3^\circ \cdot 18 \quad \text{absolute.}$$
>
> In the region of space not in the neighbourhood of any star this constitutes the whole field of radiation, and a black body, e.g. a black bulb thermometer, will there take up a temperature of $3^\circ \cdot 18$ so that its emission may balance the radiation falling on it and absorbed by it. This is sometimes called 'the temperature of interstellar space'."

The existence of this heat bath at a temperature of 2.3K was inferred from the observed rotational spectrum of interstellar cyanogen, CN, as early as 1941, and confirmed many times since then (Sciama, 1971). It only became big news when it was linked to the big bang, although there was nothing left to explain. Fred Hoyle (1997) lamented:

> "How, in big-bang cosmology, is the microwave background explained? Despite what supporters of big-bang cosmology claim, it is not explained. The supposed explanation is nothing but an entry in the gardener's catalogue of hypotheses that constitutes the theory. Had observation given 27 kelvins instead of 2.7 kelvins for the temperature, then 27 kelvins would have been entered in the catalogue. Or 0.27 kelvin. Or anything at all."

The expansion timetable of particle physics predicts that electrons and nuclei first combined to form electrically neutral atoms when the temperature dropped below 4000K at $t = 10^5$ years after the big bang. With the disappearance of charged particles that scatter photons the universe became transparent and the accumulated radiation became free to propagate through space without impediment. Electromagnetic astronomical observation therefore cannot penetrate further back in time than the so-called *recombination epoch*. As this primordial radiation spread through the expanding universe its wavelength has been stretched into the microwave region. Until the time of last scattering the younger, much smaller, universe kept all radiation confined as in a hollow cavity with the intensity distribution shown in Figure 4.1, known as a black-body spectrum. An isotropic background with exactly the predicted intensity distribution observed with high precision is generally interpreted as indisputable proof of the big bang.

A surprising property of the background radiation is its remarkable homogeneity in temperature over the entire visible universe. A signal emitted longer than time t ago cannot be observed further than a distance ct away from the source. For the same reason two particles, further than $2ct$ apart, could not have experienced contact through a common intermediary since $-t$ in the past.

In an expanding universe $2ct$ defines the particle horizon that limits the independent region over which a homogeneous equilibrium state can develop at time t. The radius of the initial homogeneous region, at a temperature of $T_{in} = 10^{28}$K, at time $t = 10^{-35}$s,

$$\begin{aligned} R_{hom} &= 2 \times 3 \times 10^{12} \times 10^{-35}\text{cm} \\ &= 6 \times 10^{-23}\text{cm}. \end{aligned}$$

The rate of expansion, as a function of temperature, has been calculated as $\eta = T_{in}/T_{fin}$. At the time of recombination, $T = 4000$K, and hence

$$R_{hom} = \frac{6 \times 10^{-23} \times 10^{28}}{4 \times 10^{3}} = 150\text{cm}.$$

The temperature of the present universe, derived from the microwave background, $T = 2.7$K, predicts that the radiation should be homogeneous over a radius of 6km. However, the radiation is observed uniform in temperature to within 10^5 on a cosmological scale of about 10^{28}cm.

A series of proposals, generically known as *inflation theory*, address this, and other related problems that appear to invalidate big-bang theory. One of these, the so-called flatness problem indicates that, as early as 10^{-10}s after the big bang, space was Euclidean and flat. General relativity, on the other hand, requires space to be highly curved in an environment of such high mass density.

6.7.4 Inflation

The flatness problem should have killed the big bang. An infinite concentration of mass implies infinitely curved space, or in Newtonian terms, an infinite gravitational gradient. This ultimate black hole must surely prevent any form of expansion. The original author[6] of the big-bang idea was aware of this problem and postulated a divine act to initiate expansion, in opposition to the law of gravity, which only kicked in later on as the divine volition, or antigravity, ceased to operate.

Standard cosmology resorted to the same device, renamed inflation, and assumed to coincide with GUT symmetry breaking at $t = 10^{-35}$s. This

[6]E.A. Poe (1848).

event caused the vacuum, which measured 1mm across, to undergo a phase transition that unleashed an irresistible force, driven by the latent heat of the transition. Conservative estimates indicate that the subsequent inflation increased the size of the universe by a factor 10^{30} almost instantaneously.

Had this process proceeded unchecked it would have by-passed several important stages in cosmic evolution between $t = 10^{-35}$ and 1s, such as baryogenesis, electroweak symmetry breaking, combination of free quarks to form hadrons and interconversion between protons and neutrons. Not to interrupt this orderly evolution it would therefore be useful to have the phase transition postponed for a while. Many phase transitions are indeed known to be delayed by the phenomenon of supercooling. Why not this one?

Why not indeed! The foregoing is exactly the common basis of the many inflationary theories that have been proposed to promote the reconciliation of the two symbiotic theories. It is pointless to discuss the subtle details of these theories. The cosmologist William Unruh has been quoted (Glanz, 1996) as stating:

> "I'll fit any dog's leg that you hand me [using] inflation".

However, a few general remarks are in order. The homogeneity and flatness problems are addressed more than adequately. The predicted homogeneous region now exceeds the observable universe by a factor 10^{23} in size. The flatness problem disappears as the rapid expansion causes space to flatten out like the surface of an inflating balloon. However, the latent heat, which acted like antigravity during inflation, is equated in particle theory with the cosmological constant Λ, at a value in total disagreement with cosmolgy, as discussed in the next section.

6.7.5 Cosmological Constant

As shown before (Vermeil, 1917) equation (6.4), with cosmological constant, is the mathematically correct form of the gravitational field equations.

Although the cosmological constant does not feature in standard cosmology any more it has a prominent meaning in the supportive field theories of physics, associated with virtual particles in the vacuum. In inflation theory it is more precisely identified as a form of antigravity – the same role initially envisaged for it by Einstein.

Another attractive feature of the cosmological constant is an ability to eliminate the gravitational singularity at zero separation, more popularly ascribed to quantum gravity. Maybe it is antigravity and maybe it is not constant, like a short-range exclusion principle, inversely proportional to distance.

The cosmological constant poses a major dilemma for standard cosmology and particle physics. In a popular article of twenty years ago (Abbott, 1988), with the byline:

> "An understanding of the disagreement could revolutionize fundamental physics,"

an interesting comparison of the different estimates of Λ is made.

In cosmological models Λ is routinely ignored because of its vanishingly small value, but in particle physics it has a non-zero value, arising from the presence of infinitely many virtual particles that constantly fluctuate in and out of the vacuum. Experimental estimates of spatial curvature predict

$$\Lambda < 1/(10^{23}\text{km})^2 \,,$$

compared to the theoretical estimate of $\Lambda > 1/(1\text{km})^2$.

A positive Λ of this order is said to have many

> "...bizarre consequences. [...] if we walked farther than a few kilometers away from home [...] the gravitational distortion of spacetime would be so great that we could never return ..."

It appears legitimate to ask:

> "Is the vacuum energy density really as large as these theories appear to suggest it is?
>
> The answer is most emphatically no."

Finally:

> "Our theoretical estimate [...] is incorrect by, at the least, an astonishing factor of 10^{46}. Few theoretical estimates in the history of physics [...] have ever been so inaccurate."

Many attempts have been made to defuse the Λ problem. A common strategy is by vague reference to unexplored, but likely, effects of otherwise well-known ideas, such as divine intervention, quantum uncertainty, many-worlds theory, wormholes or the anthropic principle, to account for any apparent theoretical inadequacies. It requires little more than some reassurance on the future vindication of a viable theory that may now appear flawed. We read (Schwarzschild, 1989):

> "In quantum mechanics almost everything is uncertain. So why should the topology of space-time be fixed?"

This means that the probabilities of a many-world wave function allows an infinitude of universes connected by wormholes that drain away the excess vacuum energy until $\Lambda = 0$, the value dictated by the anthropic principle or creationist design. The effect of the tiny wormholes, of Planck length, may be likened to osmotic seepage through a semipermeable membrane.

The problem with this handling of the cosmological constant is that it depends on too many wild ideas without common ground. On the one hand we assume a Newtonian metric to guide Einstein's equations into an exploding world, rationalized as the anthropic probability among infinitely many quantum universes, which interact through ten-dimensional wormholes. Do we forget that the small value of Planck's constant restricts quantum effects to microscopic objects?

A more logical response would be looking for a relativistic, rather than a quantum, solution of the cosmological problem in order to avoid the big-bang singularity in a natural way. Twenty years ago Maddox (1989) recommended:

> "Down with the Big Bang,"

but it still flourishes.

However, the problem of Λ in particle physics remains. It can be traced back to the concept of massive indivisible point particles with infinite self-energy and self-field – the basic premise of field theories based on the Copenhagen interpretation of quantum mechanics. As recollected by one pioneer (Heisenberg, 1955):

> 'SCHRÖDINGER atempted to deny entirely the existence of discrete energy values and quantum jumps, and to resolve quantum theory into a simple classical wave theory. [...]
>
> The months which followed [...] were a time of the most intensive work in Copenhagen, from which there finally emerged what is called the "Copenhagen interpretation of quantum theory," [...] BOHR intended to work the new simple pictures, obtained by wave mechanics, into the interpretation of the theory, while I for my part attempted to extend the physical significance ....'

This so-called orthodoxy was formulated with the express purpose of discouraging wave-mechanical interpretations of quantum phenomena, which started to appear (Gamow & Iwanenko, 1926) within the same year as Schrödinger's papers. As an alternative to the particle theory quantum objects are defined here as made up of waves, with internal structure described by a parameter of the form $\theta = \exp(2\pi ip/h)$, where p is the phase (gauge) factor, first identified by Schrödinger (1922), and which links the wave structure of an electron directly to the geometry of space-time. In the context of vacuum structure the infinite mass of particle theory does not occur and the cosmological constant becomes insignificant. Revival of the wave theory of matter in modern times is documented in detail by Milo Wolff (1995).

6.7.6 Anti-matter

To avoid the problem of CPT symmetry standard cosmology makes the empirical assumption that the universe contains no anti-matter and that all cosmological models need to explain this observation. This is clearly an unsubstantiated assumption because spectroscopic observation, on which astronomy relies, cannot distinguish between material and anti-material galaxies.

The assumed preponderance of matter over antimatter supposedly originated during the epoch of baryogenesis, soon after the big bang, as the temperature of the universe decreased to the point where heavy GUT bosons were no longer produced. These bosons and their conjugates, equal in number, started to decay and annihilate, producing baryons, antibaryons and radiation. If the decay processes, like those mediated by weak interaction, happened with CP symmetry violation, there is the possibility of an imbalance between baryons and antibaryons developing at this time. The rate of expansion at this stage should be such that the remaining antibaryons could be completely annihilated, leaving only baryons and photons behind. This is one of the reasons why $SU(5)$ symmetry breaking had to supercool before inflation could separate the chiral enantiomorphs.

It is of interest to note that electron-positron annihilation only commenced at $t = 1s$ and $T = 10^{10}$K, well into the inflation period. In the absence of CP violation and the known existence of electrons it now becomes difficult to understand the observed balance between protons and electrons or the whereabouts of the positrons.

6.7.7 Dark Matter

The idea of dark matter was first mooted to resolve Zwicky's paradox, which relates to different estimates of galactic mass, based on luminosity and rotational models respectively. Rotational analysis consistently predicts excess mass, compared to the total luminous mass within the assumed limiting size of a galaxy, as defined by its so-called Holmberg radius. The subsequent discovery of massive clouds of both atomic and molecular hydrogen around typical galaxies and clusters may well account for this missing mass.

The chemical structure of molecular hydrogen makes it extremely difficult to detect in space. Because it has no permanent dipole moment it cannot emit radio waves which could be detected by the common 21-cm receivers of radio astronomers. It therefore is a serious candidate as the major component of invisible matter that gives rise to the anomalous rotation of spiral galaxies (Rubin, 1983).

Unlike planets in the solar system the stars that rotate around the massive hub of a spiral galaxy do not obey Kepler's laws. Doppler measurements of the rotation show that stars near the galactic rims rotate just as fast as those near the hub. An attractive explanation of this phenomenon is that a halo of gas that fringes the galaxy is responsible for the distortion of the inverse-square law (Rubin, 1988). It has been argued that the invisible material need not be confined to a halo at the fringes, but could be distributed within the galaxy and even extend through all intergalactic space. Evidence of such massive molecular clouds is accumulating all the time.

The only reason why this sensible explanation is not accepted by cosmologists is that it requires a mass of hydrogen in the universe that exceeds the amount, alledgedly produced in the big bang, by many orders of magnitude. To save them this embarrassment the world has to live with the non-baryonic alternative. Apparently there is no incentive to explain the big-bang synthesis of non-baryonic matter on the required scale. The leading candidate is a swarm of neutrinos, but apart from being almost certainly massless, there is no known mechanism whereby neutrinos could be confined to orbit the galactic centre and not disappear into intergalactic space. Gravity provides this mechanism for molecular hydrogen.

Another complicating factor is the assumed stability of galaxies and clusters, as if all stars or galaxies are bound in fixed orbits. If instead, these structures can eject members into the universe at large, the estimated mass will be in error. Judging by the accepted structure of the solar system, which is surrounded by the Kuiper belt and the Oort cloud of comets, from where they escape on a regular basis, it is not unreasonable to assume that galaxies are likewise surrounded by small faint stars, solar systems, asteroids, comets, cosmic dust, molecular gas and atomic matter, at an enormous scale, but invisible from here. Even the missing matter implied by the Einstein–de Sitter model, which predicts the critical Hubble density

$$\rho_c = \frac{3H^2}{\kappa c^4} ,$$

could be accounted for by such matter. Although the proposal is largely conjectural it is sufficiently likely to avoid invoking the presence of non-baryonic matter, whatever that may be.

The demands of big-bang cosmology changed all that. The existing, but unfortunate, terminology of "dark matter" was unceremoniously adopted to satisfy the mass demands of the new inflationary cosmology. However, as these demands exceed the upper limits on total baryonic matter as dictated by big-bang nucleogenesis, it became necessary to postulate that more than 99% of the universal mass was non-baryonic. This outrageous assumption is surely enough to discredit any theory.

The reason why such a theory survives at all probably is a question of authority again. Einstein was the first to introduce hypothetical matter to support his closed-sphere cosmology. His contemporary, de Sitter (1917), comments on this assumption:

> "Einstein thus finds it necessary to add another term [λ] to his equations... Moreover it is found necessary to suppose the whole three-dimensional space to be filled with matter, of which the total mass is so enormously great, that compared with it all matter known to us is utterly negligible. This hypothetical matter I will call 'world-matter'. ... Moreover we find the remarkable result, that now no 'world-matter' is required."

Not only world-matter (dark matter) but also the cosmological constant, λ, abandoned by Einstein himself, today feature again prominently in standard cosmology, in both cases inflated by orders of magnitude, and dressed up with completely different meaning.

6.7.8 The Anthropic Principle

Like any other creation myth big-bang cosmology also needs a purpose, in this case provided by the anthropic principle. The argument goes that the laws of physics had to be fine-tuned in the beginning to ensure that the purpose of the creation would be fulfilled with the appearance of intelligent human beings at the right time. To dress up the argument as physics it is postulated that a delicate balance between a number of dimensionless constants is required as initial conditions. The crucial conclusion is that the emergence of humans retroactively finalized the initial conditions. In some cases the constants, assumed to be functions of time, are required to vary in concert, so as to peak just when mankind is ready to appear. As remarked by a sceptic, the anthropic principle has

> '...put Descartes before the horse: "I am, therefore..." (I think).'

A logical explanation is said to follow from many-worlds quantum theory.

A dimensionless constant is always obtained as the ratio of two quantities measured in the same units. A familiar example is a freely moving electron which is characterized by a de Broglie wavelength $\lambda_{dB} = h/mv$ and a Compton wavelength $\lambda_C = h/mc$. The ratio $\alpha = \lambda_C/\lambda_{dB} = v/c$ is a dimensionless quantity which depends on the velocity of the electron. For an electron in the ground-state H atom, $\lambda_{dB} = 2\pi a_0 = 2\pi\hbar^2/me^2$ and α assumes the constant value of $e^2/\hbar c$, known as the fine-structure constant.

It has been a major objective of standard cosmology to explain the fine-structure constant as dictated by the anthropic principle (Barrow & Tipler, 1988) p. 288–310, or even divine intervention. An award-winning theorist (Feynman, 1990) writes:

> It (α) has been a mystery ever since it was discovered more than fifty years ago,.... is it related to π, or perhaps to the base of natural logarithms? Nobody knows. It's one of the *greatest* damn mysteries of physics: a *magic number* that comes to us with no understanding by man. You might say the "hand of God" wrote that number, and "we don't know how He pushed His pencil".

By considering the electron as a wave structure rather than a point particle (Boeyens, 2010) the need of such speculation falls away. The same applies to the other dimensionless quantities of cosmological significance.

A gravitational analogue of the fine-structure constant is defined by

$$A = \frac{Gm_p^2}{\hbar c} = 5 \times 10^{-39}$$

in terms of proton mass. Alternatively the ratio of electrostatic to gravitational attraction between a proton and an electron is calculated as

$$N_2 = \frac{e^2}{r^2} \cdot \frac{r^2}{Gm_em_p} = \frac{e^2}{Gm_em_p} = 2.3 \times 10^{39} \simeq A^{-1}$$

The significance of this number is its close correspondence with another large number, obtained as the ratio of the Hubble radius (c/H) of the universe and the radius that encloses the rest mass of an electron, e^2/m_ec^2,

$$N_1 = \left(\frac{c}{H}\right)\left(\frac{m_ec^2}{e^2}\right) = 1.5 \times 10^{40}.$$

The Friedmann model defines the density of a closed spherical universe as

$$\rho = \frac{3H^2}{8\pi G}$$

and for a Hubble radius c/H the total mass follows as $M = c^3/2HG$, equivalent to a number of

$$N_3 = \frac{c^3}{2HGm_p} = 1.5 \times 10^{78} \quad \text{protons.}$$

The anthropic argument that ties the three large numbers together interprets the Hubble radius as increasing with time until N_1 reaches its current value, which matches N_2. It is argued that this is no coincidence and more likely an identification of the epoch during which intelligent life first became possible. This is confirmed by calculation (Dicke, 1961), based on the condition that the supply of elements required for biological life had to await the death of first-generation stars and the subsequent formation of a suitable solar system. From the calculated stellar lifespans and the current value of N_2 it is deduced that the universe has now reached precisely that epoch. The final conclusion is that the near equality $N_1 \simeq N_2$, which defines the epoch of intelligent life, is induced by the anthropic principle. The match between $\sqrt{N_3}$ and N_2 is explained by Mach's argument that the gravitational coupling constant depends on the total mass of the universe. It would be of interest to know how Dicke responds to ΛCDM cosmology.

6.8 State of the Art

Standard cosmology of the early 21st century rests on three observational pillars: gravity, spectroscopic redshift and microwave background radiation, interpreted in terms of general relativity, an expanding universe and big-bang genesis, respectively. The scientific basis of general relativity is well defined. The expanding universe is an assumption and the big bang is inspired conjecture. The same mix of fact, delusion and mysticism is recognized in all cosmological systems, starting with the most primitive. Never content to look for a rational interpretation of observed phenomena there is the constant desire to identify an underlying purpose and understand the design principles that shape the cosmos. This desire, which clearly has no scientific basis, is known as teleology, or the doctrine of final causes.

Many individuals commonly identified as the great scientists and philosophers of the past actually belong to the teleological design school – from Plato, Aristotle, the Stoics, Galen, Bœthius, Aquinas, Boyle, Copernicus, Kepler and Newton, to the big-bang cosmologists. In opposition are those who advocated a natural order of things, underpinned by universal laws – starting with the atomists Democritus, Leucippus, Epicurus, followed by the more modern thinkers like Cicero, Roger and Francis Bacon, Galileo,

Descartes, Leibniz, Kant, Laplace, Hume, Darwin, Huxley, Spinoza and Russell. Both schools claim the allegiance of Einstein.

There is no intention to belittle the scientific contributions of those with teleological convictions, but to arrive at a scientific model of the world their interpretations should be ignored. According to Bertrand Russell, "... the universe is just there, and that's all". To, therefore, infer the origin of space and time from the extrapolation of an unobserved effect to a mathematical singularity, violates all the rules of the science game. Rather than biting the honest bullet and look for an alternative interpretation of the observed red-shifts, the big-bangers look to buttress their house of cards with the standard model of particle physics, equally in need of experimental support, which they hope to find in the big bang. As another alternative they rope in the anthropic principle and large-number coincidences to end up with a cosmetic version of Edgar Alan Poe's cosmogony.

A distressing aspect of standard cosmology and the anthropic principle is the total lack of consistency that precludes any logical debate. For instance:

> What appears to be a logical sequence of events, based on uniform expansion, is interrupted, but not disrupted, by a dramatically accelerated expansion that ensures large-scale isotropy and homogeneity almost instantaneously, but without distorting the temporal evolution of temperature and density that controls events for another 10^5 years as if there were no inflation.

Inflation also distorts the mass budget of the universe so badly that less than 1% of the total mass appears visible, but this is the total mass that neatly balances the large-number coincidences of the anthropic principle. The excess must therefore be non-baryonic matter, for which there is no evidence. At the same time, the equal mass of antibaryons, implied by CPT symmetry, is declared non-existent. Antileptons are simply ignored.

The anthropic principle is by no means the only baseless theoretical concept of standard cosmology. Perhaps equally prominent and equally spurious is quantum gravity. Although nobody has formulated such a theory it is being referenced as imbued with the capacity to generate universal wave functions that solve the problems around space-time singularities and cosmic probabilities. This could be the only instance where the solutions of a non-existent equation override empirical observations.

However, most distressing is the selective use of astrophysical data in support of standard cosmology. Twenty years after the discovery of massive inhomogeneities in the form of supergalactic clusters, the mere formation of which would have taken much longer than the alleged Hubble age of the universe, their cosmological importance remains to be acknowledged. No other theory could survive such blatant falsification.

The way in which big-bang theory handles the problem is by changing the ground rules. The result, also known as ΛCDM cosmology, was introduced in section 5.2.2. It overcomes the dark-matter problem by having CDM produced in the big bang. Baryonic matter appears later as the rapidly expanding "CDM plasma" develops inhomogeneities before the onset of clumping, which results in the formation of galaxies. As baryonic matter does not feature in the theoretical simulations, the thorny question of CPT symmetry is neatly avoided. To allow sufficient time for the conversion of elementary matter into heavy elements, by circulation over three generations of galaxies, the process of early galactic evolution is accelerated by inflation and increased Λ, which supplies the required amount of dark energy; resulting in total nuclear synthesis within 500 000 years.

The success of the simulation relies on the availability of enough adjustable parameters to match any observation. It is foreseen that the observation of larger and larger redshifts could further reduce the lifespan of early galaxies to approach time zero. In principle even the existence of objects that antedate the big bang, could be simulated by introducing time inversion, like the gravity reversal of cosmic inflation, as a feature of the early universe. Without physical constraints on the simulation parameters all possibilities are equally feasible. The ultimate scenario is creation *ex nihilo (nihil fit!)*.

Progressive commentators already refer to ΛCDM as standard cosmology. To avoid confusion we shall retain this term for 20th century big-bang theory and use ΛCDM for the cosmology of the new millenium.

6.9 References

Abbott, L. (1988): *The Mystery of the Cosmological Constant*, Sci. Amer., 1988 (256) 106–113.

Barrow J.D. & F.J. Tipler (1988): *The Anthropic Cosmological Principle*, University Press, Oxford, 1988.

Boeyens, J.C.A. (2005): *New Theories for Chemistry*, Elsevier, Amsterdam.

Boeyens, J.C.A. (2010): *Emergent Properties in Bohmian Chemistry*, in M.V. Putz (ed.), *Quantum Frontiers of Atoms and Molecules*, NOVA, New York.

Born, M. (1962): *Einstein's theory of Relativity*, Revised edition, Dover, N.Y. p. 369.

Burbidge E.M., G.R. Burbridge, W.A. Fowler and F Hoyle. (1957): *Synthesis of the elements in stars*, Rev. Mod. Phys., 1957 (29) 548–650.

Capria, M.M. (2005): Chapter 6, in *Physics before and after Einstein*, IOS Press, Amsterdam.

Chown, M. (1993): *All you ever wanted to know about the big bang ....*, New Scientist, 17 April 1993, 32–33.

Coughlan, G.D. & J.E. Dodd (1991): *The ideas of particle physics* , 2nd ed., University Press, Cambridge.

De Sitter, W. (1917): *On the relativity of inertia. Remarks concerning Einstein's latest hypothesis*, Proc. Kon. Akad. Wetensch. Amsterdam, 1917 (19) 1217–1225.

Dicke, R.H. (1961): *Dirac's Cosmology and Mach's Principle*, Nature 1961 (192) 440–441.

Eddington, A.S. (1930): *The Internal Constitution of the Stars*, University Press, Cambridge.

Einstein, A. (1950): *The Meaning of Relativity*, 4th ed., Methuen, London, p. 99.

Einstein, A. & W. de Sitter (1932): *On the relation between the expansion and the mean density of the universe*, Proc. Nat. Acad. Sci., 1932 (18) 213–214.

Feynman, R.P. (1990): *QED*, Penguin Books, London, p. 129.

Fuller, F.W. & J.A. Wheeler (1962): *Causality and Multiply Connected Space-Time*, Phys. Rev., 1962 (128) 919–929.

Gamow, G. & D. Iwanenko (1926): *Zur Wellentheorie der Materie*, Z. Phys., 1926 (39) 865–868.

Glanz, J. (1996): *Debating the Big Questions*, Science, 1996 (273) 1168–1170.

Hawking, S.W. (1988): *A Brief History of Time*, Bantam Press, London.

Heisenberg, W. (1955) *The development of the interpretation of quantum theory*, in W. Pauli (ed.), *Niels Bohr and the development of physics*, Pergamon, London, 1955.

Hoffmann, B. & H. Dukas (1972): *Albert Einstein. Creator and Rebel*, The Viking Press, NY.

Hoyle, F. (1997): *Home is where the wind blows*, University Press, Oxford.

Hubble, E. & R.C. Tolman (1935): *Two methods of investigating the nature of the nebular red-shift*, Astrophys. J., 1935 (82) 302–337.

Krane, K.S. (1988): *Introductory Nuclear Physics*, 1988, Wiley, N.Y., pp. 764–768.

Kruskal, M.D. (1960): *Maximal Extension of Schwarzschild Metric*, Phys. Rev., 1960 (119) 1743–1745.

Lerner, E.J. (1989): *Galactic model of element formation*, IEEE Trans. Plasma Sci. 1989 (17) 259–263.

Maddox, J. (1989): *Down with the Big Bang*, Nature, 1989 (340) 425.

Milne, E.A. (1948): *Kinematic Relativity*, 1948, Oxford.

Narlikar, J.V. (1992): *The concepts of "beginning" and "creation" in cosmology*, Philosophy of Science, 1992 (59) 361–371.

Narlikar, J.V. (2002): *An Introduction to Cosmology*, 3rd ed., University Press, Cambridge.

Parish, L. (1981): *The Theory of Cosmic Aberration*, Cortney, Luton.

Pecker, J-C. (2001): *Understanding the Heavens*, Springer-Verlag, Berlin.

Rubin, V.C. (1983): *The rotation of spiral galaxies*, Science 1983 (220), June 24.

Rubin, V.C. (1988): *Dark matter in the universe*, Proc. Amer. Phil. Soc., 1988 (132) 258–267.

Schrödinger, E. (1922): *Über eine bemerkenswerte Eigenschaft der Quantenbahnen eines einzelnen Elektrons*, Z. Phys., 1922 (12) 13–23.

Schwarzschild, B. (1989): quoting A. Strominger, in: *Why is the cosmological constant so very small?*, Physics Today, March 1989, 21–24.

Sciama, D.W. (1971): *Modern Cosmology*, reprinted 1975, University Press, Cambridge.

Synge, J.L. (1950): *The gravitational field of a particle*, Proc. Roy. Irish Acad., 1950 (53A) 83–114.

Vermeil, H. (1917): *Notiz über das mittlere Krümmungsmass einer n-fach ausgedehnten Riemann'schen Mannigfaltigkeit*, Nachrich. Kgl. Ges. d. Wiss. zu Göttingen, 1917 (Heft 3) 334–344.

Weyl, H. (1922): *Space–Time–Matter*, translation of the 4th (1921) German edition, Dover, N.Y., Appendix II.

Wolff, M. (1995): *Beyond the Point Particle – A wave Structure for the Electron*, Galilean Electrodyn., 1995 (6) 83–91.

Chapter 7

Relativistic Cosmology

General relativity is widely accepted to have superseded Newton's laws as a model of gravitational interaction. Assuming gravity to be the major factor that decides the geometrical arrangement of heavenly bodies, general relativity is clearly the best algorithm for cosmology. The theory of special relativity, which requires a four-dimensional space-time, is contained within the general theory, which in turn dictates non-Euclidean geometry of space-time. The minimum constraint on the topology, assumed for any cosmological model, is to be consistent with these principles of special and general relativity.

These conditions are not generally appreciated. It is a common fallacy to assume that the use of Einstein's field equations:

$$R_{\mu\nu} + \Lambda g_{\mu\nu} = -\kappa(T_{\mu\nu} - \tfrac{1}{2}Tg_{\mu\nu}), \tag{6.4}$$

with or without the cosmological constant, automatically defines a relativistic cosmology. In this respect it is illuminating to re-examine the meaning of each term, following Capria (2005).

The first term, known as the Ricci tensor, is obtained from the 4-index Riemann–Christoffel tensor on contraction with the mixed fundamental tensor:

$$R_{\mu\nu} = R^{\alpha}_{\mu\nu\beta}g^{\beta}_{\alpha}\,.$$

The curvature tensor:

$$R^{\alpha}_{\mu\nu\beta} = \frac{\partial\Gamma^{\alpha}_{\mu\beta}}{\partial x^{\nu}} - \frac{\partial\Gamma^{\alpha}_{\mu\nu}}{\partial x^{\beta}} + \Gamma^{\alpha}_{\nu\tau}\Gamma^{\tau}_{\mu\beta} - \Gamma^{\alpha}_{\beta\tau}\Gamma^{\tau}_{\mu\nu}$$

describes how the manifold deviates locally from Euclidean n-space. When it vanishes everywhere the manifold is flat.

J.C.A. Boeyens, *Chemical Cosmology*, DOI 10.1007/978-90-481-3828-9_7,

The Christoffel symbols are the analogues of the functions that minimize the distance between points in a two-dimensional curved space:

$$\Gamma^{\alpha}_{\mu\nu} = \tfrac{1}{2}g^{\alpha\beta}\left(\frac{\partial g_{\beta\nu}}{\partial x^{\mu}} + \frac{\partial g_{\beta\mu}}{\partial x^{\nu}} - \frac{\partial g_{\mu\nu}}{\partial x^{\beta}}\right).$$

The important conclusion is that the Ricci curvature tensor depends entirely on the form of the fundamental tensor.

The right-hand side of (6.4) is implied by the fact that the Ricci tensor has zero divergence for any metric, which requires that the stress-energy tensor should likewise have $\nabla_{\lambda}T^{\lambda\mu} = 0$.

Equations (6.4) describe the balance between two unknowns – the fundamental tensor and the distribution of matter in a system of interest. Although the distribution function is not conditioned by the theory of relativity in any way, it assumes critical importance in deciding the appropriate space-time geometry in cosmological applications. This way the metric tensor is defined, not on the basis of relativistic considerations, but on Newtonian principles. Cosmological models arrived at in this way we consider non-relativistic, unless the metric tensor has the correct relativistic signature. To explain the reasoning we consider a few elementary models.

Newtonian Universe

A common simplifying assumption is that matter is distributed uniformly throughout the universe, subject to gravitational interaction. On this basis Newton could argue that in a universe of limited size all matter would eventually clump together in a central mass. Dynamic equilibrium, as observed, can only develop in an infinite universe. The resultant gravitational field should vanish everywhere, but local instabilities can give rise to the formation of stars and solar systems. The metric of such a system is clearly Euclidean.

Einsteinian Universe

Inspired by Riemann's ideas William Clifford proposed, as a modification of Newton's universe, a space with constant positive curvature, except for small local variations. This is the most likely precursor of Einstein's closed 3-sphere universe in which matter is modelled as a pressure-less incoherent fluid, or dust, of constant density. The closure ensures constant density in space, that remains constant by defining a time coordinate orthogonal to space. The Robertson–Walker metric is defined on the same principle of constant density in a comoving coordinate system.

De Sitter's Model

Of all the models considered in the previous chapter only the original de Sitter model meets the relativistic requirements. It is the only one to assume a space-time metric rather than a privileged time coordinate. It transcends special relativity in assuming a space-time embedded in five-dimensional space. The four-dimensional metric appears on projection along a space axis, which has no other role. An alternative would be representing the closed de Sitter space-time in projective 3-space (also known as elliptic space) rather than a 3-sphere. Other cosmologies based on five-dimensional metrics have been shown more recently to induce the physical structures observed in four-dimensional space-time. One of these is briefly discussed next.

7.1 Induced-Matter Theory

Kaluza and Klein managed to formulate a unified theory of gravitation and electromagnetism in terms of Einstein's field equations in five-dimensional space, but with the metric tensor $\eta_{\mu\nu}$ defined to be independent of the fourth space dimension. Without this restriction, solution of the equations in apparent $5D$ vacuum:

$$G_{AB} \equiv R_{AB} - \tfrac{1}{2}g_{AB}R = 0\,, \tag{7.1}$$

where the upper case Latin letters, *e.g.* $A = 0, 4$, contains, embedded, the familiar four-dimensional equations (Wesson, 1999):

$$G_{\alpha\beta} = kT_{\alpha\beta} \quad , \quad \text{Greek letters } (\alpha = 0, 3). \tag{7.2}$$

Comparing the vacuum solutions of G_{AB} and $G_{\alpha\beta}$ the former contains extra terms, which were shown (Ponce de Leon, 1988) to correlate with the mass terms of (7.2).

A typical calculation is based on a $5D$ metric like

$$ds^2 = e^{\nu}dt^2 - e^{\omega}\left(dr^2 + r^2 d\Omega^2\right) - e^{\mu}dl^2.$$

Here the time coordinate $x^0 = t$ and the space coordinates $x^{123} = r\theta\varphi$, $(d\Omega^2 = d\theta^2 + \sin^2\theta d\varphi^2)$, $x^4 = l$. The calculated $5D$ components of the Einstein tensor, G^A_A and G^0_4 are then matched with the usual $4D$ perfect-fluid energy-momentum tensor, containing density ρ and pressure p terms

$$T_{\alpha\beta} = (p + \rho)u_\alpha u_\beta - pg_{\alpha\beta}\,.$$

Explicit solutions of the field equations to replicate most of the popular $4D$ cosmological models have been obtained, subject only to an appropriate choice of parameters.

Another class of solution corresponds to massive objects without gravitational effects, also referred to as solitons. These have been suggested as viable candidates for dark matter or even black holes. The logic is not too clear.

The main conclusion is that, whereas the universe in the large has matter and structure in four dimensions, it may be empty and flat in five dimensions. With respect to the standard model this could be interpreted to mean that the big bang in four dimensions is an artefact produced by an unfortunate choice of coordinates in five dimensions.

Apart from interesting speculations the modern theory produces little more than classical Kaluza–Klein theory.

7.2 A Rotating Universe

A solution of the gravitational field equations:

$$G_{ik} \equiv R_{ik} - \tfrac{1}{2}(R - \Lambda)g_{ik} = -\frac{8\pi\kappa}{c^2}T_{ik}\,, \tag{7.3}$$

obtained by Gödel was recently re-interpreted (Boeyens, 2010) in projective space. It was shown by Gödel (1949) that elimination of the absolute time coordinate leads to the rotation of matter relative to the compass of inertia. He proposed a metric to reflect nine special properties of the assumed four-dimensional space S, which it defines:

1. S is homogeneous;
2. (Neighbouring) world lines of matter are equidistant;
3. S has rotational symmetry;
4. A positive direction of time can be introduced in the whole system;
5. The direction of the time flow is not uniquely defined for each space-time point;
6. Every world line of matter is infinitely long and never returns to the same point; but there also exist closed time-like lines;
7. There are no space-like three-spaces;
8. There is no absolute time;
9. Matter everywhere rotates relative to the compass of inertia

The Gödel (1949) metric:

$$\begin{aligned} ds^2 &= \alpha^2\left[dx_0^2 - dx_1^2 + \left(\frac{1}{2}e^{2x_1}\right)dx_2^2 - dx_3^2 + 2e^{x_1}dx_0dx_2\right] & (7.4)\\ &= \alpha^2\left[(dx_0 + e^{x_1}x_2)^2 - dx_1 - \frac{1}{2}e^{2x_1}dx_2^2 - dx_3^2\right] & (7.5)\\ &= (dx_0 + e^{\alpha x_1}dx_2)^2 - (dx_1)^2 - \frac{1}{2}e^{2\alpha x_1}(dx_2)^2 - (dx_3)^2 & (7.6)\end{aligned}$$

contains the Schwarzschild,

$$\begin{aligned} ds^2 &= c^2dt^2 - \left(dr^2 + r^2d\theta^2 + r^2\sin^2\theta d\varphi^2\right)\\ &= Adx_0^2 - Bdr^2 - r^2\left(d\theta^2 + \sin^2\theta d\varphi^2\right)\end{aligned}$$

and Robertson-Walker (6.67) metrics as special cases, with expansion factor $S(t) = 1$.

The cross term in the time interval of (7.4) is reminiscent of the metric of cylindrical space, rotating with constant angular frequency and can be shown (Adler et al. 1965) to be equivalent to that.

The stress tensor, as was done by Einstein, is obtained from the kinetic theory of galactic clusters, assumed to behave like dust in a fluid of average density ρ, internal pressure p and rms velocity $\overline{v^2}$,

$$p = \tfrac{1}{3}\overline{v^2}\rho$$

On lowering indices with the metric tensor, the contracted Riemann tensor follows as:

$$R_{ij} = -\frac{\alpha^2}{\rho}T_{ij}$$

showing that the Gödel metric is a solution of (7.3) if

$$\Lambda = -\frac{\alpha^2}{2} \qquad \frac{\alpha^2}{\rho} = \frac{8\pi\kappa}{c^2}$$

For $\alpha = 0$, both Λ and ρ must be zero, which implies flat space. The parameter α therefore measures the curvature of space. Although the solution will be shown to define an aesthetically more pleasing cosmology, it has been ignored for many years because it fails to predict a Doppler redshift, does not give a clear definition of the compass of inertia, proposed as rotation axis, and allows closed time loops.

It was shown (Boeyens, 2010) that all of these objections are eliminated on modification of assumption 6 by the addition of a point at infinity, which turns infinite space into projective space. The metric remains the same under this transformation and some of Gödel's perceived problems, which he addressed in footnote material, are also resolved. In particular [1] footnote 11:

> "There exist stationary homogeneous solutions in which the world lines of matter are not equidistant. They lead, however,into difficulties in consequence of the inner friction which would arise in the 'gas' whose molecules are the galaxies, unless the irregular motion of the galaxies is zero, and stays so".

Gödel describes his affine space-time as follows:

> "The space S [...] is the direct product of a straight line and the three-space S_0, defined by $x_3 = 0$; and S_0 obtained from a space R of constant curvature.[...] This definition of S_0 also leads to an elegant presentation of its group of transformations. To this end we map the points of R on the hyperbolic quaternions $u_0 + u_1 j_1 + u_2 j_2 + u_3 j_3$ [...] by means of projective coordinates $u_0 u_1 u_2 u_3 \ldots$".

In terms of this simple alternative geometry, geodesic transplantation fixes points on the line element to occur along the double cover of a narrow Möbius band. It is a known property of a Möbius band that a point, which moves along the double cover, close to one edge, rotates around the central line without intersecting it. This is what Gödel describes as rotation with respect to a compass of inertia.

To further elucidate this description it was shown that a Möbius strip represents a section through a closed projective plane embedded in four-dimensional space. As an example consider four-dimensional Minkowski space. A pseudocircle in this space,

$$x^2 - (ct)^2 = \pm r^2$$

consists of two hyperbolas (Jennings, 1994), shown in Figure 7.1. By following the asymptotes of the hyperbolas as an involuted closed curve *e.g.* ($A \to B \to A$), two Möbius strips occur as sections through the projective plane. Any geodesic in projective Gödel space therefore rotates around an inertial compass in the same way. The constant separation between symmetry-related points determines the number of turns before returning to the starting

[1] In view of modern observations Gödel's concerns appear less serious. Galaxies are now known to wander and even collide. There is hardly any reason why not to explore the alternative possibility.

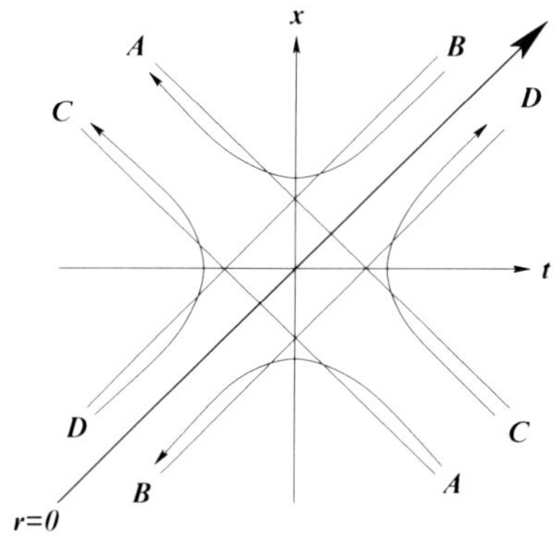

Figure 7.1: *Pseudocircles* ($r \neq 0$) *in Minkowski space*

point. To define a non-orientable plane this number must be odd. All essential properties are recovered, without loss of generality, by assuming a single Möbius twist, shown in Figure (7.2), alongside a one-sided ribbon with five twists.

On adding points at infinity to all coordinates, S transforms into the projective space, best described as the direct product of S_0 and a circle (at infinity), *viz.* the conformal compactification of Minkowski space-time. More important than the precise topology of the modified space-time is that conversion to projective space, with five homogeneous coordinates, leaves the gravitational potentials unchanged, while the extra degrees of freedom represent the electromagnetic field (Veblen & Hoffmann, 1930). The Gödel solution therefore remains valid.

Although the Gödel metric solves the field equations, large-scale astronomical observations in the system can only be understood as features of the topological space (Flegg, 2001). The real projective plane is topologically closed, multiply-connected, one-sided and non-orientable, with Euler characteristic $\chi = 1$. Moving a normal to the surface at P (*e.g.* time axis) along the continuous path to P', so that the foot of the normal remains in contact with the surface, demonstrates the one-sided non-orientable property. The fixed-point theorem for a disc shows that any continuous transformation to itself maps at least one central point to itself. For the real projective plane, which is not homeomorphic to the disc, there is no fixed point, except for a rotation of $2\pi n$. The Gödel rotation defines such a transformation. All coordinate points are equivalent and the fixed point can be placed anywhere.

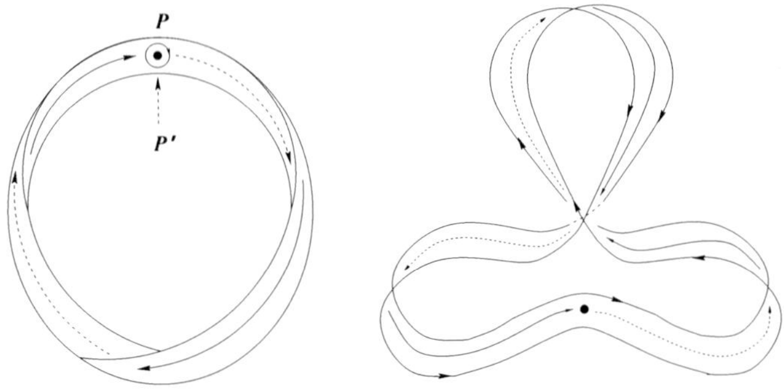

Figure 7.2: *Non-orientable surfaces with a single twist (left) or five twists (right)*

Cosmic rotation, which appeared to be in geocentric mode, has been observed in radio astronomy (Birch, 1982) and excited the comment that:

> "This would have drastic cosmological consequences, since it would violate Mach's principle and the widely held assumption of large-scale isotropy".

Topologically there is no mystery. Any large-scale rotation must appear to be centred at the observer and indeed, the identity of the Einstein and Gödel stress tensors shows that the field equations have two basically different solutions for the same T_{ij}, one rotating and the other static. The Gödel solution therefore is consistent with the general theory of relativity, but not with Mach's principle (Adler et al., 1965, p.377).

The time paradox is also resolved in projective space. The closed time loops now connect any point, such as P, through involution, to its antipode in conjugate space-time. P and P' are related by CPT symmetry and separated in time.

7.2.1 The Cosmological Model

Gödel's solution of the relativistic gravitational field equations (7.3) is compatible with an incoherent matter distribution, in uniform rotation about a compass of inertia that coincides with the involuted geodesic of high-dimensional projective geometry. In one-dimensional analogue the geodesic traverses a Möbius band, which is the interface between symmetry-related segments of the non-orientable surface that constitutes a double cover. It is natural to identify the interface with the physical three-dimensional vacuum which is interpenetrated by a pair of chiral subspaces that define the material universe. Curvature of the surface occurs with respect to the local time coordinate of the non-Euclidean manifold, and is everywhere perpendicular to the surface. The interpenetrating dual spaces are seperated in time, which prevents mutual annihilation.

This model can hardly be more unlike a universe that expands in three-dimensional Euclidean space. The special-relativistic requirement of four-dimensional space, with the curvature of general relativity superimposed, seems to demand that space-time has a minimum of five dimensions, which is equivalent to four-dimensional projective space, described by five homogeneous coordinates. Locally perceived three-dimensional space therefore is an illusion and extrapolation of local structure, beyond the Galactic borders, a gross distortion.

While the dimensionality of space-time remains largely speculative, the argument will not be pursued any further, beyond pointing out that the naïve interpretation of spectroscopic redshift as a three-dimensional Doppler effect is a futile attempt to visualize four-dimensional events. By a remarkable coincidence Segal (1976) has shown that a chronometric redshift and a Planckian microwave background occur in the same projective space as the modified Gödel solution and that it satisfies all the main cosmological criteria.

7.2.2 Chronometric Redshifts

According to general relativity the cosmos is a four-dimensional manifold, $\overline{M}$. At each point there is given a convex cone of infinitesimal future directions, in the space tangent to the manifold. Space and time coordinates are entangled in $\overline{M} = T \times S$. According to Segal (1976) space-time events in the vicinity of an observer have a linear temporal order T^1 and a three-dimensional space S^3. An observer is said to collapse the event into a stationary state by splitting space-time into *space* and *time* components. The non-Euclidean space-time model, which may be globally acausal, becomes locally Minkowskian.

The universal space-time geometry itself is not necessarily directly observed; no apparent departures from a Euclidean model have been found by classical measurements. It might be argued that curved space-time coordinates are split into flat space and time coordinates. There is no direct observational basis for asserting that the Cosmos is Minkowskian at large distances and times.

Locally, Minkowski and universal space are identical as causal manifolds, but the universal time is not equivalent to the time registered in the local Minkowski frame. Quantum mechanically temporal evolution and energy are defined by conjugate operators. The operator $-i(\partial/\partial t)$, which defines the energy (or frequency) depends on the geometry of the stationary state.

The difference between standard time and Minkowski time therefore becomes measurable in the form of alterations produced in the apparent frequency of a freely propagated photon. The relative shift can be calculated as the frequency difference as measured by two observers, one of whom is at rest relative to the other.

Minkowski space, M, is assumed embedded in a more general universal closed (compact) space $\overline{M}$, the so-called conformal space. This is the projective space proposed as a model of the universe by Oswald Veblen (1933), translated in the Appendix. Roughly speaking, $\overline{M}$ is obtained from M by adding a light cone at infinity. More precisely, it is the double cover of the space so generated. Segal (1976) refers to $\overline{M}$ as *unispace* and to the natural time τ in this space as *unitime*.

Temporal generators (*e.g.* energies) are basically different between unispace and local Minkowski space. The unispace generator is strictly greater than the Minkowski energy. An observation consists of the local decomposition of either the curved unispace or flat Minkowski space into space and time components. Physically only one of these analyses can be globally valid. Therefore, if global conservation of energy is valid in one analysis, it cannot be valid in the other, for the respective energy operators do not commute.

Conventionally flat frequency is represented by the operator $-i(\partial/\partial t)$, but in terms of the unitime it becomes $-i(\partial/\partial\tau)$. The experimental observation of redshifts indicates conclusively that local measurements correspond to flat dynamic variables, but that the universe operates on curved dynamics. In the process of measurement the frequency shifts from the actual $\nu'(\tau)$ to the stationary value $\nu(t)$. By symmetry

$$\nu' = \frac{1+\cos\tau}{2}\nu$$

which predicts a so-called *chronometric* redshift:

$$z = \frac{\nu-\nu'}{\nu'} = \frac{1-\cos\tau}{1+\cos\tau} = \tan^2\left(\frac{\tau}{2}\right)$$

The excess energy appears with redshifting, and is then diffused in space in a fashion which causes no observable local particle production.

The apparent motion between the cosmos and the stationary Minkowski frame is entirely virtual. To establish a cosmic distance scale, stationary states at two different points, r apart, are compared, choosing unit radius for unispace; leading to

$$z = \tan^2(\rho/2) = \tan^2(r/2R)$$

where R is a Minkowskian radius of the universe.

The quadratic relationship of redshift to distance of the source is a complete departure from the conventionally assumed Doppler shift and Hubble's linear law. In order to test the quadratic model it is necessary (Segal, 1980) to eliminate the distance, which is not an observable quantity, using geometrical relations with parameters such as apparent luminosity or angular diameter. These relations were tested on data available for galaxies, quasars and radio sources.

Graphical comparisons (Segal, 1976) appear to favour the square law overwhelmingly and this was further confirmed by sophisticated statistical analysis (Nicoll & Segal, 1978). Several spurious attempts to discredit this work and editorial refusal to publish Segal's rebuttals have been documented and analyzed by Daigneault (2001).

Microwave Background

As $\rho \to \pi$, at the antipodal point in the projective space, the redshift approaches totality as $z \to \infty$. In the stationary Minkowski frame the antipode is infinitely distant at an infinite time coordinate. The conditions leading to the derivation of the redshift formula are not met for radiation with a propagation interval close to a half-circuit in space, and such photons will appear entirely delocalized and severely redshifted, constituting the isotropic microwave background. Segal ascribes the Planckian distribution to the conservation of energy, which is tantamount to the fact that any closed space must eventually impose a Planckian spectrum on stray radiation.

Chiral Matter

A predicted gradual change in chirality with transplantation along the geodesic follows the gauge transformation of electric charge, such that (Boeyens, 1992):

$$\Psi' = \Psi \exp\left[-(2\pi i\hbar)\int e\phi_i \mathrm{d}x_i\right]$$

as $x_i \to -x_i$, and an involution of π turns matter into conjugate antimatter. There are no separate matter and antimatter domains as chiral forms transform smoothly into each other, preserving an element of CPT symmetry in the interface.

The secondary criteria are functions of a steady state that reflects the exchange of matter and energy across the vacuum interface.

7.3 Projective Relativity

The idea of a projective theory of relativity was developed early in the twentieth century, but never fully implemented as a cosmological model and by now is largely forgotten. I have traced the reason for this negative response to specific statements of two highly respected authorities on the theory of relativity, Pauli (1958) and Bergmann (1976). Pauli states:

> "We mention here briefly another equivalent way to formulate the geometrization of the gravitational and the electromagnetic field, namely the *projective* formulation. Many authors have contributed to it, among them Veblen and Hoffmann, Schouten and van Dantzig and myself. Bergmann has shown, however, that – in contrast to what I believed myself for a while – this formulation is *not* more general than Kaluza's, and that it is easy to pass from either of these two formulations to the other".

Another reason is the way in which Veblen introduced the topic in his monograph, which is presented here in English as an Appendix, *i.e.* as: Unified Theories for Gravitation and Electromagnetism.

In all of these statements the emphasis is on unified fields and not on cosmology. As a matter of fact, the equivalence of the projective unified model to five-dimensional spaces in general, and to that of Einstein and Mayer in particular, was first demonstrated by Veblen himself (Monograph: Chapter VIII). The crucial observation is that this equivalence mapping is done in the tangent space, without implying the equivalence of the Einstein–Mayer five-dimensional construct with four-dimensional projective space-time. The five-dimensional spaces are not projective, but affine spaces.

The purpose of projective relativity is to derive the equivalent of Einstein's field equations in homogeneous projective coordinates, which requires definition of projective scalars, vectors, displacements, connections and tensors in projective space. Such procedures are described in detail in the monograph.

The field equations in empty space are obtained as:

$$\Gamma_{\alpha\beta} - \varphi_\alpha \varphi_\beta \Gamma = 0\,,$$

where the scalar $\Gamma = \gamma^{\alpha\beta}\Gamma_{\alpha\beta}$, is related to a Ricci tensor, as in the affine formalism, and the φ_α are invariant projective vectors.

The tensor

$$\varphi_{ij} = \tfrac{1}{2}\left(\frac{\partial\varphi_i}{\partial x^j} - \frac{\partial\varphi_j}{\partial x_i}\right)$$

is of central importance in electromagnetic theory. In projective relativity it appears (Monograph p.47) in the equations of motion of an electrical particle of charge e and mass m:

$$\frac{d^2x^i}{dt^2} + \left\{ \begin{array}{c} i \\ jk \end{array} \right\} \frac{dx^j}{dt}\frac{dx^k}{dt} + \frac{e}{m}\varphi^i_j \frac{dx^j}{dt} = 0\,,$$

together with the gravitational potentials g_{ij}.

Because of the property of homogeneous coordinates to remain valid on multiplication with a randomly selected proportionality factor, which may be defined as

$$k = e^{x^0}\,,$$

by reference to the plane at infinity, there is a close connection with Weyl's definition of a gauge variable. The corresponding gauge transformation (Monograph, p. 8–10) becomes

$$\bar{x}^0 = x^0 + \log \rho(x)$$

on extension of the homogeneous coordinates by a position function $1/\rho(x)$.

Applied to the potentials of the electromagnetic field the coordinate system is determined only to within an additive gradient, which is the well-known property of the vector potential of the Maxwell field. In common practice it is necessary to *assume* the gauge invariance, which appears naturally in projective relativity.

The affine form of the projective field equations is shown (Veblen & Hoffmann, 1930) to be:

$$R - 3N^2 + 9N\Phi^{\sigma}{}_{;\sigma} + 15N^2\Phi^{\sigma}\Phi_{\sigma} = 0$$

where the projective covariant derivative,

$$\Phi^{\alpha}{}_{;\alpha} = \frac{\partial \Phi^{\alpha}}{\partial x^{\alpha}} + \Gamma^{\alpha}_{\sigma\alpha}\Phi^{\sigma}\,.$$

By substituting the projective scalar of index N,

$$\Phi = \psi^{\frac{3}{5}}$$

the equation reduces to

$$\frac{1}{\sqrt{g}}\frac{\partial}{\partial x^i}\left(g^{ij}\sqrt{g}\frac{\partial \psi}{\partial x^j}\right) - \frac{10N}{3}g^{ij}\varphi_j\frac{\partial \psi}{\partial x^i} - \frac{5N}{3}\frac{1}{\sqrt{g}}\psi\frac{\partial}{\partial x^i}\left(g^{ij}\varphi_j\sqrt{g}\right)$$
$$+\frac{25N^2}{9}g^{ij}\varphi_i\varphi_j\varphi + \left(\frac{5}{27}R + \frac{20}{9}N^2\right)\psi = 0\,.$$

This is the relativistic Schrödinger equation, also known as the Klein–Gordon equation, in tensor notation. It reduces to a more familiar form on substituting

$$\begin{cases} N = \frac{3\pi mc}{\sqrt{5}h} i \\ \varphi_j = \frac{2}{\sqrt{5}} \frac{e}{mc^2} V_j \end{cases} \tag{7.7}$$

$$\frac{1}{\sqrt{g}} \frac{\partial}{\partial x^i} \left(g^{ij} \sqrt{g} \frac{\partial \psi}{\partial x^j} \right) - \frac{4\pi i e}{h} g^{ij} V_j \frac{\partial \psi}{\partial x^i} - \frac{2\pi i e}{h} \frac{1}{\sqrt{g}} \psi \frac{\partial}{\partial x^i} \left(g^{ij} V_j \sqrt{g} \right)$$
$$- \frac{4\pi^2 e^2}{h^2} g^{ij} V_i V_j \psi \left(\frac{5}{27} R - \frac{4\pi^2 mc^2}{h^2} \right) \psi = 0 \, .$$

When g^{ij} has the Minkowski signature $\{1\ 1\ 1\ -c^2\}$ the gravitational factor R vanishes to give:

$$\left(i\hbar \frac{\partial}{\partial t} - q\phi \right)^2 \Psi = m^2 c^4 \Psi + c^2 \left(-i\hbar \boldsymbol{\nabla} - q\boldsymbol{A} \right)^2 \Psi \, .$$

This is the common form of the Klein–Gordon equation in the electromagnetic field with scalar and vector potentials ϕ and $\boldsymbol{A}$ (Bransden & Joachain, 1983). Equations (7.7) represent the transformation from projective space to an affine tangent space, which embodies two important principles. The generation of a compensating field by the gauge principle has been mentioned. By choosing the proportionality function as a gauge transformation according to Weyl and Schrödinger (1922). The transformation

$$V_j = (\tau + \tfrac{1}{2}) mc^2 \phi_j$$

demonstrates, for the first time, the appearance of the golden ratio as a descriptor of projective space-time. The ubiquitous self-similarity in Nature has often been ascribed to space-time curvature related to τ.

We have now demonstrated that projective relativity provides a complete unification of gravitational and electromagnetic fields. This unification remains valid on mapping the homogeneous coordinates to five-dimensional Kaluza–Klein spaces, but losing the projective topology that defines the gauge principle, the vacuum interface and self-similar symmetry. In both types of geometry the field equations produce fifteen variables, only fourteen of which feature as gravitational and electromagnetic potentials. The vacuum interface only occurs in projective space-time where the fifteenth parameter is required to label the time axis, and hence the difference between spin and anti-spin; space-time and anti-spacetime.

The cosmological projective equations have never been solved directly, but Gödel's solution fits the projective topology closely enough to serve as a model of the implied steady-state universe.

7.4 The Steady State

Mathematical singularities in physical theory appear for only one reason – a wrong model.

Although the Gödel solution is free of singularities the need to accommodate black holes in the cosmic model requires an interpretation of the Schwarzschild singularity which occurs with infinite curvature of space-time. A new interpretation is rather obvious. Such a high degree of curvature must clearly rupture the interface between adjacent sides of the postulated cosmic double cover. Rather than disappear into a singularity, the matter,

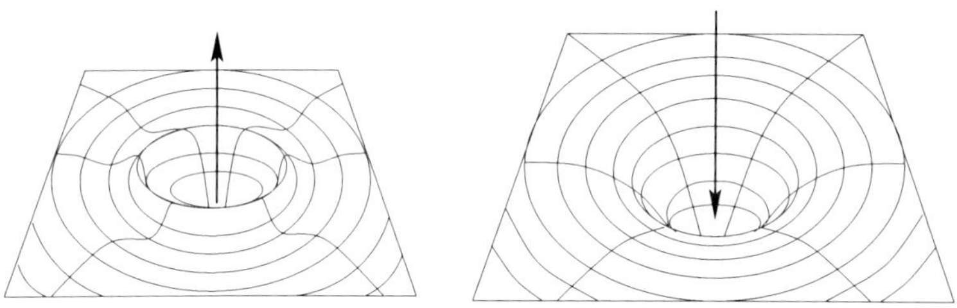

swallowed up by a black hole, therefore reappears, with inverted chirality, on the opposite side of the interface. Such a connection is known as an Einstein–Rosen bridge, never precisely localized before. A schematic drawing of the gravitational gradients suggests sufficient asymmetry to interpret the bridge as operating between a black hole and a cosmic volcano that injects matter into space.

Many sources with prominent emission spectra, such as Seyfert galaxies and quasars, which could be of this type, have been observed. According to this interpretation cosmic matter is neither dispersed nor created in time, but recycled. The constant two-way flow across the interface has reached a steady state which gives the universe the appearance of being static.

The observed periodicity in the cosmic abundance of atomic nuclides (Boeyens, 2005) finds a ready explanation in this steady-state recycling. Under simulated high pressure all matter collapses into elementary forms with inversion of chirality. On emergence, through the throat of the black hole, complex nuclides are reconstituted by the fusion of α-particles in an equilibrium process involving all compositions with the proton:neutron ratio of unity. On release into free space only those isotopes with appropriately adjusted composition survive radioactive transformation, giving rise to a numerically precise range of nuclides. The nucleogenesis of standard cosmology cannot reproduce the observed mix.

7.4.1 Universal Self Similarity

As recently shown (Boeyens, 2009) the Bode–Titius law, which hints at some harmonious regular organization of planetary motion in the solar system, is dictated by a more general self-similar symmetry that applies from sub-atomic systems to galactic spirals. The common parameter is the golden ratio, $\tau = 0.61803\ldots$. Any such cosmic symmetry should be dictated by a successful cosmological model.

The golden ratio is superimposed on a logarithmic spiral, $r = \mu^\theta$, by setting $\mu = \tau^{2/\pi}$ to produce the golden spiral, $r = \tau^{2\theta/\pi}$ that leads to the Bode–Titius law. The orbital radii of planets and moons in the solar system are characterized correctly by divergence angles of $n\pi/5$ ($\simeq n\tau$) on the spiral that fits into a golden rectangle, as shown diagrammatically in Figure 7.3 for $n = 2$.

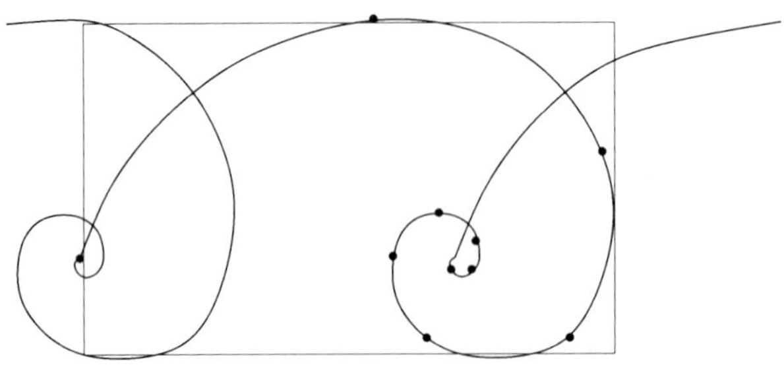

Figure 7.3: *A set of logarithmic spirals, such as the golden planetary spiral with divergence angle 2τ, may serve as a model of Gödel's compass of inertia, going through an odd number of involutions*

It is now conjectured that the Gödel compass of inertia has the same structure with five involutions, which divide the universe into ten segments with alternating chirality. The topological structure remains the same as for the single involution considered before, but the five-fold self-similarity is now imprinted on all space.

The consequences of rotating space with periodically alternating chirality have never been contemplated before. If it has some dynamo effect it could explain the appearance of Einstein's cosmological constant Λ, which stabilizes Gödel's solution.

The idea of a closed chiral universe in rotation is an irreducible concept which is only defined in hyperspace. Unlike the concept of an expanding universe, which is routinely projected into infinite three-dimensional Euclidean space, there is no incentive to perform such a visualization here. The possibility of observing multiple images of astronomical objects are obvious and real. With a regular scattering of black holes and their conjugates, concepts such as gravitational lensing and astronomical measures of distance and time need serious reconsideration. Rescaling by many orders of magnitude may not be unreasonable. The Hubble telescope is like an observer, tricked into reporting from a hall of mirrors.

The volume of astronomical data has grown too large for disentanglement until a well-defined geometrical framework is in place. Further speculation about great attractors, megawalls, quantized redshifts and dark matter are doomed to remain unproductive on the hand of a quasi-three dimensional model of the universe. It is not even clear which extra-Galactic objects have real existence in geocentric projection.

7.5 References

Adler, R., M. Bazin & M. Schiffer. (1965): *Introduction to General Relativity*, McGraw–Hill, NY.

Bergmann, P.G. (1976): *Introduction to the theory of relativity*, Dover edition, NY.

Birch, P. (1982): *Is the Universe Rotating?*, Nature, 1982 (298) 451.

Boeyens, J.C.A. (1992): *The Geometry of Quantum Events*, Specul. Sci. and Techn., 1992 (15) 192–210.

Boeyens, J.C.A. (2005): *Number theory of nuclide abundance*, 5th Int. Conf. Isotopes, 2005, Brussels;
— & D.C. Levendis. (2008): *Number Theory and the Periodicity of Matter*, Springer.

Boeyens, J.C.A. (2009): *Commensurability in the solar system*, Physics Essays, 2009 (22) 493–499.

Boeyens, J.C.A. (2010): *The universe in perspective*, Physics Essays, 2010 (23) 337–343.

Bransden, B.H. & C.J. Joachain (1983): *Physics of atoms and molecules*, Longmans, London.

Capria, M.M. (2005): *Physics Before and After Einstein*, IOS Press, Amsterdam.

Coxeter, H.S.M. (1989):*Introduction to Geometry*, 2nd ed., Wiley.

Daigneault, A. (2001): *Is the Universe Expanding?*
http://www.ams.org/notices/fea-daigneault.pdf
– & A. Sangalli (2001): *Einstein's Static Universe: An Idea Whose Time has Come Back?*, Notices of the AMS, January 2001, 9–16.

Flegg, H.G. (2001): *From Geometry to Topology*, 2001, Dover, NY. (Reprint of 1974, Crane & Russak edition.)

Gödel, K. (1949): *An Example of a New Type of Cosmological Solutions of Einstein's Field Equations of Gravitation*, Rev. Mod. Phys., 1949 (21) 447-450.

Jennings, G.A. (1994): *Modern Geometry with Applications*, Springer-Verlag, NY.

Narlikar, J.V. (2002): *An Introduction to Cosmology*, 3rd ed., University Press, Cambridge.

Nicoll, J.F. & I.E. Segal (1978): *Statistical Scrutiny of the Phenomenological Redshift–Distance Square Law*, Ann. Phys., 1978 (113) 1–28.

Pauli, W. (1958): *Theory of Relativity*, Translated from the German by G. Field, Dover edition, N.Y.

Ponce de Leon, J. (1988): *Cosmological models in Kaluza–Klein theory with variable rest mass*, Gen. Rel. Grav., 1988 (20) 539–550.

Schrödinger, E. (1922): *Über eine bemerkenswerte Eigenschaft der Quantenbahnen eines einzelnen Elektrons*, Z. Phys., 192 (12) 13–23.

Segal, I.E. (1976): *Mathematical Cosmology and Extragalactic Astronomy*, 1976, Academic Press, NY.

Segal, I.E. (1980): *Time, Energy, Relativity and Cosmology*, in B. Grüber & R.S. Millman (eds.), *Symmetries in Science*, Plenum, N.Y.

Veblen, O. (1933): *Projektive Relativitätstheorie*, Springer-Verlag, Berlin.

Veblen, O. & B. Hoffmann (1930): *Projective Relativity*, Phys. Rev., 1930 (36) 810–822.

Wesson, P.S. (1999): *Space–Time–Matter*, 1999, World Scientific, Singapore.

Chapter 8

Reasoned Alternatives

8.1 Alternative Perspectives

The proliferation of textbooks over the last half century has proceeded at such a pace that the burden of keeping up with new developments forces many a new author to repeat the conclusions of their immediate predecessors without thought or question. In the process good faith leads on to dogmatic belief in certain fundamentals, never to be questioned again.

One of the most glaring examples is seen in the theory of chemical bonding which now enters science curricula at the primary-school level. When reaching graduate level, chemists with this background experience, not surprisingly, may react with bewildered disbelief should anybody venture to query the inviolate truths of their science. Even experienced scientists often fail to recognize the most elementary inhereted delusions.

The present author is no exception, having spent many hours in fruitless pursuit of the elusive quantum/classical limit, commonly defined by the unphysical condition $h \to 0$. Stepping back from the traditional perspective, in this case, suggests some unexpected aspects of vacuum structure.

8.1.1 Quantum Potential in the Vacuum

The most convincing analysis of the quantum limit relies on the quantum potential which is given in a system with wave function

$$\Psi = Re^{iS/\hbar} \quad \text{as} \quad V_q = -\frac{\hbar^2\nabla^2 R}{2mR}. \tag{4.27}$$

Systems with vanishing V_q (*i.e.* $\hbar/m \to 0$) must clearly behave classically. A photon of zero mass is a pure quantum; an electron less so and a marble is classical. There is no discontinuity from classical to non-classical behaviour,

J.C.A. Boeyens, *Chemical Cosmology*, DOI 10.1007/978-90-481-3828-9_8,

which clearly continues through the electromagnetic spectrum as shown in Table 8.1.

Table 8.1: *Frequency and Wavelength of selected quantum objects*

Item	Mass(kg)	$\nu(s^{-1})$	$\lambda_C(m)$
TMV*	6.6×10^{-20}	10^{31}	3×10^{-23}
Hemoglobin	10^{-22}	1.3×10^{28}	2.6×10^{-20}
C_{60}	1.2×10^{-24}	10^{26}	3×10^{-18}
U	4×10^{-25}	5×10^{25}	6×10^{-16}
p	1.67×10^{-27}	2.2×10^{23}	1.4×10^{-15}
γ		10^{22}	10^{-14}
e	9.1×10^{-31}	1.2×10^{20}	10^{-14}
X-ray		10^{18}	10^{-10}
UV		10^{16}	10^{-8}
Vis		10^{14}	10^{-6}
MW		10^{10}	10^{-2}
Radio		10^{6}	10^{2}
		10^{4}	3×10^{4}

*TMV – Tobacco Mosaic Virus

The frequencies and wavelengths of massive objects derive from their rest mass through

$$E = mc^2 = h\nu\,, \quad i.e.\ \nu = mc^2/h\,, \quad \lambda = h/mc\,,$$

which is recognized as the Compton wavelength, λ_C.

It is of interest to examine how the quantum potential varies as a function of λ_C and to what extent extrapolation to higher values may identify such properties of the vacuum that may be of importance in intergalactic space.

A quantum object confined to an impenetrable spherical enclosure of radius r has quantum-potential energy

$$V_q = \frac{\hbar^2}{2mr}. \tag{4.29}$$

Substituting $m = h/c\lambda_C$, gives $V_q = \hbar c\lambda_C/4\pi r^2$, anywhere within the sphere. This corresponds to the potential energy of a spherical electronic wave packet

(Figure 4.6) of radius r, as calculated from its interaction with a second electron:

$$V = \frac{e^2}{4\pi\epsilon_0(2r)}.$$

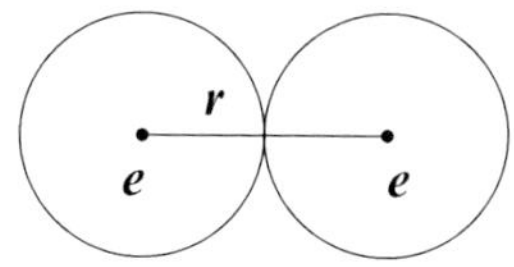

Putting $V = V_q$, one has $e^2 = 2\epsilon_0\hbar c\lambda_C/r$. If the radius of this sphere satisfies de Broglie's condition, $n\lambda = 2\pi r$, then

$$e^2 = 4\pi\epsilon_0\hbar c(\lambda_C/n\lambda_{dB}).$$

As all other parameters are constants, the ratio $\alpha = \lambda_C/\lambda_{dB}$ is also constant for $n = 1$.

By a second de Broglie condition, the wavelength $\lambda_{dB} = h/p = h/mv$, depends on the velocity of the electron and can vary continuously. This means that α is a constant only for special quantum states of the electron, in which case

$$\alpha = \frac{e^2}{4\pi\epsilon_0\hbar c} = 7.3 \times 10^{-3}.$$

One such state is of the electron in the hydrogen ground state, where

$$\frac{\lambda_C}{\alpha} = \lambda_{dB} = 2\pi r$$

predicts the correct value of $r = a_0$.

Returning to Table 8.1 it is inferred that the quantum potential increases in the same order as λ_C at a given radial distance and in each case has a maximum at $r = \lambda_C/2\pi\alpha$. If we interpret the quantum potential and the associated wave motion with virtual photons we therefore find no evidence of infinite self-energies or infinite mass in the vacuum – only waves.

Although there is also no evidence of a quantum limit there is a discontinuity between massive and massless objects to be considered. Why, for instance, does a γ-ray have a shorter wavelength than an electron, but remains massless? Although this question cannot be answered there are clear differences between electrons, protons and γ-rays. The first two are fermions with respective negative and positive charges, while γ-rays are neutral bosons. The difference lies in spin, electric charge and mass, as shown in Table 8.2.

We conjecture that the mass difference between leptons and baryons is due to a different internal wave structure, related to the $SU(2)$ and $SU(3)$ gauge symmetries of quantum field theories. Antispin is responsible for the appearance of antimatter. The annihilation of matter and antimatter occurs

Table 8.2: *Spin, charge and mass of elementary waves*

	Mass	Spin	Charge
Proton	H	$\overset{\rightarrow}{0}$	+
Electron	L	$\overset{\leftarrow}{0}$	-
Positron	L	$\underset{\leftarrow}{0}$	+
Anti-proton	H	$\underset{\rightarrow}{0}$	-
γ-ray	0	1	0

when the chiral fermionic spin states are racemized and no longer able to contain the energy of the wave packet as mass:

$$(m)\text{atter} + (m)\text{atter}^* \rightarrow 2 \times \gamma(energy) = 2mc^2,$$

to create energetic bosons. The inverse process is also known to occur.

The inference from all this is that the vacuum is a universal wave field of long wavelength. In an open universe the wavelength becomes infinitely long, resulting in a quiet flat space-time. In a universe of closed topology there is endless wave motion, which we identify here with the quantum potential and which Milo Wolff (1990)describes[1] as the solution of a universal scalar wave equation. Our interpretation of cosmic physics differs from Wolff's model only in his proposal (p. 171)

> "that matter and antimatter are [...] *gravitationally repulsive.* This makes it possible for equal amounts of matter and antimatter to exist together in the universe and accounts for only matter being found in our galaxy."

No such assumption is needed in a chiral universe wrapped around an involuted vacuum interface.

The scalar wave equation only refers to long wavelength vacuum undulations and cannot be linked to the wave function of the universe, postulated

[1]This book is recommended to readers less familiar with new interpretations of modern physics.

by some cosmologists in a theory of quantum gravity. It is evident from Table 8.1 that quantum behaviour gradually disappears towards the top of the table as $V_q \to 0$ and $h/m \to 0$, as gravity gains importance.

The wave function of the universe should also not be confused with wave properties of the gravitational field (Einstein & Rosen, 1937). Gravitational waves may be emitted when large masses interact. The most likely source that has been mooted is a binary pulsar system, or on a much smaller scale, even an orbiting planet. The existence of gravitational waves is inferred from the fact that electromagnetic and gravitational disturbances follow the same null-geodesic paths. However, unlike the electromagnetic field the gravitational field equations for free space are not linear, but by approximation can be linearized (Adler et al. 1965), such that the Riemann tensor obeys a wave equation:

$$\Box^2 R^{\alpha}_{\eta\beta\gamma} = 0.$$

According to this equation gravitational waves resemble electromagnetic waves in polarization properties and speed of propagation.

The interaction between two electrical charges has been described (section 3.5.1) as mediated by a standing wave, which we called a photon, or a virtual photon in a balanced situation. Gravitational interaction can now be viewed in the same way. Two masses at the same potential exchange virtual gravitons – if they are at different potentials, transfer of energy takes place and the mass lost at the higher potential is gained at the low potential. This mass transfer further distorts the gravitational field and the feedback enhances the radial attraction. In fact, Einstein and Rosen pointed out that, like any wave equation, the gravitational equation is satisfied by both retarded as well as advanced potentials, or a mixture of the two, which ″...corresponds to the case without damping, in which a standing wave is present″. This result is directly in line with Wolff's (1990) proposal.

A planet in orbit therefore looses energy, and like a classical orbiting electron (section 3.4.2), spirals in towards the nucleus of the system. In the case of an atom this is prevented by the quantum potential and there is no reason why a cosmological quantum effect could not be responsible for the stabilization of satellite orbits. In fact, there is the evidence of planets and moons in the solar system on orbits characterized by integers, as discussed in section 5.3.1. The cosmological term, Λ, which Einstein included in the gravitational field equations (6.4) describes exactly such an effect.

8.1.2 The Vacuum Interface

The idea that the vacuum represents an achiral interface that separates two space-time segments of opposite chirality developed from the notion that mass-dependent quantum effects arise from a field in the vacuum which affects the smallest of objects most prominently. The original argument (Boeyens, 1992) was that quantum behaviour results from feeble interactions through the interface, which create the impression of random wave-like perturbations imposed on the regular motion of small particles.

Without the particle concept the same argument is restated by postulating a general wave pattern, which cannot be uniform everywhere in non-Euclidean curved space, giving rise to elementary wave packets that behave like particles with mass, spin and charge. The uniform curvature necessitates a closed topology. The most economical arrangement, respecting Occam's razor, occurs for closure with an involution, which results in projective space with a double cover that meets in an interface, recognized as the vacuum. The involution inverts the chirality of objects (wave packets) in the double cover as well as the time coordinate, which is everywhere perpendicular to the interface. The resulting space-time corresponds to a Riemannian manifold with the local tangent spaces of projective relativity.

The result is a universe that consists of exactly fifty percent antimatter, which however, can never be detected in convential observations – only when the interface is penetrated. Although matter and antimatter therefore occupy the same space, there is no possibility of direct interaction as the two antipodes of the double cover are at different time coordinates. It is important to realize that transportation along the double cover, through the involution, gradually converts matter into antimatter.

An age-old argument about the heat-death of the universe is also settled by the interface model. It relates to the problem that the second law of thermodynamics is time-irreversible, but based on time-reversible laws of physics. It has been argued (Boeyens, 2005) that, because the world lines in neighbouring tangent spaces of the curved manifold are not parallel, a static distribution of mass points must be inherently unstable. As systems with non-parallel world lines interact a chaotic situation such as the motion in an ideal gas occurs, which means that time flow generates entropy.

Because of spatial curvature an initially stationary array of non-interacting particles (ideal gas) spontaneously generates relative internal (zero-point) motion. This intrinsic microscopic instability is responsible for the dispersal of energy and the source of entropy. Transportation along the interface inverts, not only the time coordinate, but also the entropy production. Integrated over the entire closed universe the total entropy production is zero and the heat death is averted.

8.1.3 Cosmic Dispersal of Matter

The near-equality of cosmic and solar abundancies of the chemical elements points at either a single synthesis event, starting from elementary matter, or a common mechanism of nucleogenesis, wherever it happens. The first possibility is the one originally preferred in big-bang cosmology, but later abandoned as there was considered not to be enough time available for this process in the early universe.

Small, but significant, differences between solar, meteoric and cosmic-ray abundances imply that environmental effects, which vary between different space-time regions, are also at play in nucleogenesis.

We consider black holes as the most likely source of reconstituted atomic nuclei. All forms of gravitating matter are concentrated in the black hole, subject to an ever increasing pressure gradient, which inverts the electronic configuration of atoms and eventually produces an equilibrium plasma of neutrons, antineutrons and elementary nuclides. The outlet for matter accumulated in a black hole is provided through an Einstein–Rosen bridge, in which the radial coordinate becomes time-like, as for a stationary particle in Minkowski space. The residence time becomes indefinite and the time-constraint for equilibrium synthesis in a big bang no longer applies. Inversion of the time coordinate at $r = 0$ implies that incoming matter is converted into antimatter.

As the plasma moves through the centre of maximum curvature it is forced through the vacuum interface into a space-time domain of inverted chirality and decreasing pressure. New matter (antimatter) emerges beyond the interface as an equilibrium mixture of cosmic rays, in a soup of α-particles, as it is squirted out from the black hole into free space. Some of the newly formed nuclides decay radioactively on moving into regions of lower curvature and a set of nuclides, characteristic of local curvature, survives. In the solar system the set consists of the 264 stable isotopes of 81 elements described before. The closed periodic sytem of nuclides and anti-nuclides, related by inversion, is consistent with the proposed mode of cosmic circulation of mass.

This scenario explains the periodic abundance of atomic nuclides and the origin of cosmic rays, which consist of the bare nuclei of all natural elements moving at relativistic speeds through the galaxy. By comparison, the standard model of stellar nucleosynthesis cannot explain the matching abundances of elements in the solar system and in cosmic rays.

Astronomical Evidence

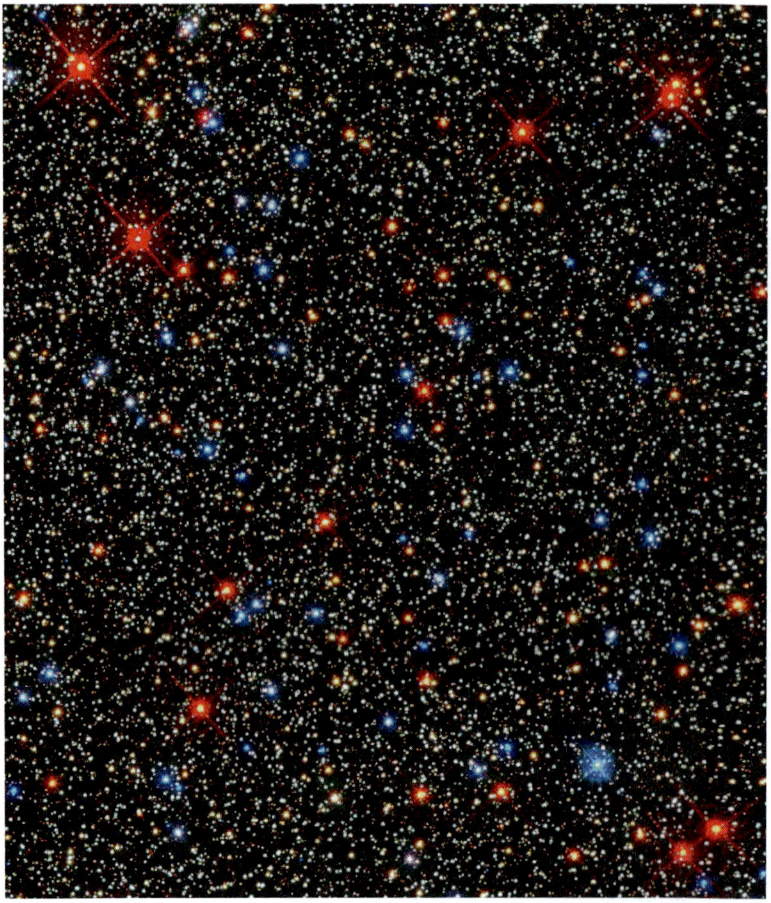

Figure 8.1: *Hubble picture of the globular cluster M3 NGC5272 in Omega Centauri*

The conjecture that matter, which emerges from an Einstein–Rosen outlet, is visible as a globular cluster, is supported by a recent Hubble picture (Figure 8.1) of stars in the Omega Centauri globular cluster. It shows a myriad of blue and red giants beside yellow, sunlike stars, all of various sizes and arranged in conspicuously regular geometrical patterns. The most obvious pattern, displayed by the prominent red stars, defines a golden logarithmic spiral, as shown in Figure 8.2.

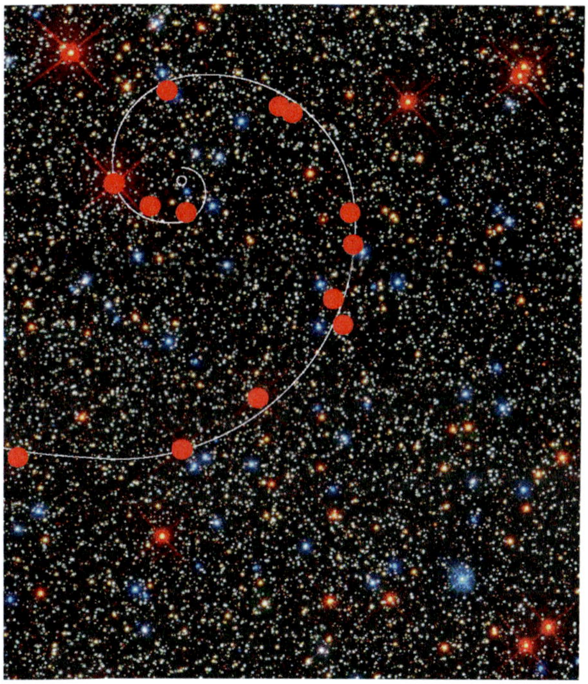

Figure 8.2: *A spiral set of red stars*

Also readily distinguished are the large number of blue stars and those that appear as yellow dots with red halos. By superimposing logarithmic spirals, either as coupled pairs, as in Figure 5.8, or as single spirals, where appropriate, most of these images are seen (Figure 8.3) to fit the same pattern as the prominent red stars.

It would appear that each of the many types of star that appears in Figure 8.1 is arranged along a number of logarithmic spirals. Because of the self-similar symmetry of a golden spiral, groups of any size must follow the same spiral curve, if this conjecture is valid. To test the proposal the readily noticeable circular arrangement of small white dots towards the lower right-hand corner is shown, on larger scale, in Figure 8.4, with a golden spiral superimposed.

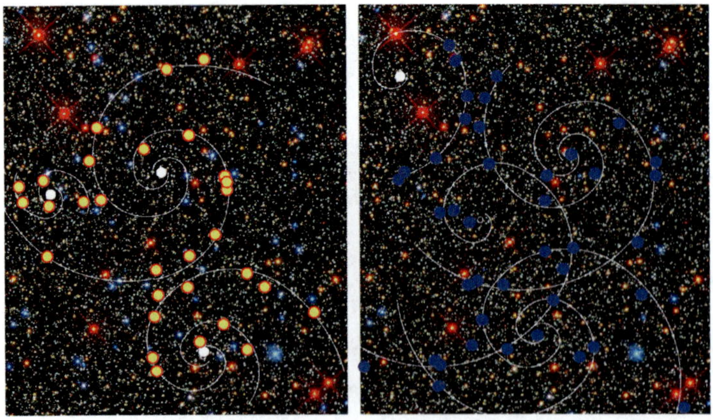

Figure 8.3: *Seperate logarithmic spirals account for most of the yellow stars with red halos (left) and the blue stars (right) in the cluster*

It is considered quite likely that a computerized pattern search would arrange most spots in the picture on seperate self-similar golden spirals. Whatever it signifies it certainly militates against a random distribution of stars in the cluster. The star cluster is clearly not what it appears to be at first glance. Rather than a single conglomerate of stars, it appears to represent a superposition of many galaxies.

All of the spirals identified here are of the same chirality and appear in projection along a line close to their plane normals. The undisputable fact is that the surprising internal structure of the globular cluster, whatever its size, is intimately related to the golden ratio and verifies the notion of self-similarity at all levels of the physical world. With hindsight spiral structures become obvious in the earlier picture of the globular cluster, shown as Figure 5.12.

A notable feature of the spiral structures is the lack of a heavy core as seen in spiral galaxies, such as M51, shown in Plate 5.1. A reasonable explanation of this and of the uniform alignment and chirality of the spirals is that they originate in turbulent flow from a central point, and therefore seen oriented in the same radial projection. In hydrodynamics such structures are known as vortices or eddies. Each of the eddies generated by the turbulent expansion is the potential nucleus of a new galaxy.

This pattern of expansion, driven by the time-inverse of a black hole, is the complete opposite of the commonly accepted structure of globular clusters, associated with the Milky Way, as ancient conglomerates, all of the same age,

Figure 8.4: *Spiral through small white spots of similar size*

and in a state beyond gravitational collapse, described as 'reincarnation'.

The origin of cosmic rays presents an even more difficult problem for standard cosmology, with supernovae as the most likely source. However, a supernova origin cannot account for the isotropic distribution of the high-energy particles, which suggests an extragalactic origin. Another, more fanciful, model (Ramathy, R. et al. 1999) proposes the accumulation of cosmic rays in superbubbles, almost the size of the entire galaxy, from where they are accelerated into the "interstellar medium". To quote:

> "Cosmic rays [...] come from the matter in what are known as superbubbles, tremendous cavities blasted out of the interstellar medium by the winds of giant stars and the explosions of supernovae. Typically, a superbubble is formed over millions of years, not by one star, but by the winds and supernovae of many. The explosions fill the superbubbles with hot, rarefied plasma, or in other words ionized gases. Compared with the surrounding interstellar medium, the gases in superbubbles are rich in metals (*sic*) such as carbon and oxygen, which are cooked up inside the stars and in the supernovae that created the superbubbles in the first place.
>
> ... a typical superbubble can last for tens of millions of years, during which it swells with the matter of hundreds of supernovae. New explosions within the superbubble, roughly every 100,000

> years, continually accelerate more cosmic-ray particles out of the mix and into the interstellar medium.
>
> A superbubble can be 2,000 light-years across, and the bubbles merge and overlap to fill roughly half the volume of our galaxy.
>
> ... supernova explosions are the major source of refractory dust. [...] We think those grains of dust are the ultimate source of cosmic rays. As the grains speed through a superbubble, they collide with atoms of gas. The collisions knock loose, or sputter, atoms in the dust grains, and the new atoms emerge as high-velocity nuclei of refractory elements [....] the force of the collisions scatters the volatile gas atoms and brings some of them up to speeds comparable to the speeds of the sputtered atoms. They, too, can be preferentially accelerated to relativistic speeds, becoming cosmic rays".

It is concluded that the study of cosmic-ray origins remains a field in ferment, and the dust grains in superbubbles are just one of several scenarios under investigation.

This explanation is hardly convincing and difficult to assess without better characterization of the superbubble and interstellar medium. Apart from producing cosmic rays the high-energy collisions, which occur all the time, must surely cause enough excitations to make the bubble glow. There is no evidence of this.

8.1.4 Quasars and Radio Sources

The idea of a black hole finding an exit for accumulated mass through an Einstein–Rosen bridge, into a conjugate region of space-time, has been considered as a mechanism for nucleogenesis. Considered in conjunction with the Naan model of parallel universes, and the Bohr–Lüders theorem of CPT symmetry, a new angle on the nature of quasars opens up. The scenario envisioned here is a build-up of mass on opposite sides along the vacuum interface, as shown schematically in Figure 8.5.

In the same way that the excessive curvature of space-time, generated by a growing black hole, eventually penetrates the vacuum interface, a pair, consisting of a sub-critical black hole and a symmetry-related anti-black hole would coalesce as the interface between them collapses. The effect is catastrophic. Two nearly infinite masses of matter and anti-matter, on mutual annihilation, must generate more than enough energy to produce the type of fireworks observed as quasars, Seyfert galaxies, γ-ray bursts or radio (X-ray) galaxies, depending on the size of the event.

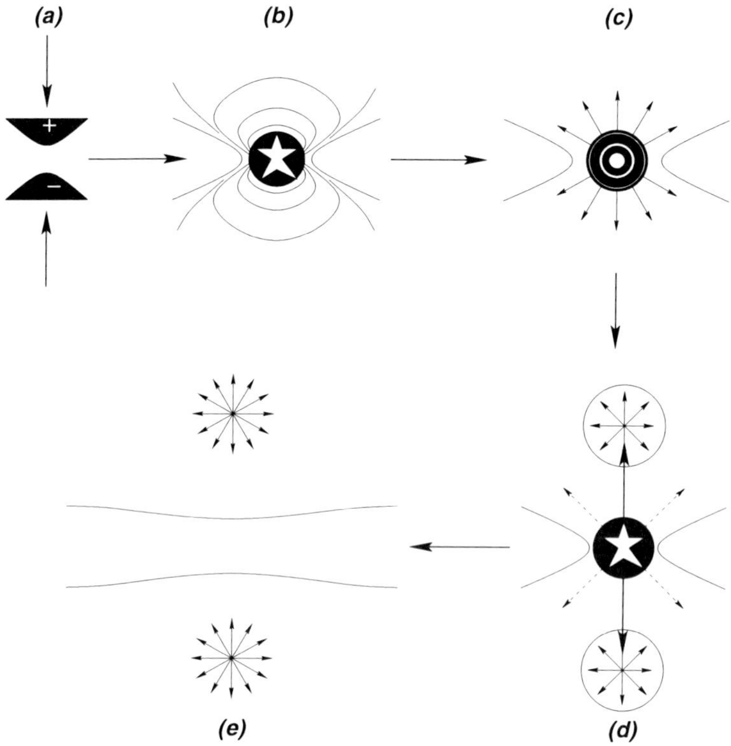

Figure 8.5: *Birth of a quasar and a radio galaxy between two black holes*

As the pair of black holes (a) break the vacuum interface the annihilation explosion (b) between matter and antimatter appears as a point-like energy source surrounded by a massive magnetic field. As in an oscillating chemical reaction radiation is emitted in a periodic fashion (c), commonly observed as a variable quasar.

Relativistic leptons and cosmic plasma excite synchrotron radiation in the environment (d) with the appearance of a typical radio galaxy. As the interface repairs itself isolated radio sources in a blank field are left behind (e).

The interacting matter within the quasar is under monstrous pressure and emits extremely redshifted radiation. These chemical redshifts are not distance indicators, but as black holes are more likely to develop in regions of high mass density, quasars are likely to occur intragalactic. The net effect would be the creation of isotropic radiation fields such as the X-ray background radiation and cosmic rays in the Milky Way. This model explains the variability over minutes in the radiation output of quasars, commonly assumed to occur simultaneously over the entire extent, of light-year order, of the quasar (Gaskill, 1983). We postulate here that the variability is confined to the active centre, which so dominates the energy production that the spectral features of the extended background is effectively masked.

The total background X-ray flux points at some 100 quasars per square degree, amounting to more than 4 million quasars in the sky (Powell, 1991). Even more enigmatic are the isotropically distributed sources of more than 150 γ-ray bursts per year, lasting about half a minute each, without evidence of repeaters. The amounts of energy (Schwarzschild, 1992) released are indicative of the "annihilation of infalling chunks of antimatter", the existence of which is ironically denied by standard cosmology. As already indicated however, such annihilation can occur through the vacuum interface, but for these events it should be at a much smaller scale than coalescing black holes.

A possible explanation of the bursters is suggested by the gravitational field, postulated in Figure 5.8c, representing a galaxy centred on a black hole. The gravitational profile on the black-hole (space) side of the interface has a shape, self-similar to that of a solar system, but not on the antipodal side. Towards the edge of the galaxy there is a region where the interface could be readily penetrated by approaching objects of moderate mass and where chunks of antimatter could penetrate into familiar space. γ-ray bursts occur when these wanderers encounter matter of comparable mass. However, it does not mean that such events are confined to the edge of the galactic disc.

Space-time is not three-dimensional as suggested by the diagram, but four dimensional and the interface is three dimensional. The proposed activity therefore occurs, not in a two-dimensional circular region in the galactic plane, but in a three-dimensional shell that surrounds the galactic nucleus, and appears isotropic from within the Galaxy. The Milky Way does not have the three-dimensional structure suggested by the gravitational field of Figure 5.8c because it is flattened into a disc by the angular momentum created by the vortex which is generated by the matter, which is sucked into the black hole.

The proposed (Manuel et al., 2006) nuclear cycle that powers the cosmos has many elements in common with some of our arguments. Not unlike the periodic model of stable nuclides and the notion of cosmic self-similarity these authors suggest that stars are subject to the same types of interaction that occur in radioactive nuclides, which depend on the relative amounts of nucleons defined by the numbers A, Z and N. Because of chemical layering an accumulation of neutrons that resembles a neutron star develops at the core of an ordinary star. This core is left behind as the remains of a supernova.

The analogy is extended to galaxies, but avoids the postulate of a black hole at galactic centres. The proposal of a nuclear cycle is intended as a substitute for an initial big bang or the disappearance of matter into black holes. What they offer instead as a driving force is the energy generated during prevalent galactic collisions, giving rise to the formation of quasars, γ-ray bursts and active galactic nuclei. The definition of an AGN remains problematical, with neutron repulsion as the main source of energy for such products of gravitational collapse.

Most of their interesting conclusions find a simple explanation in matter-antimatter annihilation through the vacuum interface, as proposed here.

8.1.5 Redshifts Revisited

Modern cosmology is dominated by the Doppler interpretation of cosmological redshifts (Section 6.6.1) and the assumed expansion of the universe. It is therefore of interest that several alternative explanations of redshifting have been proposed. These proposals are essentially of two types, predicting redshifts that are either distance dependent, or not. Of those already discussed in these pages chronometric redshifts (7.3) are distance related while chemical shifts (5.1.2) are not. Observed redshifts are most likely due to more than just one of the factors discussed below. Not surprisingly, anomalies, like discordant redshifts observed from physically connected objects, are frequently observed.

A comprehesive review (Reboul, 1981) of 772 reports, dealing with anomalous redshifts published before 1980, provides an interesting insight into an ongoing redshift controversy. A total of 216 of these were found to be explained adequately by standard cosmology while 33 more were explained with bias and in 8 cases there were not sufficient data for useful comment.

The remaining 515 instances were adjudged as unexplained by the standard approach. 91 of these clearly depend on some form of intrinsic redshift and 146 more could be accounted for by more than one alternative explanation. A further 57 reports dealing with multiple redshifts, superluminal velocities, energetics and other general problems remain uncoded and for 71 no preferred interpretation could be identified. Other interpretations relate to gravitational effects (8), motion of the sun (7), variation of physical constants, in large-number hypotheses (21), photon-boson interaction (31), superclusters (23) and the rest (60) were ascribed to alternative cosmological factors.

These anomalies are rarely addressed, if not ignored altogether, in the cosmological literature, but they certainly show that no single model is likely to explain all observed redshifts. The Doppler interpretation, in particular, assumes a simple linear law which is statistically testable. Such testing could either confirm or reject the assumed distance dependence and/or the linearity. Alternative explanations should, in any case, be considered for outliers. Possible alternatives are discussed in the following.

Doppler Shifts

The reigning cosmological paradigm is based on a Doppler interpretation of galactic and quasar redshifts. Verbal disagreement about this interpretation leads nowhere. The only decisive test must be an unbiased statistical test of the basic assumption, which is a linear relationship between measured redshifts and some dependent variable, such as distance to the source or rate of recession. Such a test was conducted by Troitskij (1996). Distance, and hence velocity, estimates are derived from observable parameters such as apparent luminosity $m(z)$, and angular size, $\theta(z)$. To avoid possible bias the dependence of these parameters on the redshift z was tested, using standard methods of regression analysis, applied to all available data sets.

To perform a regression analysis on a set of data points (x_i, y_i), $i = 1 \rightarrow n$, it is assumed that the x values are fixed and by least-squares fit find the best values of m and b that describe the linear relationship:

$$y = mx + b\,.$$

This is done by minimizing the squared vertical distances between the data points and the regression line. The best values of m and b are calculated as

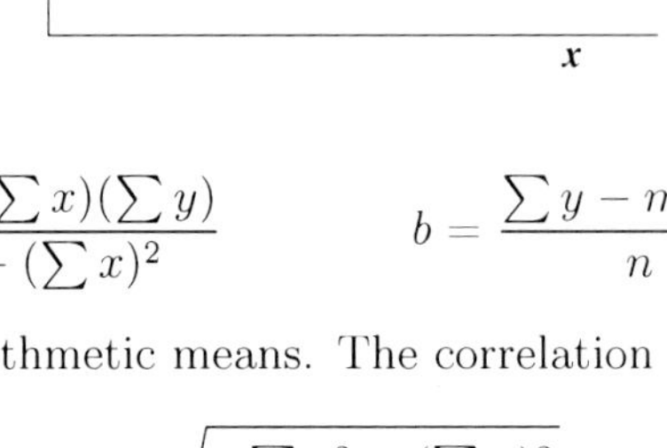

$$m = \frac{n\sum xy - (\sum x)(\sum y)}{n\sum x^2 - (\sum x)^2} \qquad b = \frac{\sum y - m\sum x}{n} = \bar{y} - m\bar{x}\,,$$

where $\bar{y}$ and $\bar{x}$ are arithmetic means. The correlation coefficient

$$r = m\sqrt{\frac{n\sum x^2 - (\sum x)^2}{n\sum y^2 - (\sum y)^2}} \leq 1$$

measures the goodness of fit.

In Troitskij's test the astrophysical observable parameters $m(z)$ and $\theta(z)$ are derived to reveal the distance function $R(z)$, starting from absolute luminosities $M(z)$ and linear sizes $l(z)$ of the galaxies. Whereas $m(z)$ and $\theta(z)$ are random values at any z this complication was overcome by demonstrating that $\overline{M(z)}$ and $\overline{\log l(z)}$ are constant values independent of z. Many checks were introduced to ensure statistical uniformity in the sampling process.

In standard cosmology galactic and quasar redshifts do not obey identical Hubble relationships. The larger intrinsic luminosity of quasars is said to be inappropriate in the $m(z)$ test of uniform galaxies, and hence not taken into account. In Troitskij's procedure this discrepancy does not occur and the $m(z)$ data for quasars and galaxies join up smoothly to yield a single global statistical dependence with demonstrable Gaussian distribution. This is the first indication that the linear Hubble law is not upheld statistically.

The inadequacy of Hubble's law becomes abundantly clear in an effort to fit the statistical data to the astrophysical predictions of the standard model which requires $R(z) = R_0 z/(z+1)$. Instead it is found that $R(z) = R_0\sqrt{z}$, with $R_0 = 600\text{Mpc}$. Figure 8.6, redrawn from Troitskij's original shows how this new relationship accords with experimental data.

To quote:

> "The statistical uniformity of galaxy characteristics in the universe space testifies to the stationary state of its processes whence it follows a conclusion on its considerable age which at least by an order or two has to exceed the age of galaxies estimated as 15–20 billion years".

Troitskij claims rediscovery of a quadratic law, first identified by Lundmark (1925) and later confirmed by Segal (1976). This is not strictly correct. From the available data Lundmark established a clear non-linear relationship of the form

$$z \simeq k + lr - mr^2$$

where k, l and m are constants, such that a maximum redshift is predicted for galaxies at a distance of 10^8 light-year.

Troitskij's analysis is based on more than 37 000 galactic and quasar redshifts. By comparison, the meagre sample that originally established the linear Hubble law is rather crude. However, as defenders of universal expansion freely use data conditioned by Hubble's law, further debate of the issue becomes pointless.

An interesting example of data processing is outlined by Rowan-Robinson (1988) who published the same diagram as Troitskij (Figure 8.6), with the

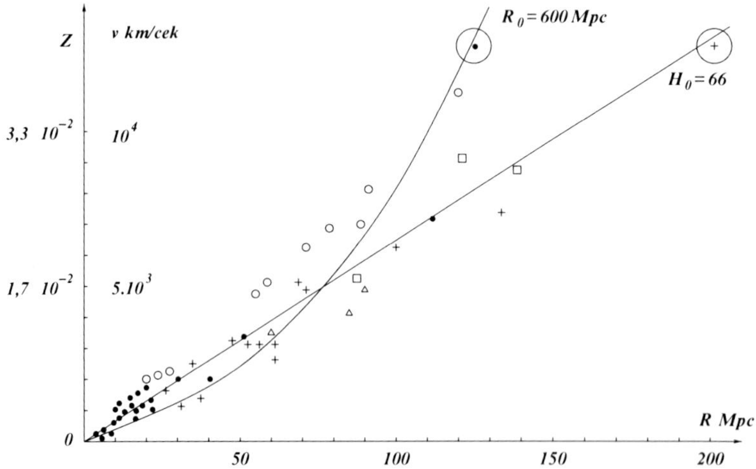

Figure 8.6: *Redshift-distance dependence taken from Rowan-Robinson (1988). As in Troitskij's original, different presentations of experimental points correspond to different methods of distance measurement. +: Supernovae; ○: Tulley-Fischer; Δ: Cepheids; □: Dispersion of star velocities; •: Two or more methods; Solid curve is $R = 600\sqrt{z}$ Mpc. The point, circled on the left, is replaced by R-R with the circled point close to his straight line for* $H_0 = 66$

linear Hubble curve for $H_0 = 66$ superimposed. By noting that the Tulley-Fisher method defines a Hubble constant of 89 and the Type Ia-Supernova method a value of 58, the empirically established parameters of both methods are adjusted to produce a common $H_0 = 66$. These adjustments result in a Hubble diagram, said to be "...indicating the extreme linearity of the Hubble flow".

The evidence in favour of a quadratic dependence of redshift on distance cannot be ignored and the Doppler interpretation of cosmological redshifts needs revision.

Troitskij's analysis found immediate application to another nagging problem of standard cosmology. Superluminal expansion of quasars and galaxies has been reported for decades, assuming a cosmological (*i.e.* Doppler) origin for the redshift used as distance indicator, (Cohen et al. 1977). A separation angle, θ is measured at intervals of many months or years by very-long-baseline interferometry that involves several widely-spaced radio telescopes.

The standard response of cosmologists is to treat the observed rate of expansion, often exceeding $10c$ as an apparent velocity "caused by relativistic motion of the emitting material towards the observer" (Moore & Readhead, 1983). The weakness of this argument is that it only applies in the rare event when the actual expansion happens on a line almost directly towards the observer. A more convincing explanation would be that the formulae of standard cosmology overestimate the distance to the quasar.

Troitskij demonstrated a simple direct solution of the problem. By plotting $\log\dot{\theta}$ as a function of $\log z$ for 32 independently reported angular rates of expansion, $d\theta/dt = \dot{\theta}$, he obtained a regression function by least-squares methods. The derived value of $\dot{\theta}(z)$, the apparent increase in angular size of the source, is then equated to

$$\dot{\theta}(z) = r_0 \sin\alpha / R(z) \, ,$$

where α is the angle between the directions of expansion and observation. By averaging $\dot{\theta}(z)$ over α within the limits $\alpha = \pi/2 \pm \pi/3$, the theoretical value of

$$\dot{\theta}(z) = 0.85 r_0 / R(z)$$

is obtained, compared to the general case, $R(z) = R_0\Psi(z)$. From the experimental value of $\dot{\theta}(z)$ then follows:

$$\Psi(z) = \sqrt{z}; \qquad 0.85 r_0 = 0.5 \times 10^{-9} R_0 \, .$$

The previous result, $R(z) = R_0\sqrt{z}$ therefore follows directly from the data, without further assumption. Setting $r_0 = c = 0.31\text{pc.year}^{-1}$, the calculated value of $R_0 = 530\text{Mpc}$ is equally dramatic.

An unbiased statistical analysis of the relationship between observed redshifts and the rate of quasar expansion, when ascribed to a light beam or other relativistic motion in the source frame of reference, leads directly to the quadratic redshift–distance relationship and predicts an eminently reasonable radius of the universe.

Gravitational Redshift

The idea of spectroscopic redshifts was inferred for the first time from the equations of general relativity. Writing the curved-space line element in the from

$$ds^2 = -dS^2 + dT^2$$

the temporal part, according to Einstein's (1950) equation

$$dT = \left(1 - \frac{\kappa}{8\pi}\int \frac{\sigma dV_0}{r}\right) dl \tag{106}$$

defines the time element, dT as measured by a clock at rest relative to a gravitating mass of radius r. In this σ is the density of ponderable material, dV_0 a volume element and dl is the time differential. It is emphasized that, despite appearances, the space coordinates are non-Euclidean. With κ defined as $8\pi K/c^2$ equation (106) reduces to

$$t \simeq \left(1 - \frac{\Phi}{c^2}\right)\tau$$

in terms of the gravitational potential Φ, proper time τ and coordinate time $t = x_0/c$. Rearranged into

$$\frac{t-\tau}{\tau} = \frac{\Delta t}{\tau} = \frac{-\Phi}{c^2}$$

the equation shows that a clock which is kept in the gravitational field lags behind a moving clock, considered to be in rythm with the vibration of a light ray. The frequency measured by this clock is independent of position whereas the frequency measured in terms of the proper time is a function of position.

A ray of light moving from the gravitational field of the sun therefore appears with a frequency shift

$$\frac{\Delta\nu}{\nu} = \frac{\Phi_E - \Phi_S}{c^2}$$

when measured on earth. As the sun is at a large negative potential relative to the earth, $\Delta\Phi$ is negative and the frequency decreases en route to the earth towards the red end of the spectrum. The important physical argument is that the frequency, expressed as a function of the proper time, is the same as the coordinate frequency at the surface of the sun, but gets out of step with increasing distance. This is essentially the same argument that led to the recognition of chronometric redshifts as arising from two independent units of time at the sites of emission and absorption respectively.

Calculation shows $\Delta\nu/\nu = 2.12 \times 10^{-6}$, which, interpreted as a Doppler shift, amounts to 0.63 kms^{-1}. This is orders of magnitude lower than the speeds of recession derived from cosmological Doppler shifts. However, as an average galaxy contains more than 10^{11} stars, the associated gravitational field must be enormous. Still, gravitational redshifts are said (Narlikar, 2000) to be generated only by compact massive bodies. There is no justification for this assumption in the relativistic equations and gravitational redshifts must be fairly common.

Tired Light

The first explanation of galactic redshifts became known as the tired-light explanation, which, for reasons unknown, was completely abandoned in favour of the Doppler interpretation. The well-known Robert Millikan is quoted by Marmet and Reber (1989) from a letter of 1952:

> "Personally I should agree with you that this hypothesis (tired light) is more simple and less irrational for all of us."

The hypothesis was resurrected by Marmet (1988). In this model the redshift is produced by inelastic collisions of photons on atoms and molecules. The argument recalls the common experience that light is transmitted through 100m of air without noticeable dispersion or visible fuzziness. However, the index of refraction of air ($n = 1.0003$) shows that collisions on air molecules are such that, on the average, the photons are delayed by as much as $100 \times (n-1) = .03$m during the 100m journey, compared to the vacuum[2]. This delay is ascribed to an estimated 10^6 photon–molecule collisions, without any angular dispersion.

In space, where the gas density is lower by 20 orders of magnitude there is about one such interaction per week. The calculated energy loss during such an inelastic nondispersive encounter amounts to a fraction 10^{-13} of the photon energy, which therefore produces a redshift, $\Delta\lambda/\lambda$ =constant. The secondary photon emitted in the process to carry away the lost energy has a wavelength of several thousand kilometers; too long for detection by radio astronomers.

Like the Doppler model the redshift predicted here is a distance indicator, but unlike the Doppler model it depends on the concentration of intergalactic gas, rather than rate of recession. Given all the other alternative redshift mechanisms there is no need to accept the tired-light proposal to explain all observed redshifts, but undeniably, it must make a substantial contribution especially in the light from far-away sources.

Curved Space

The simplest explanation of astronomical redshift, apparently the explanation also favoured by Hubble, is the effect of curved space (Boeyens, 1995). As the curvature of a two-dimensional surface requires a third dimension non-Euclidean three-dimensional space must curve into a fourth (time) dimension.

[2]Index of refraction is defined as if the air is a continuous medium, which it is not.

Any large displacement in curved space therefore contains an undetected displacement in time.

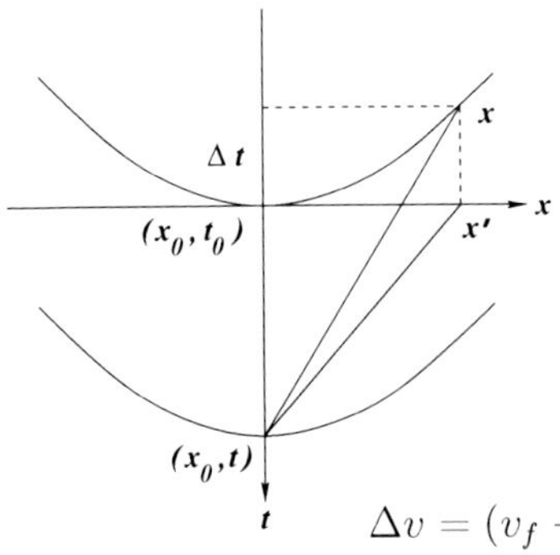

Consider a light signal that travels along the null geodesic from the point x towards an observer at (x_0, t_0). The observer, unaware of the time difference Δt interprets the signal as coming from the Euclidean point x'. On arrival at (x_0, t) a flat-space observer, calculates the velocity $v_f = x/t$, which is different from the velocity calculated by a four-dimensional observer as $v_c = x/(t + \Delta t)$. The difference

$$\Delta v = (v_f - v_c) = x[1/t - 1/(t + \Delta t)]$$

is naturally interpreted as due to recession of the source. Since v cannot exceed c the difference Δv appears as a frequency shift, interpreted as a Doppler effect, *i.e.* $z = \Delta\lambda/\lambda = v_r/v_c$. The measured shift and hence the assumed rate of recession, v_r is observed, or assumed to be proportional to the distance between receiver and source, *i.e.* $v_r = Hx$. This is Hubble's law.

An equivalent linear relationship $\Delta v = Hx$ may be assumed in curved space. In the limit $\Delta v \to \infty$, $H \to 1/t_0$ as before, but now $1/H = r_0/c$ is interpreted to define an upper bound to some radius of the curved universe, rather than its age. Unlike the Hubble age, the predicted radius is not in conflict with any known observation.

Red shifts in curved space occur by the need of a light signal to overcome a natural time lag that increases with distance, while the universe remains static.

Optical Redshifts

The prediction by Emil Wolf (1987), from the theory of optics, that a correlation of source fluctuations may cause the redshift of emission lines has been confirmed experimentally (Morris & Faklis, 1988) and provides the basis of an alternative explanation of cosmological redshifts. In later work (James & Wolf, 1994) it was further shown that scattering on fluctuating media can generate redshifts of spectral lines which are indistinguishable from those due to the Doppler effect, even though the source of the radiation, the scatterer and the observer are at rest with respect to each other. For instance, the frequency shift is the same for every line present in the spectrum of the incident light, just as in the case when the shifts are generated by the Doppler effect.

This equivalence is demonstrated specifically for the α, β and γ Balmer lines of H.

The principles of statistical optics at the root of non-Doppler spectral shifts are reviewed and evaluated by Emil Wolf (1998). It is of particular importance to note that all spectral changes of relevance could be demonstrated to occur in the laboratory. With the appropriate orientation of a beam splitter it was possible to generate both red and blue shifts in the spectra.

With respect to shifts in absorption spectra it was demonstrated that all characteristics of Doppler shifts could be reproduced by dynamic scattering on suitably correlated random media.

Too little is known about the internal structure of quasars to claim a detailed exposition of their emission spectra, but this alternative mechanism of redshifting provides an immediate rationalization of wildly different redshifts observed for physically connected galaxies (Arp, 1983), which clearly has nothing to do with velocities of recession. In the case of NGC 7603 fluctuations within the larger galaxy must be correlated differently compared to its smaller companion.

Information Theory

According to the information theory of Stonier (1990) the universe is made up of three components: matter, energy and information. In a sense, what mass is to the manifestation of matter, and heat is to energy, organization is to information. Heat is the form of energy that lacks information. All other, more structural forms of energy contain an information component.

Like mass and energy, information may also be assumed quantized as *infons*. In view of the interconvertibility of energy and information, there is a close relationship between photons and infons. The relativistic momentum equation

$$p = \frac{m_0 v}{\sqrt{1 - v^2/c^2}} \tag{8.1}$$

shows that a massless particle has zero momentum, unless it travels at speed $v = c$, in which case the momentum

$$p = \frac{0}{0}$$

becomes indeterminate but finite.

Equation (8.1) describes a photon only if $v = c$, otherwise it is postulated to describe an *infon*, which has no momentum, mass or energy, but wavelength

$$\lambda = \frac{h}{p} = \frac{h\sqrt{1 - v^2/c^2}}{m_0 v} .$$

Only at $v = c$ the wavelength, although indeterminate, is finite and the resulting photon has energy $E = h\nu$. For $v \neq c$, $\lambda \to \infty$. Thus, an infon is a photon whose wavelength has been stretched to infinity, readily identified with the quantum potential, and providing a simple interpretation of cosmological redshifts.

The photons emitted by a star that recedes at velocity v, may be assumed to approach the earth as infons with velocity $c - v$. Unlike photons, these quanta, without momentum, frequency or wavelength remain undetected by conventional receptors. In order to become detectable it is necessary for these infons to be accelerated to reach receptors with speed c. Such acceleration is experienced as it enters the earth's gravitational field. This acceleration from $c - v$ to c must increase the wavelength, causing a redshift.

All other things being equal, the greater the velocity at which a star recedes, the smaller will be the fraction of quanta that can be accelerated to reach velocity c. The effect of this is that stars with large redshifts must appear faint, not by virtue of their distance from earth, but because of their high speed of recession. Many galaxies may therefore not be as remote as suggested by Hubble's law and the universe much smaller than commonly thought.

Infons that originate at stars that move towards earth, arrive with velocities $c + v$ and there is no device to slow them down to the observable velocity of c. Such stars remain invisible and hence, all galactic objects appear to be redshifted. The light from several near-by galaxies, including M31 in Andromeda, is actually blue shifted. Like universal expansion the infon model cannot account for this observation, but without relative motion in the local group no spectral shift is predicted and the blue shift could be induced by statistical optics.

Stonier offers a novel explanation of the commonly observed dark (Bok) globules as approaching stellar objects, rather than contracting gas clouds, which occur in the early stages of star formation. He even has an explanation for the invisibility of Nemesis, the legendary companion of the sun.

The Arp Age Model

Halton Arp is best known for documenting large numbers of anomalous redshifts, mainly in the form of physically connected objects with widely different redshifts. The fraternity in which he operated persistently refused to accept his data, which cannot be explained in terms of cosmological Doppler-generated redshifts, as credible.

While looking for an alternative interpretation of observed redshifts Arp (1998) noticed

> "...that there was a continuous physical transition between the compact, high-redshift quasars through the high surface-brightness, active companion galaxies, and finally down to the more relaxed, normal appearing galaxies. This, empirically, was also a continuous sequence in age from the youngest with the highest redshift, to the oldest with lowest redshift".

The observation is rationalized in terms of the gravitational theory of Hoyle and Narlikar (1964), as summarized in equation (4.31). The new interpretation that Arp gives to this theory is that elementary matter first appears in the vacuum as massless and gradually acquires mass from existing matter on aging. As an electron ages, its increased mass results in an increased Rydberg constant, $R = 2\pi^2 me^4/h^3c$, such that the frequency of an emitted photon, $\nu = Rc(1/n_1^2 - 1/n_2^2)$, must also increase. On these grounds it is proposed that the emission line of a younger atom occurs at lower frequency and hence appears to be redshifted. New matter is postulated to emerge from *white holes*, simply defined as time-reversed black holes.

In essence, the Arp model argues that observed redshift is a measure, not so much of the distance to the source, but the age of the matter it contains. It is noted in passing that the Arp age factor is directly related to the relative mass of a source (*e.g.* the young supergiants in the Magellanic clouds), such that chemical redshifts would define the same desired trend.

8.1.6 Hubble's Constant

There is no more important concept in standard cosmology than Hubble's constant, which relates spectroscopically measured redshifts in galactic light to the estimated distance of the source and its inferred rate of recession. This definition specifies a proportionality constant only for Doppler shifts. It also occurs for redshifts with a linear distance dependence due to other causes, in which case it has no connotation to an expanding universe. It is undefined for redshifts, independent of distance.

Despite its fundamental importance Hubble's constant is also the most fiercely disputed concept in standard cosmology because of its decisive role in predicting the age of the universe. The very first version of expanding-universe cosmology was widely rejected as the original value of the constant,

$H_0 = 530\text{kms}^{-1}\text{Mpc}^{-1}$, derived from Hubble's linear plot, predicted an age of only 1.2×10^9 years, which is demonstrably less than the age of the Earth.

Later studies reduced this value to the current range of $50 < H_0 < 100$. An extensive international effort, involving 12 of the world's leading observatories, with the purpose of fixing H_0 to within an error of less than 10% produced (Freedman et al., 1994) $H_0 = 80 \pm 17$, revised to 70 ± 10 within two years (Glanz, 1996), barely overlapping the rival value of 55 ± 10. An independent survey (Rowan-Robinson, 1988) from the same era arrived at $H_0 = 66 \pm 10$.

The basic reason for the persistent discrepancy is the obstinate refusal to admit that cosmological redshifts could arise from anything but the Doppler effect. As in the Stonier (1990) model many observed redshifts may well be of the Doppler type, but unrelated to distance. In other cases the redshift may be distance dependent, but caused by curvature and not expansion. It would clearly be impossible to find a common proportionality constant for two such unrelated linear relationships. In one instance maybe $H_0 = 100$ compared to $H_0 = 50$ in the other. Who knows?

Should the observed redshifts have a variety of causes it is most likely that any effort to relate them to a single parameter such as distance or luminosity must produce a non-linear function with large scatter. Such behaviour was first observed by Lundmark (1925) and repeatedly since then by others. The situation is illustrated particularly well by Figure 8.6.

8.1.7 The Distance Scale

Each of the many interpretations of spectroscopic redshifts can probably be developed into a unique cosmology. Since there is no scientifically reliable means of identifying the correct one, a model, which is compatible with all of the likely interpretations, should be favoured. However, in current usage the term, cosmological redshift, is used interchangeably with Doppler shift, which relates distance to rate of recession by Hubble's law.

Neither distance nor speed of recession is directly measurable and it has become standard procedure to convert measurable parameters, such as apparent luminosity, into a distance measure, often relying on vulnerable assumptions. The familiar concept of *distance modulus*, $\mu = m - M$, describes the difference between apparent and absolute magnitude for an object of luminosity L:

$$m = -2.5 \log\left(\frac{L}{4\pi d^2}\right) + \text{constant} \tag{8.2}$$

$$M = -2.5 \log L + \text{constant} \tag{8.3}$$

The constants in (8.2) and (8.3) are fixed in such a way that the resulting distance scale agrees with some existing scale in a common region of overlap. In the case of galaxies the scale should smoothly extend the Cepheid scale by definition of a suitable standard candle. To this end it is assumed that the brightest star in any galaxy always has the same intrinsic luminosity. In explanation it is assumed that optimal brightness of any star occurs at commencement of helium fusion, creating a short-lived, so-called helium flash, which only depends on the critical mass that triggers the process.

Currently agreed values of the constants reduce the relationship between redshift and distance in parsec to:

$$\mu = m - M = 5\log d - 5 \tag{8.4}$$

Elimination of d, using Hubble's law: $cz = H_0 d$, gives

$$5\log h_0 = 42.38 - \mu + 5\log z$$

where $H_0 = 100h_0$. Measurement of μ and z therefore determines H_0 in what appears to be a straightforward procedure.

In the case of Cepheids the period of variability is the measurable property. In order to derive a distance modulus, using (8.4), the period is converted into an absolute magnitude, specified by two arbitrary constants:

$$\langle M\rangle = a + b\log P$$

Smart definition of the constants ensures smooth extension of the parallax distance scale (Section 2.6).

In his pioneering use of Cepheids as distance indicators Hubble treated all Cepheids alike to derive H_0. This assumption needed drastic revision when it was noticed that Cepheids occur as two types which differ in intrinsic luminosity. Although redefinition of the four new constants could rescue the form, if not the gradient of the revised scale, the possibility of the distance scale being disrupted once more by further subdivision of the Cepheids cannot be excluded. However, the only effect would be a modified Hubble constant and a new set of apologies to defend the big-bang dogma.

Extension of the distance scale to, presumably more remote objects, like galactic clusters, requires equally precarious assumptions. Like the brightest star in each galaxy, each cluster is assumed to harbour a dominant galaxy that shines as a standard candle. Finally the only criterion left to measure larger distances is Hubble's law and the redshift.

The sobering conclusion is that using Lundmark's (1925) non-linear law as the criterion in extension of the distance scale would have perpetuated an entirely different picture of the cosmos. By the same argument, the statistical preference for a square relationship, if ever accepted, would completely revolutionize most of astronomy and all of cosmology. Even worse, there is no fundamental theory that requires redshifts to be distance related at all. The prospect of giving up this belief is perhaps too painful to contemplate, but hardly as painful as living with a science dominated by medieval superstition.

8.1.8 Quasar Redshifts

The problem with Hubble's law was complicated out of all proportion by attempts to specify the observed redshifts of an increasing number of quasars by the same linear formula used for galactic redshifts. Some theoretical ideas to explain quasar redshifts within standard cosmology were reviewed by Segal et al. (1991), and shown to culminate in the theory of luminosity evolution of quasars, designed to force the definition of intrinsic luminosity as a function of distance.

Already in 1972 a quasar of redshift 3.53 was calculated to be 10 000 times as bright as the Milky Way and receding at the velocity of light. This record had long since been smashed (Peterson, 1989). A quasar with a redshift of 4.73 was calculated to have existed 10 billion years ago with the same chemical composition as much younger quasars. This creates the problem to explain how all elements were already present in their present-day abundances a billion years after the big bang.

Standard cosmology requires heavy elements to be synthesized in stars and released into space only at the end of the star's lifespan of more than 5 billion years. This means that either the Doppler-expansion theory or the big-bang nucleogenesis model must be wrong – most likely both of them. Currently reported measurements with the Hubble telescope at infrared wavelength have now reduced the period between big bang and heavy-element quasars to half a billion years and counting. Frantic adjustments to H_0 are not very likely to widen this gap again.

A graphical summary of quasar redshifts as a function of the apparent magnitude of all quasars (Hewitt & Burbidge, 1981) is shown in Figure 8.7, together with theoretical predictions according to standard cosmology and Segal's chronometric model respectively. Whereas the standard model fails to match observation, the chronometric model gives an almost perfect match, without any adjustable parameters. It is of interest to compare this result with Troitskij's corresponding curve (Figure 8.6) for galactic redshifts. Neither the linear nor the quadratic curve matches the observed as well as the quadratic quasar curve.

The fundamental difference between redshifted emission and absorption

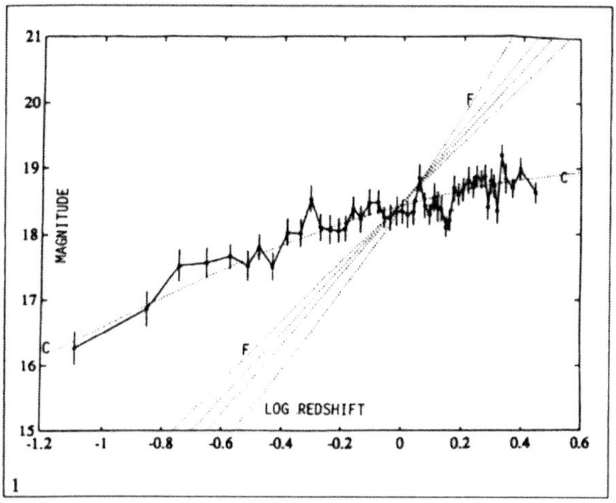

Figure 8.7: Fig. 1. (From: Segal et al. 1991, with permission). The mean apparent magnitude of quasars as a function of redshift; 2023 radio, optically and X-ray-selected quasars, 40 in each redshift bin. The *solid broken line segments* connect the means of each bin. The *dotted curves* labeled C and F are the predictions of the chronometric and Friedmann–Lemaître cosmologies for a representative sample

spectra, mentioned before in section 5.1.2 is in line with this result. The fact that chronometric redshift is a manifestation of energy conservation means that it applies directly to emission spectra in the same way as chemical redshifts.

The time discrepancy of chronometric theory translates directly into a frequency shift through the energy operator $-i(\partial/\partial t) \rightarrow h\nu$, which is related quadratically to luminosity (Segal, 1976). In the final analysis the chronometric effect, like the perturbation of electronic energy levels that causes chemical redshifts, therefore depends on the relative curvature at the sites of emission and absorption of the radiation. Although there is a distance component associated with the curvature gradient it is not the primary cause of the observed redshift.

8.2 Alternative Cosmologies

Narlikar (2002) devotes a chapter with the same title as this section to a number of variations on the theme of an expanding unverse, while specifically omitting most cosmologies that we consider to be genuine alternatives. Even Hoyle's steady-state cosmology, and later variations thereof differ from standard cosmology only in avoiding the big bang. The real malaise of Newtonian space-time and expansion of the universe remains.

8.2.1 Size of the Cosmos

The adjective 'astronomical' acquires new meaning as more powerful telescopes reveal millions of new stars and galaxies at ever larger distances. When optical probes fail to penetrate any further into the cosmic dust near-infrared devices (Cowen, 1999) show up a hundred times more stars beyond the optical horizon.

The expanding universe is believed to stretch the wavelength of long-distance photons to the point where only the faint microwave echo from the big bang remains detectable. Then comes the news of a powerful γ-ray source at the edge of the visible universe.

In order to rationalize this observation (see section 5.2.2) a simulation based on cosmic inflation, cold dark matter and dark energy (called ΛCDM cosmology) is invoked to replace classical standard cosmology and speed up stellar evolution in the early universe by many orders. The knock-on effect of this assumption on the present size of the universe is not explained and the new standard cosmology will be treated here with the scepticism it deserves.

The term light-year has become so commonplace that even with the prefix Mega or Giga it causes no real excitement. A single structure, the so-called Great Wall (Geller 1989) measures 5×10^8 light-year in length, stretches over the stupendous volume of more than 1.5×10^{24} cubic light year, consisting of some 5000 individual co-moving galaxies. The co-discoverer, Margaret Geller remarked:

> "...there is something fundamentally wrong in our approach to understanding such large-scale structures – some key piece of the puzzle that we're missing."

Astrophysicist M. Davis is quoted (Wilford, 1990) as follows in this regard:

> "...it is safe to say we understand less than zero about the early universe."

Such clustering of galaxies is not a new idea, as mentioned by Hoyle (1955), p. 272:

> "...the Local Group of galaxies is a bound system. Just as the planets are bound to the Sun and move together with the Sun

> around the centre of our Galaxy, so the galaxies of the Local Group are to be considered as a single connected physical system. [...] The Local Group is not expanding."

The problem is to explain how such clumps of matter could evolve from the homogeneous distribution assumed by the big bang. The standard answer is that tiny clumps in the early universe gradually grew bigger through gravitational attraction to form stars, galaxies and clusters. Of course, the bigger the clump, the longer the time needed for its formation. For stars a few million years are enough, for galaxies a period of one or two billion years is needed. Clusters take even longer to develop. The discovery of superclusters presented an obvious difficulty. They are just too big to form in 20 billion years.

Conservative estimates indicated times in excess of 100 billion years for the great wall and other, more extended superclusters subsequently discovered, to form spontaneously. And these are only the nearby clusters. The great wall is at a distance of only 200 million light year. The problem is not unlike that faced by early geologists when trying to fit the earth's history into a biblical few thousand years. There is simply not enough time allowed to build up mountain ranges by the known geological processes. Likewise, the big bang does not allow sufficient time for the formation of superclusters. ΛCDM cosmology could presumably be parametrized to speed up the formation of superclusters.

On weighing the impartial evidence it becomes obvious that standard cosmology underestimates the age of the world by the same margin that it overestimates the size of the universe. Both factors relate to the Doppler interpretation of redshifts, which should be re-examined, and to geometry and the topology of space-time. The obvious alternative to an infinite Euclidean universe is a world closed in space-time. As originally pointed out by Weyl (1922)(p.278):

> "If the world is closed, spatially, it becomes possible for an observer to see several pictures of one and the same star. These depict the star at epochs separated by enormous intervals of time (during which light travels once entirely round the world)."

or remarked by Segal (1974):

> "In view of the apparent transparancy of intergalactic space, the residual radiation should typically make many circuits of space before being ultimately absorbed by matter."

and summarized by Daigneault (2005):

> "Could it be that the night sky were a family album of the living and dead celestial objects, each of them being depicted a large number of times?"

Multiple images occur, not only because of light rays that circle the cosmos, but also as a consequence of multiply-connected topology, that causes a single object to be visible in different directions. If our proposition, of a black hole as a hole in the vacuum interface, holds, each such object increases the connectivity of space-time and further multiplies the possibility of multiple imaging.

The weakest link in the size argument is the large-scale distance criterion that singles out the Doppler interpretation of measured redshifts and boils down to Newton's infinite universe. This assumption simply ignores the large number of alternative effects that may cause redshifts in galactic light, each one as plausible as the next. Although some of these, like the Doppler interpretation, relate redshift to distance, they allow repeated orbiting of a topologically closed universe to account for extreme values.

The common imponderable is the topology and dimensionality of space-time, not very likely to be resolved in the foreseeable future. The smart way out is to look for the simplest solution to the problem – a closed universe with an illusionary size, suggested by an infinitude of multiple images of celestial objects. This way many exotic objects such as black holes, supernovae, quasars, radio sources and the like may be located either in the Milky Way or its immediate environs of sibling galaxies.

With all manner of gravitational lensing, revised distances and multiple imaging, also of the Milky Way itself, we may be back to Hubble's original estimates of cosmic size, without any chronological crisis. Even superclusters may turn out to be magnified images of smaller structures, much closer than imagined.

8.2.2 Plasma Cosmology

A serious defect of standard cosmology is the total neglect of electromagnetic effects, especially in view of the evidence that virtually all matter in the universe occurs in the plasma state.

The most familiar contact with the plasma state in everyday experience is in the form of lightning, produced by thunderstorms, sandstorms, snowstorms and volcanic eruptions, ball lightning, welding arcs and fluorescent lights. In his fine summary of plasmas and plasma physics Liley (1977) defines a plasma as any electrically conducting medium, including metallic solids.

A fluid plasma is produced in the laboratory by heating a gas to a temperature, high enough to cause ionization into electrons and positive ions. At sufficiently high temperatures atoms, stripped of all their electrons, are known as cosmic rays. Above 10^6K the bare nuclei undergo nuclear reactions. Thermonuclear fusion leads to the formation of heavier nuclei with the release of excess mass as kinetic energy.

Equilibrium in a fluid plasma at constant temperature defines a complex state of interactions between electrons, ions and photons, involving dissociation and recombination between all different species and the creation of equally complex electromagnetic fields. The study of conducting fluids in electric or magnetic fields is known as magnetohydrodynamics.

Radiation from a plasma consists mainly of *Bremsstrahlung*, emitted as a continuum by deceleration of electrons in collision, synchrotron radiation caused by rotation in a magnetic field and atomic emission radiation, better known as line spectra associated with transition between electronic energy levels. The equilibrium between all modes of radiation in a large plasma causes the emission of black-body radiation. The interaction between plasma particles and electromagnetic fields causes internal wave motion of both longitudinal and transverse types.

The phenomena observed in laboratory studies of plasmas also occur in extraterrestrial plasmas such as stars, interstellar and interplanetary media and the outer atmospheres of planets. The fact that cosmological implications of plasma interactions can be simulated in the laboratory opens up the possibility of studying astrophysical phenomena in the plasma laboratory, as pioneered by Hannes Alfvén.

Noting that the electromagnetic force between two currents is inversely proportional to the distance between parallel conductors (Starr, 1957):

$$F = \mu_0 \frac{i_1 i_2}{2\pi d},$$

compared to the inverse-square law of gravity, suggests that electromagnetic effects are seriously underestimated as a cosmological factor. Stars are gravitationally bound plasma and the energy that they emit originates in interior thermonuclear fusion reactions. The interstellar medium near a hot star consists mainly of hydrogen, ionized by the ultraviolet from the star. This, socalled H II region is surrounded by vast clouds of neutral H I regions. These clouds contain photoionized atoms, cosmic rays and grains of dust. Such plasma regions, which spread, at lower concentration, into intergalactic space, contain magnetic fields, which are no doubt, responsible for astrophysical phenomena.

As first proposed by Alfvén, sheets and ropes of electric current criss-cross the universe, giving rise to the formation of large-scale cellular and filamentary structures. The increasing number of galactic clusters and superclusters that show up on maps of deep space may be evidence of precisely such filamentary connections (Lerner, 1991).

Considered with an open mind plasma physics has the potential to develop into an extremely powerful guide to the physical structure of the cosmos, depending on the amount of spectroscopically invisible plasma in intergalactic space. The most sensitive method of detection of such plasma might be by looking at very long wavelength radio waves. In observations by Reber (1986) at a wavelength of 144m, galaxies are not transparent but show up as shadows on an illuminated cosmic background. Radiation observed in regions where there is no resolved galaxy was ascribed to hot intergalactic plasma at an average density of 0.01 atom per cm^3.

Before one can seriously consider an alternative plasma cosmology the ideas of Alfvén and others need to be integrated with a sensible alternative to universal expansion and the topology of space-time. Instead of chasing after non-baryonic dark matter the role of hydrogen in that regard should be explored and the interaction between matter and antimatter, an important argument in the current theories (Lerner, 1991), must be rationalized.

8.2.3 The Self-Similar Cosmological Model

The self-similar cosmological model (SSCM) was first proposed and developed by Robert Oldershaw who has been ploughing a lone furrow since 1978. It is largely in line with the ideas on self-similarity developed in this work, but based exclusively on the assumption of a universe with infinite fractal geometry. A detailed exposition of the theory appeared in two papers of Oldershaw (1989a,b). Later papers elaborated on several aspects, without adding anything essentially new.

The basic premise is that the fractal nature of the universe leads to self-similarity, or invariance with respect to scale transformation, in which small parts of the structure have geometrical properties that resemble the whole structure or large parts thereof. This relationship implies that Nature has a nested hierarchical organization. Although the overall mass range of objects that reflect the cosmological hierarchy may appear to be continuous, relatively few classes of objects account for large percentages of the mass of the observable universe, whereas objects at intermediate levels make an infinitesimal contribution. This means that the cosmological hierarchy is highly stratified.

At least 90% of the total observed mass is bound up in stellar and substellar objects ranging in mass from about 10^{-5} $M_\odot$ to $8M_\odot$ ($M_\odot \equiv$ one solar mass). This level is referred to as the stellar scale. Levels above the stellar scale do not involve appreciable mass until one reaches the galactic scale ranging from $10^7 M_\odot$ to $10^{12} M_\odot$. At least 95% of observable matter is tied up in objects that populate this level. By comparison, the atomic scale of the cosmological hierarchy is represented by atomic and subatomic matter in the mass range of about 9×10^{-31} kg for an electron, to about 4×10^{-27} kg for heavy atoms. Again, there is a cosmologically insignificant percentage of cosmic matter that appears in the interval between atomic and stellar scales.

Comparison of typical objects at the atomic and stellar scales inspired formulation of the SSCM, which assumes that atomic, stellar and galactic scale systems are rigorously self-similar, meaning that specific systems on a given cosmological level have specific analogues on all other cosmological scales, and that the properties of analogues from different scales are geometrically related by the simple scale-transformation equations:

$$\begin{aligned} R_n &= \Lambda R_{n-1} \\ T_n &= \Lambda T_{n-1} \\ M_n &= \Lambda^D M_{n-1} \end{aligned}$$

where R, T and M are length, temporal interval and mass values pertaining to analogue systems on the neighbouring cosmological scales n and $n-1$, Λ is a simple scale factor and D a fractal dimension.

An empirical value for Λ was established by comparison of objects at atomic and stellar levels, likely to be correlated, such as the radius of a Rydberg atom and the orbit of the planet Jupiter. To correlate mass scales the hydrogen atom and an M dwarf star of mass $0.15 M_\odot$ are assumed analogous. The derived values of

$$\Lambda = 5.2 \times 10^{17} \qquad \text{and} \qquad D = 3.174$$

are then used to find the analogues of 20 atomic scale objects on both stellar and galactic scales. A convincing match between the three levels is retrodicted in this way.

A conspicuous prediction of SSCM is the galactic analogue of a proton, identified as a black hole of mass $0.145 M_\odot$ and Schwarzschild radius of 20 cm. Whereas 90% of atomic matter occurs as protons, the same percentage of black-hole analogues must represent all of the dark matter predicted by astrophysicists.

If the assumed strong principle of self-similarity is valid, the cosmological hierarchy must be completely unbounded, since only in the case of an infinite hierarchy can a given system and its lower-scale analogues be totally equivalent, except for relative scale.

The SSCM makes some interesting predictions, for instance on the structure of a galaxy:

> "Virtually all of the galactic mass is in the form of a singularity at the galactic center. The bulge, disk, and halo of stars represent an infinitesimal fine 'mist' of stellar-scale objects within the galaxy's Schwarzschild radius".

Also:

> "Because of their enormous energy output and mass ejection, the SSCM proposes that supernovae are stellar scale analogues of highly excited atoms undergoing radioactive decay events".

and:

> "Here it is proposed that globular clusters may represent galactic scale analogues of free electrons. [...] Globular clusters are very numerous galactic scale constituents and they cluster around galaxies (analogues of positive ions)".

As presented here SSCM is a highly idealized system, as also recognized by Oldershaw (1989b). If the self-similarity of analogue systems on different cosmological scales is approximate (but nontrivial), then a transfinite number of cosmological scales is no longer assured. There is also the real problem of scaling that becomes dubious when applied to systems which rotate in gravitational and electromagnetic fields, respectively.

My assessment is that Oldershaw has identified a major symmetry of the cosmos, but like most symmetries in Nature, it is not perfect. Extrapolation to infinities and singularities is therefore not valid. A major deficiency is the failure to recognize CPT symmetry and the role of antimatter which, in our view, implies a closed universe.

The author himself regards SSCM in its present form as natural philosophy rather than proven science, but its potential to elucidate cosmic phenomena is enormous. Cosmological redshift is a relevant example. As observed it is a galactic-scale phenomenon, which should correlate with an atomic-scale counterpart. The proposed chemical redshift (5.1.2) is the most likely candidate for this role. The theory predicts an enormous number of small black holes, which, re-interpreted as penetrating a vacuum interface, may lead to the recognition of new sources of astronomical luminosity.

8.2.4 Population III Cosmology

Although, strictly speaking, not an alternative cosmology, the notion of Population III stars assumes increasing importance in the resolution of anomalies associated with nucleogenesis, dark matter and 3K background radiation in standard cosmology. The subject is reviewed by Carr et al. (1984), with many references.

Population III stars correspond to the missing supergiants, high up the main sequence, mentioned by Hoyle (1955). The comment that stars above the main-sequence limit of a cluster

> "...have by now completed their evolution and disappeared..."

is apparently interpreted to mean that very massive stars (VMS) in the mass range 10^2–$10^5 M_\odot$ formed in the pregalactic period 10^6–10^9 years after the big bang and disappeared soon after. Carr et al. (1984) argue that although

> "...the evolution of such stars is not well understood [.and.] theoretical reasons have been voiced for why they could never form [...], in the pregalactic context at least, these reasons no longer apply: the stars would have burnt out long ago (so their present absence need cause no embarrassment)..."

Since the nuclear-burning time of a VMS is of the order 10^6 yr for $M > 10^2 M_\odot$ several generations could exist pregalactically, having decayed into black holes which now provide the missing mass in galactic halos. The original Population III starlight is now the microwave background.
As

> "...the pregalactic nucleosynthesis constraint could be potentially embarrassing to the Population III star hypothesis..."

two generations of such stars are postulated.

> "One of these generations could contain a large fraction of the mass of the universe and provide the dark matter; the other could contain only a small fraction but comprise stars whose mass is such that they produce a large individual heavy-element yield".
>
> "If Population III remnants are to provide the missing mass, one may need to give up the conventional cosmological nucleosynthesis picture anyway [.as.] the dark matter problem cannot be solved by black holes or low-mass stars of nonprimordial origin unless one modifies the conventional nucleosynthesis scenario".

Still

> "... the circumstances under which such stars could generate the "primordial" helium abundance..."

requires strict limits on their mass

> "... in a way which avoids overproduction of heavy elements".

Although it remains unclear how to fine-tune the nucleosynthesis without overproducing either helium or heavy elements, these proposals have become incorporated into ΛCDM cosmologies (section 5.2.2), which do not rely on Population III to provide the dark matter.

8.2.5 Conclusion

All that is left of big-bang theory is the name. Even cosmologists have quietly abandoned the idea, which is no longer needed to produce either cosmic rays, primordial helium or background radiation. Some say it produced non-baryonic dark matter and molecular hydrogen, others say it marks the beginning of time. There is no consensus any more. The original purpose of the big bang, envisioned by Gamow, has been usurped by the equally mysterious Population III stars.

Standard cosmology in its new guise is little more than a computer game with the purpose of simulating the collapse of primordial molecular clouds into a first generation of VMS's, which rapidly evolve through pure He (Population III) stars towards supernova explosions (Heger & Woosley, 2002). Nucleogenesis is determined by coprocessing the stellar evolution model, with a network of nuclei representing all the necessary isotopes of elements from hydrogen to ruthenium, to produce a mass function whose "distinctive signature is a composition almost solar-like in the ratios among elements..."

Clearly, the big bang has no relevance in this simulation, which could refer equally well to the nucleogenesis proposed in section 5.2.3. The only reason why it is linked to hypothetical Population III stars is to account for the observed metal content revealed by high redshift ($z \sim 10$) events. Without an expanding universe there is no need of Population III and we consider it more likely to interpret the simulated nucleogenesis with the black-hole penetration of the vacuum interface, which is an on-going event that produces copious amounts of cosmic rays, in what Heger & Woosley (2002) describe as "the most energetic and brightest thermonuclear explosions in the universe". In a projectively closed universe there is no need to relegate such events to a hypothetical lost epoch – they happen around us all the time.

8.3 References

Adler, R., M. Bazin & M. Schiffer (1965): *Introduction to General Relativity*, McGraw-Hill, NY.

Arp, H.C. (1983): *Related Galaxies with different Redshifts?*, Sky and Telescope, 1983 (65) 307–309.

Arp, H. (1998): *Seeing Red: Redshifts, Cosmology and Academic Science*, Apeiron, Montreal.

Boeyens, J.C.A. (1992): *The Geometry of Quantum Events*, Specul. Sci. Techn., 1992 (15) 192–210.

Boeyens, J.C.A. (1995): *Red shifts in a curved space*, S.A. Jour. Sci., 1995 (91) 220.

Boeyens, J.C.A. (2005): *New Theories for Chemistry*, Elsevier, Amsterdam.

Carr, B.J., J.R. Bond & W.D. Arnett (1984): *Cosmological consequences of Population III stars*, Astrophys. J., 1984 (277) 445–469.

Cohen, M.H., K.I. Kellermann, D.B. Schaffer, R.P. Linfield, A.T. Moffett, J.D. Romney, G.A. Sielstad, I.I.K. Pauling-Toth, E. Preuss, A. Witzel, R.T. Schilizzi & B.J. Geldzahler (1977): *Radio Sources with Superluminal Velocities*, Nature, 1977 (268) 405–409.

Cowen, R. (1999): *The Universe en Rose*, Science News, 1999 (155) 172–174.

Daigneault, A. (2005): Invited lecture – *Irving Segal's Axiomatization of Spacetime and its Cosmological Consequences*, Budapest, August 2005.

Einstein, A. (1950): *The Meaning of Relativity*, 4th ed. with appendices, Methuen, London.

Einstein, A. & N. Rosen. (1937): *On Gravitational Waves*, J. Franklin Inst., 1937 (223) 43–54.

Freedman, W.L. et al. (14 authors) (1994): *Distance of the Virgo cluster galaxy M100 from Hubble Space Telescope observations of Cepheids*, Nature, 1994 (371) 757–762.

Gaskill, C.M. (1983): *Spectra that defy Explanation*, Nature (304) 212.

Geller, M.J. & J.P. Huchra (1989): *Mapping the Universe*, Science, 1989 (246) 897.

Glanz, J. (1996): *Debating the Big Questions*, Science, 1996 (273) 1168–1170.

Heger, A. & S.E. Woosley (2002): *The nucleosynthetic signature of population III*, Astrophys. J., 2002 (567) 532–543.

Hoyle, F. (1955): *Frontiers of Astronomy*, Heinemann, London.

Hoyle, F. & J.V. Narlikar (1964): *A new theory of gravitation*, Proc. Roy. Soc. 1964 (A282) 191–207.

James, D.F.V. & E. Wolf (1994): *A class of scattering media which generate Doppler-like frequency shifts of spectral lines*, Phys. Lett., 1994 (188) 239–244.

Lerner, E.J. (1991): *The big bang never happened*, Random House, N.Y.

Liley, B.S. (1977): *Plasma State*, The New Encyclopædia Britannica, Chicago, 1977 (14) 505–510.

Lundmark, K. (1925): *The Motions and the Distances of Spiral Nebulæ*, Mon. Not. Roy. Astron. Soc., 1925 (85) 865–894.

Manuel, O., M. Mozina & H. Ratcliffe (2006): *On the Cosmic Nuclear Cycle and the Similarity of Nuclei and Stars*, J. Fusion Energy, 2006 (25) 8 pages, [om@umr.edu].

Marmet, P. (1988): *A new non-Doppler red shift*, Physics Essays, 1988 (1) 24–32.

Marmet, P. & G. Reber (1989): *Cosmic matter and the nonexpanding universe*, IEEE Trans. Plasma Sci., 1989 (17) 264–269.

Moore, R.L & A.C.S. Readhead (1983): *Superluminal acceleration in 3C345*, Nature, 1983 (306) 44–46.

Morris, G.M. & D. Faklis (1988): *Spectral shifts produced by source correlations*, Optics Letters, 1988 (13) 4–6.

Narlikar, J.V. (2002): *An Introduction to Cosmology*, 3rd ed., University Press, Cambridge.

Oldershaw, R.L. (1989a): *Self-Similar Cosmological Model: Introduction and Empirical Tests*, Int. J. Theor. Phys., 1989 (28) 669–694.

– (1989b): SSCM: *Technical Details, Predictions, Unresolved Issues and Implications*, Int. J. Theor. Phys., 1989 (28) 1503–1532.

Peterson, I. (1989): *Quasar illuminates the most distant past*, Science News, 1989 (136) 340.

Powell, C.S. (1991): *X-Ray Riddle*, Sci. Amer., 1991 (264) 26.

Reber, G. (1986): *Intergalactic Plasma*, IEEE Trans. Plasma Sci., 1986 (PS-14) 678–682.

Ramaty, R., J.C. Higdon, R.E. Lingenfelter & B. Kozlovsky (1999): *Rain of Fire*, The Sciences, 1999 (39) 24–29.

Reboul, H.J. (1981): *Untrivial redshifts: a bibliographical catalogue*, Astron. Astrophys. Suppl. Ser., 1981 (45) 129–144.

Rowan-Robinson, M. (1988): *The Extragalactic Distance Scale*, Space Sci. Rev., 1988 (48) 1–71.

Schwarzschild, B. (1992): *Compton Observatory Data deepen the γ-ray Burster Mystery*, Physics Today, 1992 (45) 21–24.

Stonier, T. (1990): *Information and the Internal Structure of the Universe*, Springer, Berlin.

Segal, I.E. (1974): *A variant of special relativity and long-distance astronomy*, Proc. Natl. Acad. Sci. U.S., 1974 (71) 765–768.

Segal, I.E. (1976): *Mathematical Cosmology and Extragalactic Astronomy*, 1976, Academic Press, NY.

Segal, I.E., J.F. Nicoll, P. Wu & Z. Zhou (1991): *The Nature of the Redshift and Directly Observed Quasar Statistics*, Naturwiss., 1991 (78) 289–296.

Starr, A.T. (1957): *Applied Electricity*, Pitman, London.

Troitskij, V.S. (1996): *Observational test of the cosmological theory testifies to the static universe and a new redshift-distance relation*, Astrophys. and Space Sci., 1996 (240) 89–121.

Weyl, H. (1922): *Space–Time–Matter*, translation of the 4th (1921) German edition, Dover, N.Y.

Wilford, J.N. (1990): *Unexpected order in the universe confuses scientists*, Pittsburgh Post Gazette, May 28.

Wolf, E. (1987): *Non-cosmological redshifts of spectral lines*, Nature, 1987 (326) 363–365.

Wolf, E. (1998): *The redshift controversy and correlation-induced spectral changes*, in R. Pratesi & L. Ronchi (eds.): *Waves, Information and Foundations of Physics*, Italian Phys. Soc., Conference Proc., 1998 (60) 41–49.

Wolff, M. (1990): *Exploring the Physics of the Unknown Universe*, Technotran, Calif., ISBN 1-56072-040-9.

Chapter 9

The Big Picture

9.1 Introduction

Anybody has a sub-conscious understanding of its environment. When shared within a homogeneous society this understanding becomes a common world view, which, expressed in words, represents a cosmography, or description of a communal world. Such a world view incorporates the totality of the collective wisdom of the community. The more individuals that contribute to the common pool of knowledge, the more difficult it becomes to define the common ground. As the well-being of a larger community, or tribe, depends on collective action, consensus cosmography is of benefit, and if not spontaneously developed, needs to be formulated and enforced by the tribal chiefs. With growing sophistication, as elements of religious and social rules and dominance creep into the dogma, intolerance of dissident views increases. For lack of better terminology, such a knowledge system defines the cosmology of a civilization.

The latest version of such a system to develop in the Western World has become known as *standard cosmology*. It shares with its predecessors the hallmark of intolerance and dogmatic certainty and does not approach the ideal of an objective collection of pertinent facts that summarizes a scientifically valid desription of the world. It relies on the belief that the history of the cosmos can be traced back to a fateful moment when everything came miraculously into being and that subsequent development can be studied scientifically to within infinitesimal fractions of a second after this big event.

Without denying creation scientists the luxury to indulge in their fantasies, their intolerant suppression of dissident views through the sympathetic mass media is unfortunate. Most human beings have a natural desire to understand the world in which they live. Also, not surprisingly, the most

J.C.A. Boeyens, *Chemical Cosmology*, DOI 10.1007/978-90-481-3828-9_9,

difficult question of when, how and where it all began, always seems to be their first. Given the present state of scientific knowledge there is no hope of ever answering this question. Although this has been the case for millenia, there has never been a lack of belief that satisfactory answers had already been found. The problem is that these answers were invariably based on something other than science, and therefore untestable.

This enquiry is less ambitious in the sense of being aimed only at finding a cosmic structure which is consistent with current scientific knowledge. It is more ambitious in not relying on the intervention of supernatural agents or other miraculous events for a kickstart. It is not cosmology in the modern sense of the word, although many topics such as spectroscopic and gravitational red shifts, the microwave background, nucleogenesis and cosmic chirality are major aspects of the argument. The emphasis is on empirical evidence, which implies direct observation. The same exercise has, of course, been repeated over and over since time immemorial, giving a refined picture whenever new evidence emerges. For millenia this evidence came from patient and careful observation of celestial objects and their motion through the skies.

For many, the impact of scientific evidence, after invention of the telescope, was too sudden to shake their traditional beliefs, established over millenia. Even today there are those in the western world who feel more comfortable with flat-earth cosmology and who regulate their lives around magical practices. There is a psychological reluctance to give up private beliefs, based on personal observation of the world, in favour of unfamiliar ideas, derived from poorly understood experiments, performed by strangers.

While proposing a different view of cosmic reality the danger exists that the new ideas may simply be too radical to be taken seriously by the masses. Even established scientists, with some intellectual commitment to standard cosmology, may prefer to ignore these ideas. Such conservatism is readily understood against historical records of the slow progress from primitive observation to advanced scientific models.

9.2 Magic, Religion and Science

The myths and legends of the original inhabitants of Southern Africa, known as Bushmen and Hottentots, provide valuable insight into the cosmological thinking of primitive societies. To account for the observed sun, moon and stars in the sky the Bushmen recall the existence of a man who lived among them and from whose armpits there shone a light. So that all could share the light, the man was thrown into the sky from where he shines by day. At night he covers himself with a blanket. Related myths account for the moon and the milky way.

Closely related to the Bushmen and sharing their habitat were the Hottentots. They enjoyed a more pastoral existence and while their myths had the same solar and celestial bias, they became linked to supreme beings or deities, who created the world.

The religious practices of Bushman and Hottentot differ significantly. The Bushmen have no gods and rely on magic to assist them in the hunt. The practical ceremony is led by a shaman – a ritual practitioner who enters into a trance to cure the sick, control the weather, guide the movements of animals, foretell the future, and so forth (Lewis-Williams, 1992).

The Hottentots, in contrast, rely for their material well-being on the intervention of a god, the creator. It is informative to note their belief that this deity first existed many generations before as a renowned sorcerer of great skill (Miller, 1979). We witness here the transition from magical to religious practice. In many primitive societies these practices co-exist. What originates as magical control of natural phenomena, such as rain or thunder, gradually assumes a religious aspect linked to the appearance of a god, associated with each phenomenon. At its most elaborate form of pantheism the sum total of gods becomes identical with everything in Nature. We refrain from discussing the development of devotional religions, confining interest to those aspects of magic and religion which can be recognized as precursors of scientific enquiry.

Creation myths, based on magic and religious beliefs, define the elementary science of primitive people. Most creation myths seem to share a common core, with one significant exception. Buddhist cosmology maintains that the present world system, like all others throughout space, goes through an inevitable process of creation, duration and destruction. The universe itself is uncreated, without beginning or end while origination, duration and annihilation succeed one another in recurrent cycles.

For myths to develop into cosmology it is necessary that all related knowledge be integrated into one coherent picture. A recurrent theme of most creation myths is the struggle between antagonistic forces – wet against dry, cold against hot, up against down, *etc.* The final outcome is a separation of unlike qualities. Dry land separates from the wet ocean; solid earth settles below the ethereal heavens above. What belongs with the earth moves down and what belongs with the heavens floats up. In line with such considerations a common shamanistic world view developed among diverse peoples across the globe. It was commonly believed that the universe is full of heavenly bodies populated by vaguely perceived spiritual beings. The central world, or Earth, is saucer-like and floats in water on the back of a giant turtle, whose movements cause earthquakes. The earth has an opening in the middle, leading to a netherworld where the sun spends the night to get food and rest and hides from the hostile earthquake god.

9.2.1 The Babylonian World

An advanced form of this cosmology is found in a classical Babylonian picture of the universe, shown in Figure 9.1. The earth is enclosed beneath the dome

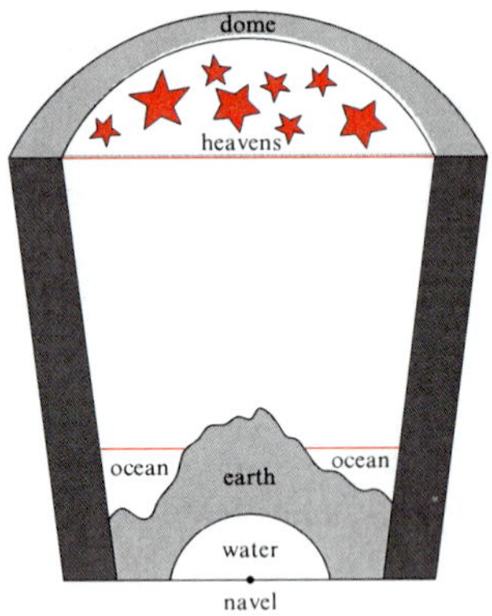

Figure 9.1: *Babylonian model of the cosmos*

of the heavens, surrounded by oceans and drifting on a water-filled cavity, without any evidence of a fixed geometrical structure. This view was shared by most nations of the Mediterranean world and the Middle East, where commercial and military contact had occured for centuries.

This Babylonian view of the world was strictly two-dimensional. The third dimension of up and down was reserved for an approach to the unknown – heaven and hell. In this model the earth is flat and of infinite extent in all directions. To make logical sense the inhabitable land was surrounded by the infinite ocean. They found this picture of the earth intellectually satisfying and the only mystery to ponder was the nature of heaven. Here, to them was a higher dimension. They made a study of heavenly phenomena like the eclipse of sun and moon, and the progress of the wandering planets through the zodiac, because these portents had a powerful effect on earthly events, such as the failure of crops. The organization of human affairs was under the control of an extensive pantheon of deities who inhabited both heaven and the dread netherworld.

It seems remarkable that the magi were well informed about the high-dimensional domains of the gods, but ignorant about the third dimension in which they lived. Like the Babylonians, modern cosmologists are fluently conversant with the twenty dimensions of string theory, but fail to appreciate that astronomical observations should be interpreted in four, not three, dimensions. As the large-scale curvature of three-dimensional space is perceived to be zero, space-time is assumed to be flat. The analogy with the flat Meditteranian world is complete. To interpret the distorted picture created by this blinkered view of the world, it is necessary to invoke the workings of unseen mythical agents. As the Babylonians relied on gods and demons to explain natural phenomena like flowing rivers and rain, the modern cosmologist invents dark matter and energy, without appreciating the importance of antimatter.

The world view of the Greeks was notably different from the Mesopotamian. It has been said (Mac Neice, 1964) that whereas science in Babylon never got out of the temple, in Greece it never got into it. Despite their unshakable confidence in the oracles of Delphi, the Greeks managed to divorce their metaphysics from magic and religion. The purpose of their metaphysics was to establish and criticize the first principles of scientific knowledge.

To forge a geocentric cosmology it was only necessary to rotate the Babylonian universe around its navel and to assign each heavenly body to an appropriate circular orbit. Even the geography of the known world (Figure 9.2) was shown to confirm this pattern.

9.2.2 The Modern World

After two thousand years we are back in a similar situation. While the evidence for an expanding universe was accumulating and the quantum theory was being formulated, theoretical advances with far-reaching consequences for cosmology were made elsewhere. Advances which led to the formulation of the theory of relativity, the only theory that enquires into the nature of matter and its distribution through space.

The theory of relativity dictates that we live in a curved space of more than three dimensions. There is probably no concept in science that the layman finds more perplexing and, not surprisingly, constant efforts by anti-scientists to discredit the theory of relativity have continued for a century. The purpose of these efforts is not to gain an improved understanding of the theory or to propose an alternative interpretation thereof, both legitimate scientific pursuits, but rather to deny the complexity of the world by simply closing their eyes to reality. The reason why anti-science enjoys some respectability is because rigorous science often appears as counter intuitive

Figure 9.2: *World map of Hecatæus of Miletus 517 BCE, redrawn*

to the uninitiated. The assumptions of Aristotelian physics demonstrate the situation well. What is more obvious and logical to the casual observer but, that heavier objects should fall down faster than lighter ones? For Galileo to assert otherwise clearly identifies him as a rascal and an heretic. Aristotle's simple prescription was to elevate empirical awareness into the only validation of physics.

Anti-science at its most effective invokes religious dogma to stifle scientific enquiry. A familiar example is Aristotle's prime mover, the only source of mechanical interaction, associated with a perfect god. In this case the philosopher came to share in the divine authority, which for centuries defied the onslaught of experimental physics.

In the case of relativity and cosmology unreason prevents the spread of scientific understanding. Based on irrelevant empirical awareness, however naïve, relativistic phenomena are discussed pseudo-scientifically in non-relativistic context, for instance, to demonstrate the impossibility of time dilation if, by definition, universal time flow is empirically known to be absolute and invariant.

Many others find it equally hard to stomach the notion of curved space. Imagine those in a two-dimensional world with all motion restricted to a flat surface. Immediately we understand how such beings must have a mental block against picturing their surface as being curved. Only being aware of forward-backward and sideways motion, we can understand their consternation on returning to their point af departure after a long trek in a fixed direction. The more enlightened ones would conclude that their world was bounded and embedded in more than two dimensions, despite a pathological inability to envisage the third dimension. We have the same dilemma. Having mastered the idea of existing on a curved two-dimensional surface, the next step is to contemplate the possibility that our three-dimensional space may also be bounded and curved, rather than extending infinitely in three orthogonal directions.

The antidote to anti-science is to find a scenario in which relativity can be perceived as both logical and comprehensible. The major coup of standard cosmology has been to create such an illusion for the masses. It found the winning recipe by offering an ideologically attractive scenario and, at the same time, claiming it to be based on the theory of general relativity. The pièce de résistance surely was the pronouncement of a Nobel laureate that the COBE data on microwave-background radiation provided decisive final proof of the big bang as scientific fact and was "like looking at the face of God."[1]

To comfort the faithful and protect them against deviant astronomical observations that may contradict the big idea, the guidance of the anthropic principle comes to the rescue. In the spirit of religious freedom we could accept standard cosmology as a belief system. As a by-product of Intelligent Design it could even be science – creation science. It certainly is the dominant cosmology of the Western World, but its status as real science must be challenged one more time.

The decisive argument against big-bang cosmology is the fact that it is not a relativistic cosmology, as claimed. Ironically, many opponents of the theory, notably Halton Arp (1998), fail to appreciate this criticism and in order to refute the big bang, reject the theory of general relativity.

The Scientific Status of Standard Cosmology

There is a popular perception that the pursuit of knowledge should be aimed at finding a theory of everything (Barrow, 1992). In mathematical terms it may be thought of as a differential equation with solutions that predict all details of the knowable universe. Conceptually it implies a fundamental thought which may be extended logically into an explanation of all phenomena.

[1]The San Francisco Chronicle, May 25, 1992, p.35.

As a model of the material universe it promises the detailed characterization of any object as predicted by the properties of less complex entities. The feasibility of this process seems to be supported by the reductionist argument whereby the properties of any complex system can be logically reduced to those of a more elementary system. In this way the properties of a complex molecule can be shown to be consistent with the properties of its atomic constituents, which in turn, can be reduced to those of their sub-atomic building blocks. To the frustration of many a philosopher the inverse, constructionist procedure, after many attempts, appears not to be feasible. At each level of higher complexity unpredictable, so-called emergent properties, invariably show up.

Once we accept the impossibility of constructionism, the dream of predicting reality from perceived origins of life and the universe, has to be abandoned, unless supported by direct scientific evidence. Such evidence does not exist and the structure of the universe must be interpreted as an emergent property.

The reductionist concept carries the germ of its own failure. It arises as an abstraction from holistic reality by separating the parts considered essential to an argument, from those considered irrelevant[2]. It only operates correctly in isolation and lacks the information to anticipate effects that may occur through interaction with a wider environment, considered as irrelevant before. Emergent properties are generated on interaction with this environment and hence unpredictable by definition.

The big bang is the ultimate in isolated events. There is nothing around to provide a frame of reference against which to specify where, how or when it happens. There is no environment in which to generate emergent properties like space or time. It cannot produce motion, let alone mass. Even a Higgs boson needs a broken symmetry, which presupposes symmetry – of what?

The momentous notion of inflation is likened to a chemical phase transition, which is a prime example of an emergent event, induced by the thermodynamic environment. There is no such an environment for an expanding embryonic universe and nothing to initiate a phase transition or define a supercooled state. Inflation is no more than idle speculation.

The theoretical basis of modern big bang is the observed frequency shift in the spectra of galactic light. Despite a large volume of counter evidence and a host of alternative explanations, all redshifts are dogmatically ascribed

[2]This is what Jan Smuts (1926) referred to as an error of abstraction.

to a distance-related Doppler effect. The theory comes into serious conflict with astronomical observation of large-scale structures, age of the universe, apparent superluminal rates of expansion, rates of luminosity fluctuation, stellar evolution and the topology of space-time. It seems that the politically correct response is to ignore all discordant data until they can be explained by, and incorporated into, a suitably modified theory.

As the source of all matter in the universe the big bang is an abject failure. By a series of convoluted assumptions and arguments it "predicts" that all matter came into being in the form of H and He, without interacting with an equivalent amount of antimatter in the mix, and overestimates the relative He/H ratio and the background temperature by an order of magnitude.

To save the day the well-known black-body radiation that pervades the vacuum is re-interpreted as a unique product of the big bang. The isotropy, homogeneity and spatial origin of the radiation is described by any of several scenarios, depending on the nature of the question.

Redshift, nucleogenesis and microwave background constitute the observational basis of big bang theory. By our analysis they amount to nothing. Still – by definition no argument prevails against dogmatic truth.

In cosmology dogmatic truth has now taken the form known as the ΛCDM model, which is claimed to eliminate the last remaining inconsistencies in the big-bang theory and being completely in line with all astronomical observations. This rapidly changing cosmology is reminiscent of 14th century alchemy, in the words of Geoffrey Chaucer (1951):

> "Their scientific jargon is so woolley
> No one can hope to understand it fully,
> Not as intelligence goes nowadays".

When the results of yet another probe (dubbed WMAP) to examine the cosmic microwave background were announced in 2003, the chairman at the news conference, John Bahcall, effused:

> "...the announcement today represents a rite of passage from speculation to precision science".

It is constructive to examine what this science consists of. As before, it starts with the assumption of a *hot* big bang, now interpreted to produce, not only space and time, but also all of the matter (as primordial molecular hydrogen) and energy observed in the universe today, as well as an excessive amount of non-baryonic *cold* dark matter and inflationary energy[3].

[3]In the most recent version of inflation the phase transition no longer features.

The main purpose of inflation is to get the expansion going. It is not too difficult to understand that a large mass of infinite density must generate an irresistible gravitational force, characterized by infinitely curved space-time. It is therefore necessary to postulate an even larger anti-gravitational force, simply known as inflationary energy. Not only was it necessary to scatter the new matter (ordinary and dark), but also to get rid of the curvature. To make ends meet, expansion by a factor of 10^{100} within 10^{-30}s has been agreed as sufficient to make the universe smooth and flat within a second.

Not to blow the new creation out of existence it is necessary to assume that after this brief period, activity of inflation energy had to be arrested, so as not to interfere with the subsequent evolution of the universe. Quantum mechanics is invoked to explain the conversion of inflationary energy into radiation, by a chance process, likened to radioactive decay, and which creates exactly the required random fluctuations to seed the growth of stars and galaxies within the dark-matter plasma[4].

In order to account for the apparent age of quasars that never exceeds 5 billion years, inflationary theory was needed again, but on a smaller scale. As the proposed decay is irreversible, anti-gravity was brought back under the new guise of dark energy, which is assumed to have been lurking in the wings for 9 billion years, only to emerge when needed most.

The scenario outlined here comes remarkably close to Edgar Allan Poe's (1848) creation model, published as serious science, but now considered to be science fiction (Beaver,1976). Poe considered himself a natural philosopher, having stated:

> "What I have propounded will (in good time) revolutionize the world of Physical & Metaphysical Science. I say this calmly – but I say it".

How true.

He proposed that in the beginning the Universe of Space was created out of nothing, by dint of God's Volition. (The Big Bang). The Matter created from Nihility was in its utmost conceivable state of Simplicity. (Cold Dark Matter). A Divine act started the diffusion of the heterogeneous multitude of atoms and continued until the diffusion was complete. (Inflation). At this point a natural reaction sets in, recognized in the tendency of disunited atoms to return into One. (Gravity). As the reaction sets in there is nothing to prevent the aggregation of various unique masses, at various points

[4]It is common practice in theoretical cosmology to rationalize illogical conclusions by mentioning the alleged probabilistic nature of quantum systems, out of context.

in space: – in other words, nothing to interfere with the accumulation of masses, each absolutely One. (First stars). This is contrary to Divine purpose and to oppose the tendency an agent that prevents total coalescence is required. (Dark Energy). In drawing together of the systems into clusters, with a similar and simultaneous drawing together of the clusters themselves (Galaxies and Clusters), while undergoing consolidation, we have attained the great *Now* – the awful Present – the Existing Condition of the Universe of Stars. (5×10^9 years ago). Perceiving the separation between stars as dependent on that stage in a cycle of either expansion or contraction, it is inferred that *Space and Duration are one.* (Measured as Redshift).

There is not a shred of evidence to support either of these creation models. Why then would so many able scientists devote their careers trying to substantiate science fiction? There is a simple two-fold answer to this question. By accepting the Doppler effect as unique explanation of *cosmological redshifts* a one-to-one relationship between space and duration is established, with the promise of spectacular computer modelling of cosmic evolution. The next step is to adjust, by international agreement, a number of crucial parameters that link observed redshifts to the computer model. The challenge is to use observed redshift data to determine the age, distance and rate of recession of celestial objects and derive matching estimates of temperature, gas density, rate of cooling, state of ionization *etc.*, consistent with an assumed pattern of cosmic evolution.

This approach led to the amusing conclusion that there are no quasars less than 5 billion years old. By admitting that redshifts could be intrinsic this conclusion is automatically avoided. Alternatively however, a new explanation of why matter stopped clumping together, is needed. Universal expansion was postulated to have decreased to a value below that of dark energy, which is not affected by expansion and is gravitationally repulsive. From that point on the universe has begun to become simple and uniform again, as in Poe's model.

The simulation of cosmic evolution as a function of redshift (Stiavelli, 2009) has precious little in common with the old standard cosmology. It explores scenarios to produce cooling phases by the production of molecular hydrogen, followed by reionization when higher temperatures are required. One of the main objectives is to simulate an accelerated process of metal production before nucleogenesis in normal stars became operational. With sufficient adjustable parameters and VMS's it becomes feasible to simulate any metal content for all redshifts. The power of the method lies in this flexibility which permits suitable parameters to retrodict any new observation. It clearly is an exciting exercise in computer modelling without scientific basis.

The final summary that follows should therefore not be seen as a challenge to the big bang, or even as cosmology. It is no more than an exercise in logic, aimed at an understanding of the world, based on internal evidence only. It summarizes our conclusions and a tentative reconstruction of cosmic structures without mathematical formulae or too many diagrams.

9.3 The Natural World

From any point of view, be it magic, religion or science, there is consensus about the splendid nature of the cosmic whole. One man's divine plan is another's magical harmony or consilience among the laws of Nature. Unity of knowledge about the natural world is not in dispute. The most striking manifestation of this unity occurs as self-similarity across all levels of observation, from the infinitesimally small to the astronomically large.

The likeness that floral structures bear to shells and curved horns has been noticed many times and the basis of the likeness has been known for millenia. The logarithmic spiral, which is the only smooth curve that is self-similar at all scales, features in the description of all these growth structures. Mathematical description of the spiral, shown in Figure 3.32, depends on the three fundamental constants, π, e and τ.

The demonstration that the solar system displays the same self-similar symmetry, also noticed in the image of spiral galaxies and hurricanes, makes it hard to deny that the pattern conveys more than coincidence. It probably is one of the more reliable guides towards a better understanding of the natural world.

Such an all-embracing symmetry can only reside in the one medium that pervades the cosmos under various names, including aether, vacuum and space. To specify the geometrical structure of the medium we shall use the concept of space or space-time. The physical aspects of the medium we describe as the vacuum. To avoid unnecessary assumptions we must rely on experimental evidence, however incomplete, to clarify the elusive topology of space and the vacuum.

9.3.1 Space-Time

The findings of Newton and Descartes developed into a beautiful model of space-time that served both microphysics and celestial mechanics convincingly well until the end of the nineteenth century. This scheme recognized the three-dimensional aspect of space that extends uniformly along three orthogonal coordinate axes to infinity. Independent of space, the inexorable flow of universal time kept all motion in step. All events unfolded against the background of an immovable aether with respect to which absolute motion is measured. As demonstrated by Laplace and others, this mechanical universe operates with clockwork precision, except for minor irregularities in the motion of celestial bodies that required occasional divine correction.

Discovery of the electromagnetic field created an insurmountable problem for this model. The propagation of electric and magnetic effects was found to be inhibited, not only in material media, but also in the vacuum. The result of this is that an electromagnetic disturbance, such as visible light, propagates through the vacuum at constant speed, irrespective of the motion of its source relative to an observer. The logic that describes the relative motion of material objects simply cannot accommodate this type of behaviour.

Resolution of the problem, known as the special theory of relativity, consists of a group of transformations that describe a rotation in four-dimensional space-time. The group has no meaning unless the fourth orthogonal coordinate, associated with time flow, functions exactly like the space coordinates.

The difficulty of a three-dimensional observer to envisage the effects of a rotation in four dimensions may be likened to the response of a two-dimensional being in the legendary Flatland of Edwin Abbott (1952) to a rotation in three dimensions. Unless the rotation axis lies perpendicular to the flat plane a rotating vector moves into the incomprehensible third dimension and appears to contract in two-dimensional projection. Seen from outside the contraction is compensated by an expansion (dilation) of the vector into the third dimension.

In the age of cellular telephones and the internet people should not doubt the reality and nature of electromagnetic signals and stop agonizing over the common inability to visualize the fourth dimension or understand time dilation. There is no better explanation of electromagnetic effects. Those science writers with a mission to enlighten the masses should reflect that any model of the world that ignores the theory of relativity and operates in Newtonian space is a fairy tale.

The Minkowski space-time of special relativity differs from conventional Euclidean space only in the number of dimensions and gives the correct description of all forms of uniform relative motion. However, it fails when applied to accelerated motion, of which circular motion at constant orbital speed is the simplest example. Relativistic contraction only occurs in the direction of motion, but not in the perpendicular radial direction towards the centre of the orbit. The simple Euclidean formula that relates the circumference of the circle to its radius therefore no longer holds. The inevitable conclusion is that relativistic acceleration implies non-Euclidean geometry.

It is important to note that the progression from special to general relativity does not require any new assumption. The demonstrable equivalence of mechanical and gravitational acceleration completes the principle of general relativity, the essence of which is that the geometry of relativistic space-time is non-Euclidean, which, in colloquial terminology, means that space-time is curved.

General relativity goes no further than this. The extent of curvature or the topology of space-time depends on additional observations and criteria. Assumptions to simplify the mathematics, but not supported by testable data, must inevitably produce spurios results. Examples of such simplifications include the assumption of universal time or Euclidean space-time. Einstein's thought experiment which established the non-Euclidean nature of space-time also predicts the nature of the curvature as positive, which excludes hyperbolic space.

The reasonable, but not essential assumption, that the general curvature of space-time be constant, predicts a closed topology in the form of either a hypersphere or a four-dimensional projective plane. Additional evidence is needed to decide between these possibilities.

9.3.2 The Vacuum

Relativity theory has equally dramatic implications on the nature of the vacuum, which is shown not to be a void, but a medium that supports wave motion and carries electromagnetic fields. A new perspective on the nature of the vacuum is provided by the principle of equivalence. Space-time curvature can be described mathematically by a Riemann tensor, which the principle implies, should balance the gravitational field, which is sourced in the distribution of matter. This reciprocity indicates that Euclidean space-time is free of matter, which only emerges when curvature sets in. This is interpreted to mean that the homogeneous wave field of Euclidean vacuum generates matter when curved. Like a flat sheet that develops wrinkles when wrapped around a curved surface, the wave field generates non-dispersive persistent wave packets in the curved vacuum.

These wave packets represent mass points, which may or may not carry electric charges. Charged objects are connected by standing waves along the null geodesics of space-time. These standing waves, which consist of advanced and retarded components, correspond to the virtual photons of quantum field theory or the infons of information theory. If there is a potential difference between two charges the energy flow that equilizes the potentials is known as a photon. The world collective of standing waves that pervades the vacuum corresponds to the quantum potential of Bohmian mechanics. It is important to note that all aspects of electromagnetic interaction and the properties of elementary charged objects are fixed by the gauge invariance of globally curved vacuum, which in turn depends on the topology of space-time.

There is a second mode of interaction between mass points, independent of charge and propagated along the general geodesics of space-time. This is gravitation – best understood as a purely geometrical effect in matter-free vacuum. But the vacuum is nowhere free of matter. Intergalactic space contains a uniform distribution of plasma, cosmic rays, neutral atoms and molecules, dust particles, meteorites and asteroids of unknown concentration. Like the presumed non-baryonic dark matter of inflation theory, this material has an enormous gravitational effect, not taken into account in conventional celestial mechanics. Interaction of this material with gravitational and plasma waves results in the accumulation of matter in characteristic patterns. These patterns are also conditioned by the local curvature of space-time and hence the resulting structures, such as the solar system, must be self-similar to atoms, which form by electromagnetic interaction. The symmetry properties of atoms could then presumably reveal the common features of curvature that relate to the topology of space-time.

9.3.3 The Periodicity of Atomic Matter

There are 81 naturally occuring non-radioactive elements with 264 stable isotopes. Each of the 264 nuclides is characterized by a mass number, A, an atomic number, Z, and a neutron number, $N = A - Z$, that respectively specify the number of nucleons (protons + neutrons), the number of extranuclear electrons and the number of neutrons per atom. Because Z and N are both integers the composition of the stable atomic nuclei is conveniently specified as the rational fraction, Z/N which, with one exception, is less than one and always larger than the irrational fraction, $\tau = 0.61803\ldots$, the golden ratio.

The Farey sequence of number theory, which arranges all rational fractions in an enumerable array, represents, by definition, a unique classification of all possible nuclides. The set of stable nuclides is projected as an 11×24 array from the infinite set, between the limits of 1 and τ, by the associated Fibonacci series. Compared to experimental values of nuclear binding energy the 11×24 array is found to represent a periodic function of nuclear stability in A, Z or N. The periodicity in N generates the magic numbers of nuclear physics and as a function of Z the periodic table of the elements is obtained when sampled at the limiting ratio of $Z/N = \tau$. This table corresponds to the empirically derived sequence in which electronic energy levels are occupied in heavy atoms. Remarkably, the periodicity sampled at $Z/N = 1$ would correspond to the occupation of energy levels in exactly the reversed order.

This observation is readily explained by the way in which atomic energy levels respond to applied pressure. The Z/N ratio of unity is inferred to occur at extremely high pressures that can only occur in the interior of monstrously heavy stars or space with nearly infinite curvature. As a corollary it is inferred that the conditions of high pressure and the Z/N ratio would be ideal for nucleogenesis by continued addition of α-particles ($^4_2$He) in an equilibrium process. It is readily demonstrated that the 264 stable nuclides can be generated in the form of four series $A = 4n, 4n+2, 4n \pm 1$, consisting of 81, 81, 51, and 51 members, produced by α-addition within the known field of stability.

The shift of electronic energy levels under pressure predicts a redshift in all spectra eminating from regions of high space-time curvature. This includes all galaxies and quasars and their immediate environments. Consistent interpretation of observed redshifts on these grounds would drastically modify the picture currently based on cosmological Doppler shifts.

The graphical representation of the way in which chemical periodicity varies continuously as a function of the limiting ratio (Figure 5.3), $1 \leq Z/N \leq 0$, appears strangely unsymmetrical, despite perfect symmetry at the extreme values. By adding an element of mirror symmetry a fully symmetrical closed function, that now represents matter and antimatter, is obtained. To avoid self overlap the graphical representation of the periodic function is transferred to the double cover of a Möbius band, which in closed form defines a projective plane.

This observation is no proof of the topology of space-time but it shows how matter and antimatter can coexist, without mutual annihilation, in projective space-time.

9.3.4 The Topology of Space-Time

The topology of space-time is not as arbitrary as often made out. The theory of relativity unequivocally specifies a manifold of more than three equivalent coordinate axes with positive curvature. Infinite universes with negative or zero curvature are therefore excluded immediately. This limitation was recognized early on by Clifford (1955):

> "We may postulate that the portion of space of which we are cognizant is practically homaloidal [*i.e.* flat], but we have clearly no right to dogmatically extend this postulate to all space. A constant curvature, imperceptible for that portion of space upon which we can experiment [...] would seem to satisfy all that experience has taught us to be true of the space in which we dwell."

Like ignoring the irregular surface features on a planet, symmetry suggests that space-time must be of constant positive curvature in the large. Only two topologies meet this requirement: the so-called 3-sphere or hypersphere and projective space. The first of these, with privileged time coordinate, was considered as cosmological model by Einstein. It also occurs as spherical *space-time* in de Sitter's universe. The projective 3-sphere (or elliptic space) is shown in this work to be consistent with Gödel's rotating universe.

Gödel's rotating-universe solution of Einstein's gravitational field equations can be interpreted in real projective space. The compass of inertia is shown to coincide with the involuted geodesic of the closed manifold and predicts a cosmological redshift for distant events. Other cosmological criteria pertaining to antimatter imbalance, the microwave background and nucleogenesis are satisfied in a natural way; also the role of black holes in steady-state circulation and universal self-similar symmetry.

Like any hypersphere the projective plane is hard to visualize. The common device of projection on to a two-dimensional surface, embedded in three, is perhaps as misleading as it is helpful. The result is the well-known Möbius band. Unlike a sheet of paper this non-orientable surface is of zero thickness, which means that antipodal regions of the double cover occupy the same space. The four-dimensional analogue consists of a three-dimensional interface, separating enantiomorphic antipodal regions that occupy the same three-dimensional space. In four-dimensional perspective the interface and the separate surfaces of the double cover are resolved, like a Möbius strip viewed in three dimensions. The illusion of two sheets with opposite curvature, inferred from the paper model, has no mathematical meaning. The constant positive curvature of the interface that generates involution of the covering surface creates this illusion.

Another way of looking at four-dimensional projective space is by adding a point at infinity on each coordinate axis of Minkowski space. Any displacement implies a change in all coordinates. The only difference between stationary antipodal points is an inversion of local chirality, which includes the direction of time flow. There is no possibility of communication between such points which move apart on their respective world lines.

It is evident from Figures 4.2 and 7.1 that in projective Minkowski space there are no separate time and space cones. Timelike and spacelike motion therefore occurs throughout all space-time. The involuted interface coincides with the null geodesics of the manifold and carries the electromagnetic (and possibly other) fields. It separates conjugate domains, identified as matter and anti-matter respectively.

This is the topology visualized for hypothetical matter-free space-time. The proposition is neither in conflict with general relativity nor with Einstein's strict interpretation of Mach's principle. Superimposed on the uniform curvature of the real projective plane is the local gravitational field that obeys Einstein's field equations. The mass points responsible for the local distortions are the products of non-Euclidean geometry. Because of the human inability to visualize such a structure the mathematical representation is doubly important.

9.3.5 Mathematical Model

All attempts to explain the beginning generate more questions than answers. Wielding Occam's razor results in the Buddhist description of the world as uncreated, or as summarized by Bertrand Russell[5]:

> "...the universe is just there, and that's all."

It is the same Russell (1918) who stated that:

> "Mathematics, rightly viewed, possesses not only truth, but supreme beauty..."

and could have added, "...it is just there [*etc.*]". We shall try to find a mathematical model of the observable details of the cosmos, with due care not to extrapolate the interpretation of local observations, such as the apparent Euclidean nature of the local environment, to all space-time. We shall argue that the global projective geometry is of Euclidean appearance only in our local tangent space.

On closest scrutiny the universe appears to be periodical, just like the natural number system. In Kepler's metaphysics this consilience was interpreted as a design principle, rooted in the golden ratio. Although the fundamental link between the world of mathematics and the observable universe remains hidden, there is no better guide to understanding of the physical world. Like the number system, physical space-time approaches *terra incognita* on extrapolation towards infinity. The geometrical remedy consists of closing the loop at infinity to create a closed system, which is periodic by definition.

Until the parallel between number and cosmos is demonstrated to be an illusion we shall use this idea to model the universe. The power of this approach lies therein that all regularities in the physical world can be reduced to the same mathematical rules as the commensurable relationships in the solar system. The same mathematics that optimizes the distribution of matter in spiral galaxies and solar systems, shapes the growth of nautilus shells and sunflower heads. This ubiquitous symmetry, known as self-similarity is

[5]From his debate with Copleston, quoted by Hick (1964).

closely associated with the golden ratio. Self-similar growth is conditioned by the exponential factor e and the metric of space-time depends on π. The interplay between the three irrational numbers, τ, e, and π is poorly understood and its resolution promises a major advance in theoretical cosmology. At this stage it is safe to assume that it represents a fundamental attribute of space-time topology.

All available evidence points at projective topology. According to a modern encyclopaedic compendium of mathematics (Gowers, 2008), which features a nautilus shell on the dust cover,:

> "Projective geometry is regarded by many as an old-fashioned subject, and it is no longer taught in schools,..."

It gives biographical profiles of 95 great mathematicians, without even mentioning the name of Oswald Veblen. Such attitude makes it difficult to write about projective relativity in a meaningful way without providing some background material. Hence the emphasis on projective geometry in chapter 3 and the translation of Veblen's monograph in the Appendix.

Addition of ideal points at infinity results in the definition of n-dimensional projective space by $n+1$ homogeneous coordinates, which remains valid on multiplication by an arbitrary gauge factor, the fundamental operation in field theory and wave mechanics. This property disappears on mapping to affine space where it is the subject of a special assumption. The unification of the electromagnetic and gravitational fields appears naturally only in projective space.

In projective relativity the field equations contain, in addition to the gravitational and electromagnetic fields, also the relativistic wave equation of Schrödinger and, as shown by Hoffmann (1931), are consistent with Dirac's equation, although the correct projective form of the spin operator had clearly not been found. The problem of spin orientation presumably relates to the appearance of the extra term, beyond the four electromagnetic and ten gravitational potentials, in the field equations. It correlates with the time asymmetry of the magnetic field and spin.

The concept of cross ratio that enables the definition of involution and the equivalence of all conics, is another unique feature of projective geometry, which may be of cosmological importance. It gives new meaning to the idea of an Einstein–Rosen bridge between two parallel sheets, now interpreted as antipodal regions of an involuted projective surface, separated in time.In projective geometry the description of a conic is a matter of perspective (Section 3.2.9). Rotation at the galactic level, best described by a four-dimensional conic, when observed on the stellar scale, appears as an irregular three-dimensional conic that creates the illusion of dark matter in galactic halos.

The isotropic lines of a projective identity mapping define the local complex Minkowski space of special relativity directly. By taking the circular points at infinity into account the global projective space of general projective relativity is obtained. No other topology reveals the transition from special to general relativity as such a simple consequence of curved space-time.

The fact that projective geometry is independent of measurement implies that size has no effect on shape. Although the curvature of space may therefore be imperceptible at the microscopic level, its effect remains geometrically the same as if expanded to astronomical scale. In the final analysis unexplained phenomena such as gauge invariance, unified fields, quantum phenomena, the golden ratio, time-asymmetry, conjugate matter, cosmological bridges (wormholes), relativistic effects, dark matter and self-similarity find natural explanations in projective space-time.

This appraisal may well contain elements of overzealous exaggeration, but the general pattern is hard to overlook and the detailed mathematical analysis is confidently anticipated to confirm most of the conclusions.

9.3.6 Cosmic Self-similarity

The number theory that underlies chemical periodicity gives a detailed account of all commensurable relationships when applied to orbital motion in the solar system. The golden mean has a pivotal role in both simulations. The second most important factor is angular momentum. In atomic systems, stabilized by electromagnetic interaction, orbital angular momentum is completely quenched to produce spherical atoms that may carry spin angular momentum. In celestial systems, stabilized by gravitation there is no such requirement and orbital angular momentum generates planar systems. However, in both cases optimized angular momentum results in quantized energy levels or orbital radii of planets, moons, rings or asteroids.

In chemical systems intramolecular interactions depend on the exclusion principle that limits interatomic electron density according to the golden ratio (Boeyens, 2008).

Photographic images of spiral galaxies point at a large degree of self-similarity between galaxies and the solar system. Instead of on a star, spiral galaxies appear to be centred on a massive black hole. The same process that shaped the solar system from a spiral vortex into an elliptical orbital system is still active in spiral galaxies. When the black hole penetrates the vacuum interface the system will presumably approach equilibrium and transform into an elliptic galaxy.

Intergalactic structures are not known in sufficient detail for comparison with molecular structure but there are indications that galactic clustering involves plasma fields in the form of filaments and ropes.

The current understanding of interaction between stellar objects is clouded by biased estimates of the cosmic distance scale and may improve after re-assessment of cosmological redshifts.

9.3.7 The Physical World

The recognition of self-similarity simplifies the description of the physical world. Everything that exists is in response of the vacuum to the curvature of space-time. Instead of the featureless uniform wave field of the hypothetical Euclidean vacuum, curved space fills the vacuum with elementary wave packets that tend to merge into larger structures, connected into a holistic whole by standing waves (called photons) in the residual wave field. The elementary wave packets are of two knotted types, distinguished physically as negatively and positively charged, each carrying half a unit of spin angular momentum ($\hbar/2$) and known as electron and proton respectively. They combine into neutral units in one of two ways:

$$e + p + \nu \rightarrow n$$

$$e + p \rightarrow H$$

Whereas a hydrogen atom, H is comparatively stable, a neutron in free space decays soon into proton and electron together with a unit of half spin, known as a neutrino. The photon that mediates interation between positive and negative charges carries one unit ($\hbar$) of angular momentum.

The half-spin of elementary waves (called fermions) appears as a chiral disturbance in the wave field and its mirror image has the opposite charge and anti-spin. An electron and its mirror equivalent, called positron, destroy each other on contact – all that remains is a high-energy photon:

$$e^- + e^+ \rightarrow \gamma\,.$$

The same applies to protons, neutrons and more complex matter. In the physical world annihilation of matter and anti-matter is prevented by separation on an interface in the vacuum. Without such an interface matter cannot exist and this mandates a space-time topology that allows the appearance of this interface.

The fusion of H atoms and neutrons into larger atoms is ascribed to nuclear interaction and further fusion of atoms into molecules and aggregates is brought about by chemical interaction. Further clumping occurs by gravitational interaction. With increasing mass such growing clumps cause a local distortion of space-time, known as a gravitational field. As the mass increases beyond a critical point nuclear reaction sets in and the cosmic cycle is restarted. Eventually another critical accumulation of matter causes rupture of the interface and newly synthesized matter is released into the vacuum for recirculation.

In some instances encounter between accumulated masses of matter and antimatter across the interface releases enormous quantities of energy by mutual annihilation. As a result of such activity intergalactic space is saturated with stray radiation, plasma and electromagnetic fields. The visible effects of such activity are difficult to interpret and often highly confusing because of multiple images created by complicated involution of space-time.

The range of events between the formation of atoms and their eventual disappearance through black holes follow a self-similar pattern that conforms to the curvature of space-time. The periodicity of atomic matter depends on the same number theory that shapes the mutual arrangement of planets, moons, rings, comets and asteroids in a solar system. By a mechanism, to be explored, solar systems are distributed along galactic spiral arms in a pattern like that which prevails in solar systems and atoms.

Apart from differences in scale, atomic systems, planets, solar systems, galaxies and clusters are arranged according to a common pattern. The interactions that generate these patterns depend on related modes of wave motion, always resulting in an inverse-square law of limited reach to avoid unphysical singularities.

9.4 A Man-sized Universe

The physical world of the previous section is not what we see when looking around. In the first instance we see the surface of the planet, which is now almost universally agreed to be spherical. Already familiar with the image of our planet, as seen from the moon, the idea of floating in space is more palatable. Space exploration has confirmed the theoretical picture of the solar system, which we now consider as home territory. We have grown accustomed to the time delay of radio signals reaching us from explorers in the far reaches of the solar system. All of this can be mapped in three-dimensional Euclidean space, allowing for a finite velocity of light.

Reaching beyond the solar system becomes more problematical. Extrapolating into the extension of familiar Euclidean space-time we are soon overwhelmed by billions of solar systems in the Galaxy, which we call the Milky

Way, followed by the appearance of billions of galaxies in an endless infinite space without limit or frontier. Something is seriously amiss. Have we ignored the admonition not to extrapolate from the known to the unknown? Can we really trust the picture presented by our increasingly powerful telescopes?

I remember a childhood fascination with a baking-powder tin, labeled with a picture of itself, in endless progression. Is that what we see through our telescopes? We are told that by looking more deeply into space we look at objects that existed further and further back in the past. Already we are looking at objects older than the universe itself and still find infinitely many objects beyond that. This is the time to reconsider our understanding of the world.

The situation in which we find ourselves was foreseen by Poincaré (1902) in his *Science and Hypothesis*, with the warning that, for the sake of simplicity, physicists would never abandon Euclidean geometry. The irony is that the theories, which have fallen into Poincaré's trap, claim to be based on General Relativity, which has no meaning in Euclidean space-time. It is equally ironic that Einstein himself reverted back to Euclidean space for his final cosmological proposal. Be that as it may, if not its author, then certainly the theory dictates a closed non-Euclidean four-dimensional space-time. Weighing the current evidence identifies the most likely space-time topology as the projective multiply-connected 3-space with constant positive curvature, also known as elliptic space.

The way in which such a topology could create the illusion of an infinite night sky is likened (Luminet, 2008) to a room panelled by mirrors on all six surfaces:

> "If we go into the room, the interplay of multiple reflections will immediately cause us to have the impression of seeing infinitely far in every direction.
>
> Cosmic space, which is seemingly gigantic, might be lulling us with a similar illusion. Of course, it possesses neither walls nor mirrors, and the ghost images would be created not by the reflection of light from the surface of the Universe, but by a multiplication of the light ray trajectories following the folds of a wraparound universe. We could live in a physical space which is closed, small and multiply-connected, yet have the illusion that the observed space is greater..."

We are beginning to see why Poincaré's prediction has been confirmed. To disentangle the composite picture generated by multiply-connected space-time will be a daunting task that promises to keep astronomers occupied for centuries. The true topology will only be recognized piecemeal as more multiple images are identified. The term *cosmic crystallography*, being heard more often, should then come into its own. The real challenge will be to generate an intelligible image of the topological space. Even the two-dimensional projection of the topology proposed here cannot be visualized in three-dimensions.

Imagine a Möbius strip centred on a great circle in a spherical surface. Suppose that the strip coincides with the surface at one (north) pole and gradually rotates around the great circle as it moves to the antipode (south pole), where it lies perpendicular to the surface of the sphere.

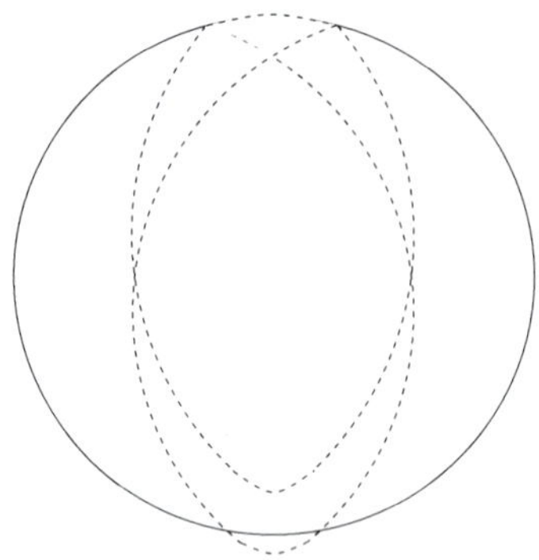

Returning in the same style along the same great circle, it joins smoothly to itself at N. A second Möbius strip along a second great circle, perpendicular to the first, intersects the first Möbius strip at S. A perspective drawing of these two Möbius strips is shown in the cover diagram.

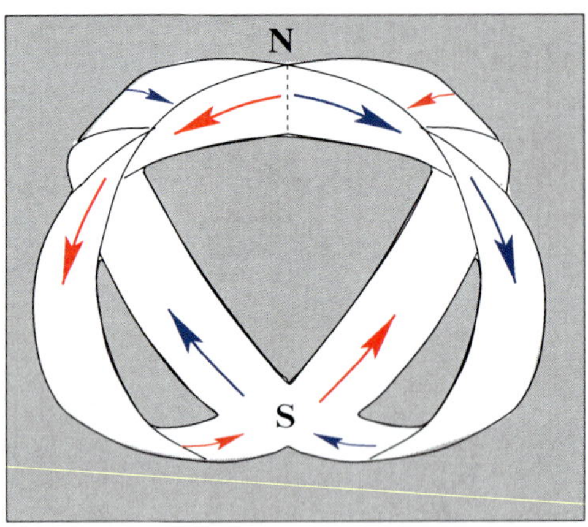

In principle, the entire surface can be covered by infinitesimally narrow Möbius strips, which are coplanar at N and intersecting at S. This set of Möbius strips joins the outside of the spherical surface to the inside, creating a single surface with two sides. This non-orientable surface is known as the real projective plane.

In four dimensions the projective plane does not intersect itself. In projective geometry there is no distinction between points, lines and planes, *etc.* and the same terminology remains valid in any number of dimensions. Hard as it is to visualize two perpendicular Möbius strip as a non-intersecting unit, it becomes impossibly hard to imagine a shape for a four-dimensional projective plane, considered the most likely representation of space-time in this work.

As projective geometry is not metrical the actual size of the universe is not determined by topology alone. A new distance scale will first have to be established by astronomical observation. To put the problem into perspective it is noted that the size of the solar system is conveniently described in units of light-hour. Extrapolation to a distance of 10 billion light year represents a scale-up of 10^{14} – like extrapolating the behaviour of an electron to the size of a marble, or the structure of the marble to the size of the solar system. Our experience with the quantum world and solid-state physics tells us that this is not feasible. It compares with the efforts of a goldfish to reconstruct the outside world, based on observations in his bowl. In the terminology of Riemann manifolds, our Euclidean tangent space differs so radically from the underlying space that linear extrapolation between these spaces is meaningless.

Newton's gravitational theory widened our understanding of the solar system. The gravitational field of Riemann and Einstein may lead to a better understanding of the Galaxy, but not without taking the electromagnetic field into account. Despite the claims of many cosmologists this has not been achieved. In fact, it was demonstrated by Milne (1935) and McCrea that the formulae of standard cosmology follow identically from Newtonian theory.

The only cosmological model that satisfies all criteria and defines a supporting topology is Veblen's projective relativity. In this geometry all points are equivalent and, as a model of the Milky Way, the galactic centre may be considered as the middle of the universe. In the same way that the Copernican model and Newton's laws of gravity replaced the geocentric Ptolemaic world by the solar system, projective relativity promises to define a galactic centred universe.

The idea of intergalactic interaction is purely speculative and probably illusory. There is no observational basis, apart from unwarrented extrapolation into infinite Euclidean space, to show that the topologically closed space-time consists of more than the Milky Way. Long before a definitive resolution of this question there may be the surprise discovery of weak fields with long-range implications, such as the vaguely perceived quantum potential, that could revolutionize our perception of space-time.

Rather than contemplating the beginning or end and infinite boundaries of the universe, enough remains unsaid about the Galaxy to inspire many generations of scientists. Instead of looking for anthropic arguments to explain the purpose of creation it may be more rewarding to study the distribution of organic matter in space (Van Dishoeck, 2008) as evidence of biological life. The developing field of astrochemistry, linked to laboratory studies, has the potential to map the evolution of intelligent life, not only on Earth, but in the entire cosmos, as an inevitable result of environmental change (Williams, 2007).

9.5 References

Abbott, E.A. (1952): *Flatland*, 6th ed., Dover, N.Y. [First published 1884]

Arp, H. (1998): *Seeing Red: Redshifts, Cosmology and Academic Science*, Apeiron, Montreal.

Barrow, J.D. (1992): *Theories of Everything*, Vintage, London.

Beaver, H. (ed.) (1976): *The Science Fiction of Edgar Allan Poe*, Penguin Books, Harmondsworth, UK.

Boeyens, J.C.A. (2008): *Chemistry from First Principles*, Springer.com.

Chaucer, G. (1951): *The Canterbury Tales*, translated by N. Coghill, Penguin Books, London. [First published ~ 1308]

Clifford, W.K. (1955): *The common Sense of the Exact Sciences*, Dover, N.Y.

Gowers, T. (2008): *The Princeton Companion to Mathematics*, University Press, Princeton.

Hick, J. (1964): (ed.) *The existence of God*, Macmillan, N.Y., p. 175.

Hoffmann, B. (1931): *Projective Relativity and the Quantum Field*, Phys. Rev., 1931 (37) 88–89.

Lewis-Williams, J.D. (1992): *Vision, Power and Dance*, 14th Kroon Lecture, University of Amsterdam, 8 May 1992.

Luminet, J-P. (2008): arXiv:0802.2236[astro-ph], *The Shape and Topology of the Universe*, in Proc. of the conference "Tessalations: The world a jigsaw", Leyden, Netherlands, March 2006.

Miller, P. (1979): *Myths and Legends of Southern Africa*, 1979, Bulpin Publications, Cape Town.

Milne, E.A. (1935): *Relativity, gravitation and world structure*, Clarendon, Oxford.

MacNeice, L. (1964): *Astrology*, 1964, Aldus, London.

Poe, A.E. (1848): *Eureka: A Prose Poem*, Putnam, N.Y.

Poincaré, H. (1902): *La Science et l'Hypothèse*, Flammarion, Paris.

Russell, B. (1918): *Mysticism and Logic*, Longmans, London, p. 60.

Smuts, J.C. (1926): *Holism and Evolution*, Macmillan, London.

Stiavelli, M. (2009): *From First Light to Reionization*, Wiley-VCH, Weinheim.

Van Dishoeck, E.F. (2008): *Organic matter in space – An overview*, Proc. IAU Symposium No. 251, p. 3–15.

Williams, R.J.P. (2007): *A chemical systems approach to evolution*, Dalton Trans., 991–1001.

Appendix A

Projective Relativity Theory

Translation of the original German text of the monograph:

PROJEKTIVE
RELATIVITÄTSTHEORIE
BY
O. VEBLEN
WITH 3 FIGURES
SECOND VOLUME OF:
ERGEBNISSE DER MATHEMATIK
UND IHRER GRENZGEBIETE

PUBLISHED UNDER THE EDITORIAL SUPERVISION OF

ZENTRALBLATT FÜR MATHEMATIK
BERLIN
VERLAG VON JULIUS SPRINGER
1933

Foreword

This booklet does not claim completeness. It rather presents a personal point of view which however, I hope will be useful in the further treatment of the geometries described here and others related to it. The same applies for the bibliography at the end. It was not attempted to make it complete. On the other hand it lists many treatises not directly referred to in the text, but which I hope, could be useful in further investigation of the subject.

I am grateful to three young German mathematicians. Mr F. John assisted me first of all, in the preparation of my lecture course in Göttingen during the summer of 1932. His composition of these lectures established a first edition of the present manuscript. For further revision I thank Mr Möbeling for his notes on occasion of the lectures that I gave in Vienna. Following my course in Hamburg Mr G. Howe finally assisted me in a comprehensive revision of the entire manuscript and in the process made many suggestions for several improvements.

Fynshaw, August 1932. **O. VEBLEN**

In conclusion of the corrections I must acknowledge my Princeton collaborators and especially Mr J.L. Vanderslice who carefully checked all details of the manuscript.

Princeton, November 1932. **O. VEBLEN**

Contents

(1)

I. Unified Theories for Gravitation and Electromagnetism

One of the many achievements of Einstein's general relativity was to geometrize gravitational theory. This geometrization consists in the first instance therein that one views the world of physical events as a space-time continuum in four dimensions. Such a continuum is, by definition, represented by a coordinate system. A coordinate system is simply a mapping of a class of world points on a class of four-fold numbers, or what may be called *number points* (x^1, x^2, x^3, x^4).

From these the first axiom or group of axioms of the relativity theory must make a statement about the existence and properties of this mapping. Although I consider it important to give a clear formulation of these axioms, I shall not give a detailed listing here, since they do not differ from the general axioms of differential geometry in the case of 4 dimensions and these axioms have been discussed by Whitehead and myself in a booklet soon to be published. (Bibliog. 1932,10).

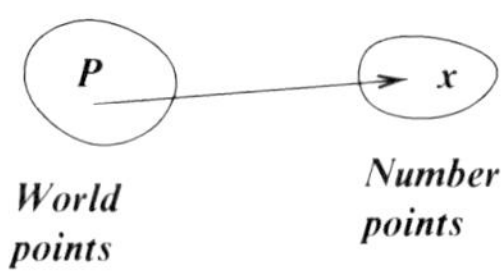

Fig. 1

Furthermore this geometrization implies the assumption of a definite space structure of the world. This structure consists in fact of nothing but the existence of ten position functions

$$g_{ij}(x^1, x^2, x^3, x^4) \quad (i, j = 1, \ldots, 4)$$

$(g_{ij} = g_{ji})$ in each coordinate system.

As these functions are clearly determined in each coordinate system we describe them as components of geometrical (or physical) objects. If these components are given in a coordinate system, they are determined in all other coordinate systems by a simple linear transformation law. Because of the special linear shape of these transformation laws this geometrical object is called a *tensor*, more precisely, the *fundamental* tensor of the Riemannian space.

It is not essential to use geometrical language. Everything proceeds smoothly logical if we define the g_{ij} as ten gravitational potentials and treat the entire theory analytical. It is not as interesting and lively (at least for the scientists of our era) as when we interpret the g_{ij} as the coefficients of a quadratic differential form

$$ds^2 = g_{ij} dx^i dx^j,$$

define the Riemannian length of a curve

$$x^i = x^i(t)$$

by the integral

$$\int ds,$$

and pronounce an entire set of propositions geometrical in this sense. One should keep in mind that in the case of Einstein's theory the geometric measurements are related to gravitational phenomena.

Each selection of fundamental tensor marks a special Riemannian space. Classes of Riemannian spaces are recognized as solutions to systems of differential equations with the g_{ij} as dependent variables. By suitable selection of such equation systems Einstein succeeded to recognize special classes of Riemannian spaces open to important physical interpretations.

To find these interpretations one uses coordinate systems (normal – or inertial coordinates) with specific geometrical meaning and allow special separation in space-like and time-like components. By this means the geometrical laws on the Einstein classes of Riemannian spaces are translated into common physical laws. In this way one finds a great deal of classical physics contained in the ten components g_{ij}, and since the gravitation phenomena play a leading role in this part of the theory, the description of the g_{ij} as gravitation potentials appears justified. The unified character of this theory finds expression in the fact that the components of the gravitational potentials are a unique geometrical object.

The essential difference between this theory and the earlier Newtonian gravitation theory that we want to raise is the following: In the old theory one assumes a Euclidean space into which a gravitation potential is introduced. However, this potential function has no influence on the space itself. The properties of space were completely independent of those of the potential. On the contrary, the properties of space are identical with those of the gravitation potentials, g_{ij}.

Against that, in the world of gravitation potentials the electromagnetic phenomena have nothing to do with the geometrical structure of space. The electromagnetic potentials are by way of speaking strangers in the Einstein space, just as the Newtonian potential functions in Euclidean space. Whether

(3) Unified Theories for Gravitation and Electromagnetism

more or less matter exists in the world, leaves the Euclidean geometry unchanged. In the same way electricity has no direct effect on general theory of relativity.

The problem to find a structure of space which not only depends on gravitational potentials but also on electromagnetic potentials was first attacked by H. Weyl in the year 1918. Although Weyl's effort did not succeed physically it had the fruit of producing a beautiful geometry. The next effort was undertaken by T. Kaluza in 1921. Kaluza replaced the four-dimensional space-time continuum by a five-dimensional continuum, followed by the introduction of a Riemannian metric into this continuum and he succeeded in the construction of field equations, which gave as first approximation Einstein's gravitation equations and Maxwell's electromagnetic equations.

The Kaluza theory was simplified by O. Klein (Bibl. 1926,5; 1927,11) in a way that the Einstein–Maxwell theory was not approximated, but came out in its exact shape. Since then several mathematical physicists investigated the theory and found its formal shape alluring. However, there remains a fundamental question without a satisfactory answer: What does the fifth dimension mean? Nobody has found convincing ground to doubt our conviction that the physical world is four dimensional. Theorists, especially Einstein himself, conducted several investigations with the intention to create a four-dimensional theory. Some of these works are referenced in the bibliography.

Five or six years ago I had the thought that a possible solution of the unity problem was to be found in terms of generalized projective geometry, the topic of many investigations over the last ten years.

Before we state the fundamental geometrical idea on which our solution of the unity problem depends, we explain some concepts from ordinary relativity theory.

Accordingly we work in the usual space-time world that depends on the coordinates x^1, x^2, x^3, x^4. We consider the differentials $dx^1, \ldots, dx^4$ of the coordinates. What do they mean geometrically? Their basic property is that they behave linearly in the transformation

$$\bar{x}^i = \bar{x}^i(x)$$

according to the formulae

$$d\bar{x}^i = \frac{\partial \bar{x}^i}{\partial x^j} dx^j; \tag{1}$$

the dx^i may also be interpreted as affine coordinates in a four-dimensional space[1].

Each point (x^1, x^2, x^3, x^4) of the underlying space has an associated affine "tangential space". The point (0,0,0,0) of the tangent space, whose coordinates remain unchanged on all transformations (1), can be associated with the point (x^1, x^2, x^3, x^4) of the underlying space and described as the contact point. Each coordinate transformation of the underlying space induces an affine transformation of each tangent space.

If our underlying space were, for instance, one dimensional, we can represent it as a curve and the tangent space in P as the usual tangent of the curve in P.

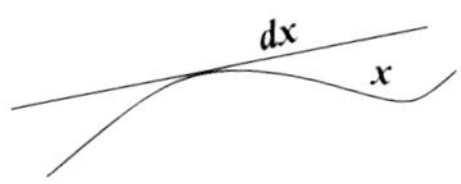

Fig. 2

The variable x is a parameter that fixes the position of a point on the curve, and the parameter dx determines the position of a point on the tangent line. Because of our limited powers of visualization such an intuitive image is not possible in the four dimensional case. Still, the corresponding geometrical expression is helpful and suggestive. We must first imagine a four-dimensional underlying space or world and then a number of tangent spaces, each tangent space connected to a specific point of the underlying space.

With respect to a fixed tangent space the $g_{ij}(x)$ are constant. Therefore

$$ds^2 = g_{ij}dx^i dx^j$$

is the square of the distance between the origin and the point dx with respect to a Euclidean metric of the tangent space in x. The points dx that satisfy the quadratic equation,

$$g_{ij}dx^i dx^j = 0$$

fill a cone through the origin – the light cone.

This cone is real in the case of relativity theory, while the quadratic form g_{ij} is indefinite. From the point of view stressed by E. Cartan (bibl. 1928,1) the Riemannian geometry of the underlying world is to be considered as the theory of these connected Euclidean tangent spaces. The generalization that we have in mind now consists of the following:

Instead of a cone, which features as light cone in relativity theory, we substitute as basis in each tangent space a completely general non-degenerate second degree surface (or quadric). Through this quadric a quadratic cone

[1]We obviously used the Einstein summation convention in this work on relativity theory.

$g_{ij}dx^i dx^j = 0$, namely the tangent cone through the origin, and a hyperplane, the polarplane of the origin, are then identified at the same time in each tangent space.

The polar plane contains the contact point of the tangent cone with the surface. The polar hyperplane should represent the electromagnetic potentials and the cone, alternatively, the gravitational potentials (Fig. 3).

Rather than a Euclidean geometry we have, a non-Euclidean geometry in each tangent space, with our quadric as absolute plane in the Cayley sense. Our new geometry therefore is the overall theory of this set of Cayley spaces, in the same way that Riemannian geometry was the theory of the Euclidean tangent spaces of the underlying space.

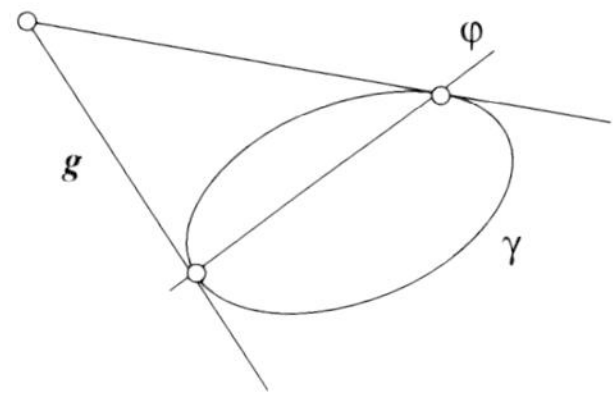

Fig. 3

It seems to me that the appropriate computational structure for our purpose was that briefly presented by B. Hoffmann and myself in the Physical Review (1930) in connection with earlier work in the Quarterly Journal of Mathematics (Oxford Series 1930). The correspondence with the formalism of the Kaluza-Klein theory is so complete that Hoffmann and I described our theory as the geometrical meaning of Kaluza-Klein. However, it must be emphasized that our theory starts from a totally different physical and geometrical point of view as Kaluza. In particular, we infer no link between electric charge and a fifth coordinate; our theory is rather four-dimensional throughout.

Independent of our investigation Einstein and Mayer (Bibl. 1930,3) have published a "unified field theory", which led to essentially the same results as ours (compare Chap. VIII). Furthermore there recently appeared several works by J.A. Schouten and D. von Dantzig (Bibl. 1931, 7; 1932, 3, 4, 8, 9; 1933, 1) in which the projective relativity theory was treated in different forms. It is remarkable that – in a mathematical sense – all these theories seem to converge together. One can therefore hope to find a final solution of the unification problem in this way. We say more about the limits of this solution in chap. VII.

II Projective Tensors

The generalized non-Euclidean geometry sketched in chapter I is the theory of a set of tangent spaces that each contains a quadric. The theory of a quadric finds its most satisfactory form in the realm of common projective geometry. Consequently it is natural to look for a representation of generalized non-Euclidean geometry in the framework of a generalized projective geometry. Such a generalized projective geometry is in fact not hard to find. It is a branch of the differential geometric researches of the last decade. The decisive idea of this geometry developed gradually from the work of a large number of mathematicians. In particular the research of H. Weyl, E. Cartan, J.A. Schouten, L.P. Eisenhart and T.Y. Thomas deserves mention. Their ideas have been presented in a variety of shapes. I shall sketch them in the form that I have internalized them myself.

Affine Tensors.

Next a few orienting words on the usual or affine tensors. In the underlying space there are many allowed coordinate systems, all related by the analytical transformation

$$\bar{x}^i = \bar{x}^i(x^1, x^2, x^3, x^4) \tag{1}$$

but in the tangent spaces only the small number of coordinate systems related to each other by the linear transformation

$$d\bar{x}^i = \frac{\partial \bar{x}^i}{\partial x^j} dx^j \tag{2}$$

We therefore call the tangent spaces "central affine spaces". Their geometry depends on their affine group and there is a special point, the contact point (0,0,0,0) whose coordinates are not affected by the transformation (2).

By this means the theory of the underlying space may be reduced to the simultaneous affine geometry of this set of affine spaces. The tensors provide a suitable aid for treatment of this simultaneous-affine theory. As first example we take contravariant vectors or contravariant tensors of first rank. That is a geometrical object that contains four components in each coordinate system

$$V^1(x), \ldots, V^4(x)$$

which are functions of x. By the equation

$$dx^i = V^i(x)$$

a specific point $(dx^1, \ldots, dx^4)$ is identified in each tangent space. While, because of the familiar transformation law of tensors the differentials transform exactly like contravariant vectors, this relationship is independent of the choice of coordinates. By contravariant vectors a specific point can be identified in each tangent space. Likewise one can, with the aid of a contravariant vector in each tangent space, identify a specific hyperplane through the origin. Indeed the points that satisfy an equation $A_i dx^i = 0$, fill a flat space in three dimensions. In general the components of a contravariant tensor of rank n are the coefficients of the equation

$$A_{ij...k}dx^i dx^j \dots dx^k = 0$$

and the points of each tangent space that satisfy this equation fill a cone of order m. In general the theory of tensors is an affine-algebraic geometry of tangent spaces with reference to their collective behaviour.

Introduction of Homogeneous Coordinates.

The question remains whether or not there are other spaces which can take over the role of the tangent spaces and a collection of which could be the subject of a new theory. This question is answered affirmatively in various ways. In our case we extend the usual tangent spaces to projective spaces. Another point of view is discussed in chapter VIII.

We know how the extension of affine tangent systems to projective spaces can be achieved geometrically. An imaginary or virtual point is added to each bundle of parallel lines. This virtual point serves as the point "infinitely far" for each bundle of lines. The points at infinity are called collinear if and only if they are the infinitely remote points of three lines in the same plane. Four points at infinity are known as coplanar if they are the infinitely remote points of four lines in the same three-dimensional space. At the same time it should be emphasized that this introduction of points at infinity was done in each tangent space.

It is appropriate to use homogeneous coordinates in the treatment of projective tangent spaces. Instead of the four affine coordinates $dx^1, \dots, dx^4$ we shall introduce five coordinates $X^0, X^1, \dots, X^4$, such that $(X^0, X^1, \dots, X^4)$ and $(kX^0, kX^1, \dots, kX^4)$ refer to one and the same point, so that only the four relationships of X are meaningful. A relation of the form

$$dx^i = \frac{X^i}{\varphi_\alpha X^\alpha} \tag{3}$$

must exist, such that the points at infinity of our space satisfy the equation[2]

$$\varphi_\alpha X^\alpha = 0. \tag{4}$$

We shall return to the relationship of this definition of homogeneous coordinates with that used in elementary textbooks ($dx^i = X^i/X^0$).

The Proportionality Factor.

The homogeneous coordinates are however, not completely determined by these formulae. It still lacks a prescription of how the X^α and φ_α should behave on coordinate transformation. To obtain a coordinate transformation that could answer this question we proceed from the fact that our homogeneous coordinates are only determined to within a proportionality factor k. It follows from this that we can relate the choice of proportionality factor to the choice of the plane at infinity.

The proportionality factor for each point of a given tangent space as well as the underlying space can be chosen freely at will. It changes in particular from tangent space to tangent space. We are not looking here for the most general possible theory, but only for one which is adequate for the purpose of our generalization of the usual theory.

In particular, nothing changes when we multiply all homogeneous coordinates[3] with a position function $\sigma(x^1, \ldots, x^4)$. To treat the extension with a position function analytically we find it useful to define the proportionality factor in the form

$$k = e^{x^0}.$$

By substituting

$$\bar{x}^0 = x^0 + \log \varrho(x) \tag{5}$$

the homogeneous coordinates are extended with a position function $1/\varrho(x)$. It is clear that geometrical or physical quantities described by a projective geometry must be invariant against (5). As in usual tensor analysis we further demand that they must also be invariant against the coordinate transformation

$$\bar{x}^i = \bar{x}^i(x) \tag{6}$$

It then raises the question about the simplest invariants against both types of transformation.

[2]Greek characters always take the values $0, 1, \ldots, 4$, Latin ones the values $1, \ldots, 4$.

[3]One could instead of this, for instance, use $\int v_i dx^i$ in place of $\log \varrho$, where $v_i dx^i$ does not necessarily imply an integrable linear differential form.

(9)	Projective Scalars.

Projective Scalars.

In order to study these invariants, we start with the scalars. An affine scalar has, in each coordinate system, only one component. The components in two coordinate systems $P \to x$ and $P \to \bar{x}$ hang together through the transformation law

$$(6') \qquad A(\bar{x}) = A(x).$$

For the definition of projective scalars we retain the transformation law

$$\bar{A} = A.$$

We further assume that x^0 enters A in the simplest manner. A should namely have the form

$$A = e^{Nx^0} f(x)$$

where N is a fixed number that obviously does not transform[4]. From(5)

$$A = \bar{A} = \bar{e}^{Nx^0} \frac{f(x)}{\big(\varrho(x)\big)^N}$$

The number N is called the *index* of scalar A. It has the approximate role of a weight. We must however reserve use of the word weight for another purpose, as further invariants with a weight in the usual sense may occur.

The part of A, which is independent of x^0, obeys the law

$$\bar{f} = \frac{f}{\varrho(x)^N},$$

while A itself is subject to the simple law $(6')$. In contrast to an affine scalar, with only one component in each coordinate system, a projective scalar has infinitely many components. Each transformation of the parameter x^0 generates a new component from a given component.

Should one however, not transform the parameter x^0 the components of a projective scalar behave under coordinate transformation exactly like the components of an affine scalar. Not transforming the parameter x^0 means, so to speak, that we keep our space fixed in a specific state. If some scalar is given, each such state corresponds to a certain component of the scalar. As we shall soon see this applies to all our projective tensors as well. Each state is specifically associated with a certain coordinate system in each tangent space.

[4] A possible generalization is for instance that one sets $N = \varphi(x^1, \ldots, x^4)$.

Gauging.

Because of the close connection of this concept with the concept of Weyl's geometry with the same name I may call it the *gauge principle.* The parameter x^0, which I simply called, following J.H.C. Whitehead, a factor, I may now call a *gauge variable.* A transformation such as (5) we call a *gauge transformation.*

Some projective scalar $e^{Nx^0}f(x)$ may be transcribed to the form e^{Nx^0}. One only needs to use the gauge transformation

$$\bar{x}^0 = x^0 + \log\left(f(x)\right)^{1/N}.$$

The gauge transformation is unambiguously established by carrying a given component of a projective scalar, with index different from zero, into another as given. This description is independent of the chosen coordinate system.

From the components of a given scalar A of index N we derive by differentiation the five functions

$$\frac{\partial A}{\partial x^0}, \frac{\partial A}{\partial x^1}, \dots, \frac{\partial A}{\partial x^4}$$

the first of which corresponds to within the factor N with the scalar itself. By any gauge transformation

$$\bar{x}^0 = x^0 + \log \varrho(x) \tag{5}$$

and a coordinate transformation

$$\bar{x}^i = \bar{x}^i(x) \tag{6}$$

these functions go over into the five functions

$$\frac{\partial \bar{A}}{\partial \bar{x}^\alpha} = \frac{\partial A}{\partial x^\beta}\frac{\partial x^\beta}{\partial \bar{x}^\alpha} \qquad (\alpha, \beta = 0, \dots, 4) \tag{7}$$

We agree to call the combination of a given gauge and a given coordinate system, a *representation.* Consequently the combined transformations (5) and (6) becomes a representation.

Projective Vectors

Equation (7) is a special case of the following:

$$\bar{\varphi}_\alpha = \varphi_\beta \frac{\partial x^\beta}{\partial \bar{x}^\alpha}. \tag{8}$$

A geometrical object with five components of the form

$$\varphi_\alpha = e^{Nx^0} f_\alpha(x^1, \dots, x^4)$$

in each representation, subject to the transformation law (8) is known as a *projective covariant vector of index N.* Each system of five components belongs to such a representation.

(11)	Projective Vectors.

Its null component φ_0 is a projective scalar for which

$$\bar{\varphi}_0 = \varphi_\beta \frac{\partial x^\beta}{\partial \bar{x}^0} = \varphi_0$$

The matrix of the transformation law is:

$$\left\| \frac{\partial x^\alpha}{\partial \bar{x}^\beta} \right\| = \left(\begin{array}{c|ccc} 1 & -\frac{\partial \log \varrho}{\partial \bar{x}^1} & \dots & \frac{\partial \log \varrho}{\partial \bar{x}^4} \\ \hline 0 & & & \\ \vdots & & \frac{\partial x^i}{\partial \bar{x}^j} & \\ 0 & & & \end{array} \right).$$

A coordinate transformation

$$x \longrightarrow \bar{x}$$

induces the transformation

$$\bar{\varphi}_i = \varphi_j \frac{\partial x^j}{\partial \bar{x}^i}$$

through the non-zero components, *i.e.*: The non-zero components of a covariant projective vector behave under coordinate transformation like the components of an affine covariant vector. However, a gauge transformation induces the transformation

$$\bar{\varphi}_i = \varphi_0 \frac{\partial x^0}{\partial \bar{x}^i} + \varphi_i$$

If φ_α is specifically of index 0 and obeys the invariance condition $\varphi_0 = 1$ the gauge transformation reads

$$\bar{\varphi}_i = \varphi_i - \frac{\partial \log \varrho}{\partial x^i}$$

We could say: The four non-zero components are only determined within a gradient $\partial \log \varrho / \partial x^i$. These properties are characteristic of the four electromagnetic potentials. In other words: The four electromagnetic potentials are the non-zero components of a projective vector whose null component is 1.

A contravariant vector is defined by analogy with a covariant one. It is only necessary to assume the particular transformation law

$$\bar{X}^\alpha = X^\beta \frac{\partial \bar{x}^\alpha}{\partial x^\beta} \tag{9}$$

instead of (8). If we split the null component from the others this law becomes:

$$\left\{ \begin{array}{l} \bar{X}^0 = X^0 + X^j \partial \bar{x}^0 / \partial x^j \\ \bar{X}^i = X^j \partial \bar{x}^i / \partial x^j . \end{array} \right. \tag{10}$$

That is: The non-zero component behaves exactly like an affine vector; the null component however, behaves like a scalar in coordinate transformation, but on gauge transformation it increases around a linear combination of the other components.

It is because of this difference in behaviour between the null component and the others in the case of a covariant and a contravariant vector, that the projective tensor calculus is non-trivial. Otherwise we could think that these tensors simply arise from the combination of affine tensors. In fact, a decomposition into affine tensors happens instead, but it happens differently for covariant and contravariant tensors.

The co- and contravariant tensors of higher rank can now evidently be defined in the same way as the corresponding affine tensors. We shall discuss some special cases in detail when the opportunity arises to use them.

Homogeneous Coordinates in Tangent Space.

We are now in a position to define the homogeneous coordinates in tangent space precisely. Consider any, but always fixed, covariant vector φ_α of index 0, with $\varphi_0 = 1$. To describe the homogeneous coordinates of a given point dx of the tangent space we take an arbitrary number k and set

$$\left\{ \begin{array}{l} X^i = k dx^i \\ X^0 = k\left(1 - \varphi_i dx^i\right) \end{array} \right. \tag{11}$$

Choosing another number k only means that one multiplies X^α with a proportionality factor. To invert these equations we note that

$$\varphi_\alpha X^\alpha = k$$

and hence

$$dx^i = \frac{X^i}{\varphi_\alpha X^\alpha}$$

which is equation (3) introduced before. A coordinate transformation

$$x \to \bar{x}$$

of the underlying space induces the transformation

$$d\bar{x}^i = \frac{\partial \bar{x}^i}{\partial \bar{x}^j} dx^j$$

and that induces, in turn the transformation

$$X \to \bar{X},$$

(13)	Homogeneous Coordinates in Tangent Space.

where
$$\bar{X}^i = Kd\bar{x}^i = \frac{K}{k} X^j \frac{\partial \bar{x}^i}{\partial x^j},$$
$$\bar{X}^0 = K\left(1 - \bar{\varphi}_i d\bar{x}^i\right) = K\left(1 - \varphi_i dx^i\right) = \frac{K}{k} X^0,$$

because the φ_i behave like the components of an affine vector. The X^i therefore transform like the components of a contravariant vector on coordinate transformation. The origin and the site opposite to it on the covering, both remain invariant. A gauge transformation induces no transformation of dx^i, but only the transformation

$$\bar{\varphi}_i = \varphi_i - \frac{\partial \log \varrho}{\partial x^i}$$

on φ_α. Hence we have the transformations

$$\bar{X}^i = \frac{K}{k} X^i,$$

$$\bar{X}^0 = K\left(1 - \bar{\varphi}_i d\bar{x}^i\right) = K\left(1 - \varphi_i dx^i + \frac{\partial \log \varrho}{\partial x^i} dx^i\right) = \frac{K}{k}\left(X^0 + \frac{\partial \log \varrho}{\partial x^i} X^i\right).$$

The X^α behave under gauge transformation like the components of a projective contravariant vector.

As we indicated before and have proven now, a gauging does not only imply a special choice of components for each projective tensor, but also a special choice of the site opposite to the origin of the coordination simplex. Each gauging therefore corresponds to a certain equation of the hyperplane at infinity. Only in this special case can we introduce the projective coordinates by the simple formula $dx^i = X^i/X^0$ or, by its equivalent $dx^i = X^i/NX^0$. That is, if a projective scalar exists for which

$$\varphi_\alpha = \frac{\partial \log A}{\partial x^\alpha}$$

we can through a gauge transformation always show that $A = e^{Nx^0}$ and $\varphi_\alpha = N\delta^0_\alpha$. Thereby we have reached the intended connection with the elementary definition of projective coordinates.

Because our homogeneous coordinates behave like the components of a contravariant vector under coordinate transformation we may interpret the equation $X^\alpha = A^\alpha$ as characreristic of one and only one point in each tangent space. The index can here be arbitrary but different from zero, *i.e.* the function A^α is of the form

$$A^\alpha = e^{Nx^0} f^\alpha(x) \quad \text{with} \quad N \neq 0.$$

Exactly as in all affine cases we can now interpret the different projective tensors geometrically. Say, for instance, that A_α is a projective covariant vector. Then

$$A_\alpha X^\alpha = 0$$

is the equation of a hyperplane. Then, and only then is it possible to choose the gauge such that the equations of all these hyperplanes reduce to the form $X^0 = 0$, if A_α is a projective gradient

$$A_\alpha = \frac{\partial A}{\partial x^\alpha}.$$

Yes, we know that A can, by gauge transformation, be brought into the form $\bar{A} = e^{N\bar{x}^0}$.

Second Rank Projective Tensors.

As the next example we take a projective covariant symmetrical tensor of second rank and index $2N$. In this case we can further clarify the picture that we sketched before as the basis of projective relativity theory.

The term $G_{\alpha\beta}$ obeys the transformation law

$$\bar{G}_{\alpha\beta}(\bar{x}) = G_{\sigma\tau}(x)\frac{\partial x^\sigma}{\partial \bar{x}^\alpha}\frac{\partial x^\tau}{\partial \bar{x}^\beta}.$$

Due to the special shape of a particular transformation

$$\bar{x}^0 = x^0 + \log \varrho(x)$$
$$\bar{x}^i = \bar{x}^i(x)$$

it follows that $$\frac{\partial \bar{x}^\alpha}{\partial x^0} = \delta_0^\alpha \quad \text{and} \quad \frac{\partial x^\alpha}{\partial \bar{x}^0} = \delta_0^\alpha.$$

Consequently $$\bar{G}_{00} = G_{00}$$

and $$\bar{G}_{0\alpha} = G_{0\tau}\frac{\partial x^\tau}{\partial \bar{x}^\alpha}.$$

Furthermore G_{00} is a projective scalar and $G_{0\alpha}$ a projective covariant vector. We write

(12) $$G_{00} = \Phi^2 = e^{2Nx^0} f(x)$$

and $$\frac{G_{\alpha\beta}}{G_{00}} = \gamma_{\alpha\beta}$$

and also $$\frac{G_{0\alpha}}{G_{00}} = \varphi_\alpha.$$

The quantities $\gamma_{\alpha\beta}$ and φ_α represent a projective tensor, or rather a projective vector, both of index 0. Φ is a scalar of index N. There are invariant conditions $\varphi_0 = 1$ and $\gamma_{00} = 1$.

(15)	Second Rank Projective Tensors.

The equations

$$\gamma_{\alpha\beta} - \varphi_\alpha\varphi_\beta = g_{\alpha\beta}$$

determine a projective tensor that satisfies the conditions $g_{0\alpha} = 0$. As a result we have the transformation law

$$\bar{g}_{ij} = g_{pq}\frac{\partial x^p}{\partial \bar{x}^i}\frac{\partial x^q}{\partial \bar{x}^j},$$

i.e. g_{ij} are the components of a second-order affine tensor.

The projective tensor $G_{\alpha\beta}$ contains, so to say, a projective scalar Φ, a projective vector φ_α, and an affine tensor g_{ij}. In fact

$$G_{\alpha\beta} = \Phi^2\left(g_{\alpha\beta} + \varphi_\alpha\varphi_\beta\right) = \Phi^2\gamma_{\alpha\beta} \tag{13}$$

The tangent space belonging to a point of the underlying space was moved to the coordinates $X^0, X^1, \ldots, X^4$.

Through the equation $$G_{\alpha\beta}X^\alpha X^\beta = 0$$

our tensor $G_{\alpha\beta}$ determines a quadric in our tangent space. The polar hyperplane of a point A^α of a tangent plane with respect to this quadric is

$$A_\alpha X^\alpha = 0\,, \qquad \text{where} \qquad \gamma_{\alpha\beta}A^\alpha = A^\beta.$$

Exactly as in the usual relativity theory we shall move covariant and contravariant indices by means of the tensor $\gamma_{\alpha\beta}$ and the associated contravariant tensor $\gamma^{\alpha\beta}$ either up or down. The tensor $\gamma^{\alpha\beta}$ is completely defined by the equations

$$\gamma_{\alpha\beta}\gamma^{\alpha\sigma} = \delta^\sigma_\beta.$$

Raising or lowering the index corresponds to a transition to polar description of the quadric. The homogeneous coordinates of the origin are

$$X^\alpha = \varphi^\alpha = \delta^\alpha_0.$$

In this we have used the relationship

$$\varphi^\beta\gamma^{\beta\alpha} = \varphi^\alpha = \delta^\alpha_0.$$

Hence the polar hyperplane of the origin has the equation

$$G_{0\alpha}X^\alpha = 0$$

or

$$\varphi_\alpha X^\alpha = 0.$$

The equation of the tangent cone with its vertex at the origin is

$$\gamma_{\alpha\beta}X^\alpha X^\beta - (\varphi_\alpha X^\alpha)^2 = 0$$

or

$$g_{ij}X^iX^j = 0.$$

The contact points of the tangent cones with the quadric are known to lie in the polar hyperplane.

The splitting of our projective tensor into the affine tensor g_{ij} and the projective vector φ_α simply means that the origin and the quadric in each tangent space geometrically determine the tangent cone through the origin and the polar hyperplane of the origin with respect to the plane.

As mentioned before it is conjectured that in projective relativity theory the coefficients g_{ij} of the conic equation are gravitational potentials and the coefficients φ_α of the hyperplane equation are electromagnetic potentials. We shall see, in fact, that the closest field equations for the $\gamma_{\alpha\beta}$ are a combination of the classical Einstein gravitation equations and the Maxwell field equations.

III. Application to Classical Projective Geometry

Projective Coordinates.

To present our point of view more clearly we shall apply it to a special case, to be precise to the case of classical projective geometry. An underlying projective space is characterized by the availability of a number of special homogeneous coordinate systems, mutually related by linear transformations

$$\bar{Z}^\alpha = p^\alpha_\beta Z^\beta. \tag{1}$$

These homogeneous coordinate systems are described as *projective*. An arbitrary permissible coordinate system, x, is associated with an arbitrary projective coordinate system through equations of the form

$$Z^\alpha = e^{x^0} f^\alpha \left(x^1, \ldots, x^4\right) \tag{2}$$

where $x^1, \ldots, x^4$ are here any point coordinates and x^0 an arbitrary parameter. The functions on the right hand side of these equations are obviously projective scalars because the choice of other coordinate systems simply means that we substitute

$$x^i = x^i(\bar{x})$$

in (2), and the meaning of (2) is not changed by a substitution

$$x^0 = \bar{x}^0 - \log \varrho(x).$$

From these reflections we conclude that the classical projective geometry is characterized by the availability of a family of projective scalars

$$Z = p_\alpha A^\alpha \tag{3}$$

where the constants p_α are arbitrary. None of these scalars is distinguished over the others. An arbitrary homogeneous projective coordinate system is fixed by the choice of five arbitrary independent scalars of the family.

Differential Equations of Projective Geometry.

For the purpose of differential geometry it is useful to eliminate the constants p_α and so remove the apparent exceptional status of the five scalars $A^0, A^1, \ldots, A^4$ in the presentation of the arbitrary scalar of the family (3):

$$Z = p_\alpha A^\alpha.$$

Next we have, due to the role of the proportionality factor,

$$\frac{\partial Z}{\partial x^0} = Z. \tag{4}$$

This equation embodies the expectation that the proportionality factor is e^{x^0}. It could just as well have been e^{Nx^0}, but a value of N, different from zero, has the same meaning as $N = 1$ in our problem.

We differentiate (3) twice to get

$$\frac{\partial Z}{\partial x^\beta} = p_\alpha \frac{\partial A^\alpha}{\partial x^\beta}, \tag{5}$$

$$\frac{\partial^2 Z}{\partial x^\beta \partial x^\gamma} = p_\alpha \frac{\partial^2 A^\alpha}{\partial x^\beta \partial x^\gamma}. \tag{6}$$

Then we define a^α_β through the equations

$$a^\gamma_\beta \frac{\partial A^\alpha}{\partial x^\gamma} = \delta^\alpha_\beta$$

and find from (5)

$$p_\alpha = a^\beta_\alpha \frac{\partial Z}{\partial x^\beta}$$

Then we substitute this expression for p_α into (6) and obtain

$$\frac{\partial^2 Z}{\partial x^\alpha \partial x^\beta} = \Pi^\sigma_{\alpha\beta} \frac{\partial Z}{\partial x^\sigma},$$

where

$$\Pi^\sigma_{\alpha\beta} = a^\sigma_\gamma \frac{\partial^2 A^\gamma}{\partial x^\alpha \partial x^\beta}. \tag{7}$$

We readily notice that these functions $\Pi^\sigma_{\alpha\beta}$ are independent of the choice of the five functions A^α. Another choice of these functions must namely be given by a linear equation

$$\bar{A}^\alpha = p^\alpha_\beta A^\beta;$$

however, by such a substitution of the variables A_α one obtains the $\frac{\partial^2 A^\alpha}{\partial x^\beta \partial x^\gamma}$ cogredient rather than the a^α_γ contragredient. This way the $\Pi^\alpha_{\beta\gamma}$ remain unchanged.

I would like to designate the differential equations

$$\text{(8)} \qquad \frac{\partial^2 Z}{\partial x^\alpha \partial x^\beta} - \Pi^\sigma_{\alpha\beta} \frac{\partial Z}{\partial x^\sigma} = 0\,, \qquad \frac{\partial Z}{\partial x^0} = Z$$

as *differential equations of projective geometry* because the entire projective geometry can be viewed as a theory of these equations.

Projective Connection.

Each coordinate system and each gauge has a specific number of 5^3 associated functions

$$\Pi^\alpha_{\beta\gamma}\,.$$

In a gauge and coordinate transformation

$$\text{(9)} \qquad \begin{cases} \bar{x}^0 = x^0 + \log \varrho(x)\,, \\ \bar{x}^i = \bar{x}^i(x) \end{cases}$$

these functions transform like the components of an affine connection

$$\text{(10)} \qquad \bar{\Pi}^\alpha_{\beta\gamma} = \left(\Pi^\sigma_{\varrho\tau} \frac{\partial x^\varrho}{\partial \bar{x}^\beta} \frac{\partial x^\tau}{\partial \bar{x}^\gamma} + \frac{\partial^2 x^\sigma}{\partial \bar{x}^\beta \partial \bar{x}^\gamma} \right) \frac{\partial \bar{x}^\alpha}{\partial x^\sigma}$$

in a five-dimensional space.

One can easily verify this by calculation. In this calculation we only use the formula (7) and not the specific form of the transformation (9). We can evidently always interpret our particular transformation as coordinate transformation in a five-dimensional space. The calculation is thereby exactly the same as by corresponding introduction of an affine connection in a smooth affine space. (Bibl. 1932, 10, p. 41–43.)

We call each invariant or each geometrical object whose components are functions of the coordinates $x^1, \ldots, x^4$ and behave like the components of a five-dimensional affine connection, a *projective connection*. The functions $\Pi^\alpha_{\beta\gamma}$ that we defined above by (7) are then components of a particular projective connection. The theory of general projective connection is a generalization of classical projective geometry.

If we only transform coordinates, our transformation law (10) reduces to

$$\text{(10a)} \qquad \bar{\Pi}^i_{jk} = \left(\Pi^s_{rt} \frac{\partial x^r}{\partial \bar{x}^j} \frac{\partial x^t}{\partial \bar{x}^k} + \frac{\partial^2 x^s}{\partial \bar{x}^j \partial \bar{x}^k} \right) \frac{\partial \bar{x}^i}{\partial x^s}\,,$$

$$\text{(10b)} \qquad \bar{\Pi}^0_{jk} = \Pi^0_{rt} \frac{\partial x^r}{\partial \bar{x}^j} \frac{\partial x^t}{\partial \bar{x}^k}\,.$$

On coordinate transformation the Π^i_{jk} behave like the components of an affine connection and the Π^0_{jk} like the components of an affine tensor.

Alternatively if we change the gauge we obtain the transformation formulae:

$$(10c)\quad \begin{cases} \bar{\Pi}^i_{jk} = \Pi^i_{jk} - \Pi^i_{0k}\frac{\partial \log \varrho}{\partial x^j} - \Pi^i_{j0}\frac{\partial \log \varrho}{\partial x^k} + \Pi^i_{00}\frac{\partial \log \varrho}{\partial x^j}\frac{\partial \log \varrho}{\partial x^k}, \\ \bar{\Pi}^0_{jk} = \Pi^0_{jk} + \Pi^i_{jk}\frac{\partial \log \varrho}{\partial x^i} - \Pi^i_{0k}\frac{\partial \log \varrho}{\partial x^i}\frac{\partial \log \varrho}{\partial x^j} \\ \quad -\Pi^i_{j0}\frac{\partial \log \varrho}{\partial x^i}\frac{\partial \log \varrho}{\partial x^k} + \Pi^i_{00}\frac{\partial \log \varrho}{\partial x^i}\frac{\partial \log \varrho}{\partial x^j}\frac{\partial \log \varrho}{\partial x^k} - \Pi^0_{0k}\frac{\partial \log \varrho}{\partial x^j} \\ \quad -\Pi^0_{j0}\frac{\partial \log \varrho}{\partial x^k} + \Pi^0_{00}\frac{\partial \log \varrho}{\partial x^j}\frac{\partial \log \varrho}{\partial x^k} - \frac{\partial^2 \log \varrho}{\partial x^j \partial x^k}. \end{cases}$$

On particular transformation $\Pi^\alpha_{0\beta}$ and $\Pi^\alpha_{\beta 0}$ behave like the components of a projective tensor. Hence the equation

$$\Pi^\alpha_{0\beta} = \Pi^\alpha_{\beta 0} = \delta^\alpha_\beta$$

is invariant. This equation is evidently satisfied by the special projective connection (7). In this special case the transformation formulae (10c) reduce to:

$$(10d)\quad \begin{cases} \bar{\Pi}^i_{jk} = \Pi^i_{jk} - \delta^i_j\frac{\partial \log \varrho}{\partial x^k} - \delta^i_k\frac{\partial \log \varrho}{\partial x^j}, \\ \bar{\Pi}^0_{jk} = \Pi^0_{jk} - \frac{\partial \log \varrho}{\partial x^j}\frac{\partial \log \varrho}{\partial x^k} - \frac{\partial^2 \log \varrho}{\partial x^j \partial x^k} + \Pi^i_{jk}\frac{\partial \log \varrho}{\partial x^i} \end{cases}.$$

Five-dimensional Presentation.

In this section we have used with advantage an interpretation essentially due to T.Y. Thomas (Bibl. 1925, 8; 1926, 13). Our entire theory finds a representation in a five-dimensional space with coordinates $x^0, \ldots, x^4$, not subject to the most general coordinate transformation, but only to

$$x^i = x^i(\bar{x}),$$
$$x^0 = \bar{x}^0 - \log \varrho(x).$$

We can also interpret our particular transformations as transformations of this five-dimensional space. The lines $x^0, \ldots, x^4 = \text{const.}$, x^0 arbitrary, play a special role because of the special form of the particular transformation.

One now obtains a precise picture of our four-dimensional projective geometry by interpreting the x^0-lines of the five-dimensional space as points of a four-dimensional space. Exactly as the straight lines in elementary geometry, through a fixed point of a five-dimensional affine space, build a four-dimensional projective space, so there also arises in the theory of a general projective connection a four-dimensional "projective" space from a five-dimensional "affine" space. The role of the straight lines through a fixed point here represents the x^0-lines. The common point of the x^0-lines are here transformed together (Bibl. 1931, 16; 1929, 2).

Projective Derivative.

With the help of a projective connection we can construct, from the components of an arbitrary projective tensor, a new projective tensor, of higher rank, through the formulae of covariant differentiation. For instance, if A_α is a covariant projective vector then

$$\frac{\partial A^\alpha}{\partial x^\beta} + \Pi^\alpha_{\sigma\beta} A^\alpha$$

become the components of a projective tensor A^α_β that we describe as the *projective derivative* of A^α. It follows without further ado from the five-dimensional affine interpretation of particular transformations that A^α_β are indeed the components of a tensor.

Correspondingly, the same law and formulae, as in affine theory, also serve for the projective differentiation of a completely general tensor. There are admittedly still further laws, not available in affine theory because they depend on the special nature of a gauge transformation. We shall develop such laws only when we need them.

Integrability Conditions.

The usual projective geometry is identifiable as the theory of systems (8) of differential equations. These differential equations are however, not the most general differential equations of the form (8). They better satisfy a row of integrability conditions. We write the differential equations in the form

$$\frac{\partial Z_0}{\partial x^\alpha} = Z_\alpha\,, \tag{11a}$$

$$\frac{\partial Z_\alpha}{\partial Z_\alpha} = \Pi^\sigma_{\alpha\beta} Z_\sigma\,, \tag{11b}$$

where Z_0 now denotes Z. The integrability conditions of these equations are known as

$$\Pi^\alpha_{\beta\gamma} = \Pi^\alpha_{\gamma\beta}\,, \tag{12}$$

$$R^\lambda_{\alpha\beta\gamma} = \frac{\partial \Pi^\lambda_{\alpha\beta}}{\partial x^\gamma} - \frac{\partial \Pi^\lambda_{\alpha\gamma}}{\partial x^\beta} + \Pi^\sigma_{\alpha\beta}\Pi^\lambda_{\sigma\gamma} - \Pi^\lambda_{\sigma\beta}\Pi^\sigma_{\alpha\gamma} = 0\,. \tag{13}$$

(21)	Integrability Conditions.

The calculation is exactly the same as for the corresponding problem of affine geometry (Bibl. 1927, 22; 1932, 10).

$R^{\lambda}_{\alpha\beta\gamma}$ is a projective tensor of fourth rank, the curvature tensor of the connection $\Pi^{\alpha}_{\beta\gamma}$. If it vanishes we call the connection *even*.

From (11a) and (11b) also follows the invariance relation

$$\Pi^{\alpha}_{\beta 0} = \delta^{\alpha}_{\beta}\,. \tag{14}$$

We now have to show that the integrability conditions are in fact met in the case of projective geometry. The functions Π have the form (7). From that one infers the existence of (12). The condition (13) can be verified by an elementary calculation based on (7). The considerations are however exactly the same as in affine theory so that we know immediately that the integrability conditions are necessary.

Alternatively we shall now prove that our integrability conditions are also sufficient. Thereby it is proven at the same time that the equations (8) together with the conditions (12), (13) and (14) are characteristic for the projective geometry.

We must also show that the availability of five independent solution functions $A^0, \ldots, A^4$ follows from the existence of the integrability conditions, and from which the additional solutions can be combined linearly with constant coefficients. On the basis of the five-dimensional theory it is further clear that because of (12) and (13) (11b) has exactly a solution vector $A^0, \ldots, A^4$ which assumes the previously given value in a specific point. If we specify five independent vectors as initial values in a point we obtain five independent solution vectors A^{α}_{β}.

From the five-dimensional representation it further follows that because of the symmetry of Π the solution gradient vectors are five affine scalars A^{α} in five dimensions.

However, we must still show that these scalars have the form

$$A^{\alpha} = e^{x^0} f^{\alpha}\left(x^1, \ldots, x^4\right) \tag{15}$$

It follows from (14) that $\quad \dfrac{\partial A_{\alpha}}{\partial x^0} = \Pi^{\sigma}_{\alpha 0} A_{\sigma} = A_{\alpha}$

or $\quad A_{\alpha} = e^{x^0} f_{\alpha}\left(x^1, \ldots, x^4\right)$.

If again, we write f for f_0, then

$$Z = e^{x^0} f\left(x^1, \ldots, x^4\right)$$

is a solution of (8), while, due to (11a) the solution vectors of (11b) are the gradients of the function Z.

So we have proven that the classical projective geometry is completely equivalent to the theory of differential equations (8) with the conditions (12), (13) and (14), in any case for a given region of the underlying space.

Homogeneous Projective Coordinates as Functions of Boundary Conditions.

We describe five independent solutions of the equations (8) by Z^α and the corresponding solution vectors by $Z^\alpha_\beta = \frac{\partial Z^\alpha}{\partial x^\beta}$. We now choose $\left(Z^\alpha_\beta\right)_q = \delta^\alpha_\beta$ as initial value in the point $x = q$. We interpret the five independent solutions determined by $Z^\alpha = e^{x^0} f^\alpha\left(x^1, \ldots, x^4\right)$ as functions of both coordinates and of the initial points:

$$Z^\alpha = Z^\alpha(x, q).$$

It now follows that

$$Z^\alpha = e^{x^0 - q^0} f^\alpha\left(x^1, \ldots, x^4, q^1, \ldots, q^4\right). \tag{16}$$

In fact (11) contains the equation

$$\frac{\partial^2 f^\alpha}{\partial x^j \partial x^k} \Pi^i_{jk} \frac{\partial f^\alpha}{\partial x^i} + \Pi^0_{jk} f^\alpha$$

We can assume as initial values $f^\alpha(q) = \delta^\alpha_0$, $\left(\frac{\partial f^\alpha}{\partial x^i}\right)_q = \delta^\alpha_i$. Then f^α has, as a function of the regarded coordinates and initial points, the form

$$f^\alpha = f^\alpha\left(x^1, \ldots, x^4, q^1, \ldots, q^4\right).$$

Because each linear combination of solutions is again a solution, we can in fact assume form (16) and confirm that Z^α_β takes the value $\left(Z^\alpha_\beta\right)_q = \delta^\alpha_\beta$ in the point $x = q$. Since the solution of (11) is unambiguous it follows that Z^α necessarily has the form (16).

If we subject x^i to the transformation

$$x^i = x^i(\bar{x}),$$

q^i behaves according to

$$q^i = x^i(\bar{q}).$$

Now again we look in the transformed coordinate system for a system of solutions with initial values $Z^\alpha_\beta(\bar{q}) = \delta^\alpha_\beta$. Because it can give only five independent solutions there must evidently exist a relationship of the form

$$\bar{Z}^\alpha = p^\alpha_\beta Z^\beta$$

between the new and the old solutions, such that the p^α_β are constants and $|p^\alpha_\beta| \neq 0$. Differentiation of these equations gives

$$\frac{\partial \bar{Z}^\alpha}{\partial \bar{x}^\beta} \frac{\partial \bar{x}^\beta}{\partial x^\gamma} = p^\alpha_\beta \frac{\partial Z^\beta}{\partial x^\gamma}$$

identical in x. For $x = q$ it follows from the use of intial values that

$$\left(\frac{\partial \bar{x}^\alpha}{\partial x^\gamma}\right)_{x=q} = p^\alpha_\gamma$$

or
$$\bar{Z}^\alpha\left(\bar{x}, \bar{q}\right) = \frac{\partial \bar{q}^\alpha}{\partial q^\gamma} Z^\gamma(x, q).$$

Z^α considered a function of q therefore is a contravariant projective vector with index -1 due to (16).

Inhomogeneous Projective Coordinates.

For our further development we use with good effect a special coordinate system. The equation $y^i = Z^i/Z^0$ determines a coordinate transformation $y^i = y^i(x)$. The coordinate system arising from this we call an inhomogeneous coordinate system.

From this inhomogeneous system we can go over again into a homogeneous system, *e.g.* through the formula

$$Z^i = \left(y^i - q^i\right) e^{y^0 - q^0}, \quad Z^0 = e^{y^0 - q^0}. \tag{17}$$

From the Z^α we retrieve the most general form of our homogeneous coordinates as we subject the y and q simultaneously to the same particular transformation.

We can calculate the value of Π in these special coordinate systems if we apply the same considerations, that we used for A^α at the beginning of this chapter, once more for the Z^α. That leads to

$$\Pi^\alpha_{\beta\gamma} = \delta^\alpha_\beta \delta^0_\gamma + \delta^\alpha_\gamma \delta^0_\beta - \delta^\alpha_0 \delta^0_\beta \delta^0_\gamma. \tag{18}$$

It means that $\Pi_{jk} = 0$ and $\Pi^\alpha_{\beta 0} = \delta^\alpha_\beta$. Using by comparison that the Z^α must in any case satisfy equations of the form (8) it follows again from

$$\frac{\partial^2 Z^\alpha}{\partial y^j \partial y^k} = 0\,, \quad \frac{\partial^2 Z^\alpha}{\partial y^\beta \partial y^0} = \delta^\gamma_\beta \frac{\partial Z^\alpha}{\partial x^\gamma}$$

that $\Pi^\alpha_{jk} = 0$, $\Pi^\alpha_{\beta 0} = \delta^\alpha_\beta$.

We still remark that the components (18) of projective connections on gauge transformation

$$\bar{x}^0 = x^0 + \log \varrho\left(x^1, \ldots, x^4\right)$$

take the form

$$\Pi^\alpha_{\beta\gamma} = \delta^\alpha_\beta \varphi_\gamma + \delta^\alpha_\gamma \varphi_\beta - \varphi_\beta \varphi_\gamma \delta^\alpha_0 + \frac{\partial \varphi_\beta}{\partial x^\gamma} \delta^\alpha_0 \tag{19}$$

in which $\varphi_i = -\frac{\partial \log \varrho}{\partial x^i}$, $\varphi_0 = 1$.

We could identify the inhomogeneous coordinate systems as those systems in which Π has the form (19). Two inhomogeneous coordinate systems are connected by a linear broken transformation

$$\bar{z}_i = \frac{p^i_j z^j + p^i_0}{p_j z^j + p_0}. \tag{20}$$

This follows immediately as two homogeneous systems are connected by a linear substitution.

From our current point of view we can say that each of two coordinate systems in which Π has the form (19) are connected by the substitution (20). If we further demand that Π has the invariant form (18) we must associate a special gauge transformation, namely

$$\varrho = k u^{-\frac{1}{5}},$$

with each coordinate transformation, where u is the functional determinant of our coordinate transformation and k is a constant. By these remarks the relationship between our present theory and the formulation in earlier work of T.Y. Thomas and others (Bibl. 1926, 13; 1928, 10; 1930, 5) is clarified.

Homogeneous Projective Coordinates as Functions of Boundary Conditions (Continued).

We have seen before that the functions

$$Z^\alpha = e^{x^0 - q^0} f^\alpha \left(x^1, \ldots, x^4, q^1, \ldots, q^4\right)$$

with $Z^\alpha_\beta = \delta^\alpha_\beta$ for $x = q$, mediate the transition between the arbitrary coordinate system x and those homogeneous coordinate systems connected through the boundary conditions. Each point q and each parameter value q^0 of the underlying space belongs to a specific coordinate system. How does Z behave as a function of q?

The answer is as follows: The Z are functions of q that satisfy the following differential equations:

$$\frac{\partial Z^\alpha}{\partial q^\beta} + \Pi^\alpha_{\sigma\beta} Z^\sigma = 0. \tag{21}$$

We know that the Z are components of a contravariant vector of index -1. We next consider the projective derivative

$$\frac{\partial Z^\alpha}{\partial q^\beta} + \Pi^\alpha_{\sigma\beta} Z^\sigma$$

of this vector. We can easily show that the projective derivative of Z^α vanishes.

(25)	Displacement along a curve

In the inhomogeneous system Π has the value (18) while the Z take the form (17). If we substitute (18) for the value of Π, the components of the second-rank tensor

$$\frac{\partial Z^\alpha}{\partial q^\beta} + \Pi^\alpha_{\sigma\beta} Z^\sigma$$

reduce to the form $\quad \dfrac{\partial Z^\alpha}{\partial q^0} + Z^\alpha , \quad \dfrac{\partial Z^i}{\partial q^i} + \delta^i_j Z^0 , \quad \dfrac{\partial Z^0}{\partial q^j} .$

Because of (17) all other components must vanish. This proves the assumed equation (21).

Whereas the left side of the equation is independent of the representation, we know quite generally that the homogeneous coordinates satisfy the equations (21). The equations (21) characterize the projective geometry just like (8).

Their integrability conditions are $\qquad R^\alpha_{\beta\gamma\delta} = 0.$

If these integrability conditions are not obeyed we are dealing with a generalization of projective geometry that we shall consider in a later chapter.

IV Projective Displacements

In the previous chapter we considered the solutions Z^α of the differential equations of projective geometry as functions of the boundary conditions. We have seen that Z^α, as a function of q, is a projective tensor of index -1 when we demand that

$$\left(Z^\alpha_\beta\right)_{x=q} = \delta^\alpha_\beta$$

It further turned out that Z_α satisfy the differential equations

$$\frac{\partial Z^\alpha}{\partial q^\beta} + \Pi^\alpha_{\sigma\beta} Z^\sigma = 0. \tag{1}$$

In this we always assume the integrability conditions.

Displacement along a curve.

We now start with a generalization somewhat analogous to the transition from Euclidean to Riemannian geometry. We no longer demand that the equations (1) be integrable; the tensors

$$\Pi^\alpha_{\beta\gamma} - \Pi^\alpha_{\gamma\beta} \quad \text{and} \quad R^\alpha_{\beta\gamma\delta}$$

are therefore not necessarily equal to zero.

Next we retain the conditions $\Pi^\alpha_{\beta 0} = M\delta^\alpha_\beta$.

Here we have written a softer generalization $M\delta^\alpha_\beta$ instead of δ^α_β. x^0 and q^0 enter Z correspondingly in the form

$$e^{M(x^0-q^0)}.$$

As in the general treatment of partial differential equations we now select an arbitrary curve

$$q^i = x^i(t) \tag{2}$$

and also set quite generally

$$q^0 = x^0(t). \tag{3}$$

If we multiply (1) with dq^β/dt, we obtain the equations

$$\frac{dZ^\alpha}{dt} + \Pi^\alpha_{\lambda\beta} Z^\lambda \frac{dq^\beta}{dt} = 0. \tag{4}$$

Then, due to the condition $\Pi^\alpha_{\beta 0} = M\delta^\alpha_\beta$

$$\frac{\partial Z^\alpha}{\partial q^0} + \Pi^\alpha_{\sigma 0} Z^\sigma = 0, \tag{5}$$

as long as the Z^α have the form $e^{-Mq^0}f(t)$ and hence, the equations (4) have the same meaning as

$$\frac{dZ^\alpha}{dt} + \Pi^\alpha_{\lambda j} \frac{dq^j}{dt} Z^\lambda = 0. \tag{6}$$

It follows from the five-dimensional affine theory that (4) is invariant against particular transformations and from that also follows the invariance of the four-dimensional equations (6). These equations only depend on the functions Π and on the curve in the parameter presentation (2) and not on (3).

We now write the equations (6) in the form

$$\frac{dZ^\alpha}{dt} + \Pi^\alpha_\lambda(t) Z^\lambda = 0 \quad \text{where} \quad \Pi^\alpha_\lambda(t) = \Pi^\alpha_{\lambda j} \frac{dq^j}{dt}. \tag{7}$$

According to the known existence theorem on linear differential equations the solutions of our equations have the form

$$X^\alpha = A^\beta p^\alpha_\beta(t), \tag{8}$$

where the A are initial values in the point $t = t_0$. The p^α_β must also reduce to the values δ^α_β in the point $t = t_0$.

(27)	Generalized Projective Geometry.

When we interpret the arbitrary specifiable numbers A as homogeneous coordinates of the tangent space in $q^i = x^i(t_0)$ and the numbers X as coordinates of the tangent space in $q^i = x^i(t)$, we are given a projective mapping of the first tangent space on the second by equation (8).

In this sense we can say that the equations

$$dX^\alpha + \Pi^\alpha_{\lambda j} X^\lambda dx^j = 0$$

represent an infinitesimal projective transformation. If we therefore connect two points a and b by a curve then (8) determines a projective transformation of the tangent space in a on a tangent space in b. These mappings are completely determined by the specification of a and b and their connecting curve. In the flat case, *i.e.* when the conditions (12) and (13) of chap. III are satisfied, the equations (1) are fully integrable and the mapping determined by two points and their connecting curve is not changed on arbitrary transformation of the curve.

In the integrable case we may consider A^α and X^α as homogeneous coordinates of the underlying space, or more precisely, the A^α as coordinates and hence the coordinate system connected with the point $q^i = x^i(t)$. Then (8) implies a transformation of the underlying space to itself.

Generalized Projective Geometry.

We formulate the facts of the case once more in a different form:

The equation $$Z^\alpha(q) = X^\alpha$$

specifies a mapping of the tangent space on the underlying space, which we can visualize intuitively as the covering of the underlying space by the tangent space. Because the differential equations of projective geometry are fully integrable, all the tangent spaces, so to say, coincide with the underlying space.

However, should the differential equations not be integrable, we have no projective coordinate system $Z(q)$ and hence no such cover of the underlying space by tangent spaces either. In this case we only have displacements along arbitrary curves. The tangent spaces are connected by the displacements, but this relationship is not as close as in the integrable case where we could describe it as coincidence. By analogy with the flat image we can say: The tangent spaces fall apart when the differential equations (1) are not integrable.

We also see geometrically how the theory of non-integrable projective connection can be considered as the generalization of ordinary projective geometry. Our more general point of view is as follows:

A geometry is the theory of a geometrical object. If this object is a projective connection we have the generalized projective geometry. If the projective connection satisfies the integrability conditions considered before the classical projective geometry results, albeit in a small way.

If the condition $\Pi^{\alpha}_{\beta 0} = M\delta^{\alpha}_{\beta}$ is fulfilled, the generalized projective geometry contains a theory of the projective displacements of tangent spaces along arbitrary curves, as we indicated above. We shall explain that a little further.

As in affine theory it is seen how to define the projective displacement of a hyperplane $B_\alpha X^\alpha = 0$ by the differential equations

$$\frac{dB_\alpha}{dt} - \Pi^{\sigma}_{\alpha j} B_\sigma \frac{dx^j}{dt} = 0. \tag{9}$$

Likewise we also explain the displacements of higher algebraic images in tangent space, *i.e.* of projective tensors of higher order. We obtain, especially for the displacement of a quadric the equation

$$\frac{dG_{\alpha\beta}}{dt} - \Pi^{\sigma}_{\alpha j} G_{\beta\sigma} \frac{dx^j}{dt} - \Pi^{\sigma}_{\beta j} G_{\alpha\sigma} \frac{dx^j}{dt} = 0. \tag{10}$$

Displacements in Inhomogeneous Coordinates.

To describe the displacements precisely we need the introduction of projective coordinates as given in chap. III, *i.e.* by using the relation

$$\frac{X^i}{\varphi_\alpha X^\alpha} = dx^i. \tag{11}$$

The actual form of the transformation of tangent spaces must depend on the projective derivative of φ. We shall calculate that now. From (11) we obtain

$$X^i = \varphi_\alpha X^\alpha V^i, \tag{12}$$

by substituting $V^i = dx^i$. We differentiate (12)

$$\frac{\partial X^i}{\partial x^j} = \varphi_\alpha X^\alpha \frac{\partial V^i}{\partial x^j} + \varphi_\alpha \frac{\partial X^\alpha}{\partial x^j} V^i + \frac{\partial \varphi_\alpha}{\partial x^j} X^\alpha V^i.$$

We now multiply with dx^j/dt. By use of (1) then follows:

$$-\Pi^{i}_{\lambda j} X^\lambda \frac{dx^j}{dt} = \varphi_\alpha X^\alpha \frac{dV^i}{dt} - \varphi_\alpha \Pi^{\alpha}_{\lambda j} X^\lambda \frac{dx^j}{dt} V^i + \frac{d\varphi_\alpha}{dt} X^\alpha V^i$$

(29)	Paths.

or after division by $\varphi_\alpha X^\alpha$

$$\frac{dV^i}{dt}+\varphi_{\alpha;j}\frac{X^\alpha}{\varphi_\sigma X^\sigma}V^i\frac{dx^j}{dt}+\Pi^i_{\lambda j}\frac{X^\lambda}{\varphi_\sigma X^\sigma}\frac{dx^j}{dt}=0. \tag{13}$$

With the help of (12) and

$$X^0 = X^\alpha\varphi_\alpha\left(1-\varphi_i V^i\right)$$

we finally obtain:

$$\begin{cases} \frac{dV^i}{dt} & +\Pi^i_{kj}V^k\frac{dx^j}{dt}+(1-\varphi_j V^j)\Pi^i_{0k}\frac{dx^k}{dt}+\varphi_{k;j}V^kV^i\frac{dx^j}{dt} \\ & +\varphi_{0;j}(1-\varphi_k V^k)\frac{dx^j}{dt}V^i=0. \end{cases} \tag{14}$$

in which $\varphi_{\alpha;\beta}$ represents the projective derivative of φ_α. Thus we have derived the inhomogeneous form of a projective displacement.

Paths

Given a curve in parameter presentation, $x^i = x^i(t)$, its "velocity vector"

$$V^i=\frac{dx^i}{dt} \tag{15}$$

determines a point in each tangent space belonging to a point on the curve.

We may ask if there exist such curves in which these points go over in themselves by displacement along the curve. In answer we must replace (14) by dx^i/dt. In case that

$$\Pi^\alpha_{0\beta}=M\delta^\alpha_\beta\,,$$

we get

$$\frac{d^2x^i}{dt^2}+\Pi^i_{jk}\frac{dx^j}{dt}\frac{dx^k}{dt}+\frac{dx^i}{dt}\left[\varphi_{k;j}\frac{dx^k}{dt}\frac{dx^j}{dt}+M\left(1-\varphi_k\frac{dx^k}{dt}\right)^2\right]=0\,.$$

As the expression in square brackets is independent of i, this equation can be cast into the form:

$$\frac{\frac{d^2x^i}{dt^2}+\Pi^i_{jk}\frac{dx^j}{dt}\frac{dx^k}{dt}}{\frac{dx^i}{dt}}=\frac{\frac{d^2x^l}{dt^2}+\Pi^l_{jk}\frac{dx^j}{dt}\frac{dx^k}{dt}}{\frac{dx^l}{dt}}\,. \tag{16}$$

It is remarkable that these are not dependent on φ at all. Equation (16) is a known expression for a system of paths. Under a system of paths we understand a system of curves with the property that, within an appropriately small region, one and only one curve of the system goes through two specified points. The paths are a generalization of the geodesics of Riemannian geometry.

The form of equations (16) does not depend on the parameter presentation of the paths, as is easily demonstrated. Likewise one shows on the basis of equations (10a) of chap. III, that equations (16) are also invariant against coordinate transformation. It finally follows from (10c) that they are invariant against gauge transformation as well.

The expression

$$\frac{\partial \log \varrho}{\partial x^k}\frac{dx^k}{dt} - \frac{\partial \log \varrho}{\partial x^j}\frac{dx^j}{dt}$$

indeed enters the left side of (16) in gauge transformation. However, as this expression is independent of i it cancels out against the corresponding expression on the right of (16), whereby the gauge invariance is proven.

So we reach the proposition that our connection in the case

$$\Pi^\alpha_{0\beta} = \Pi^\alpha_{\beta 0} = M\delta^\alpha_\beta$$

produces an unambiguous system of paths. In the case of classical projective geometry the paths are straight lines.

General Projective Connections.

Up to now we have only considered the special case of a projective connection for which

$$\Pi^\alpha_{\beta 0} = M\delta^\alpha_\beta$$

is valid. We are now trying to establish a displacement, also when this condition no longer holds.

We cannot use the equation

$$\overline{(17)} \qquad \frac{\partial X^\alpha}{\partial x^\beta} + \Pi^\alpha_{\sigma\beta}X^\sigma = 0$$

because it would contradict the assumption that for the case $\beta = 0$, x^0 enters X in the form e^{-Mx^0}. However, we must keep this assumption because the displacement only depends on the curve and its parameter representation

$$x^i = x^{i(t)}\,,$$

but not on the special choice of the parameter

$$x^0 = x^0(t)\,.$$

We note that, in order to obtain a suitable definition of a displacement, the choice of the tensor 0, on the right of $\overline{(17)}$, is equally as arbitrary as the choice of some other arbitrary tensor which is invariantly connected with Π.

(31) The Associated Projective Connections of the Tensors $G_{\alpha\beta}$.

We now define a displacement by the equations

$$X^{\alpha}_{;\beta} = X^{\alpha}_{;0} C_{\beta} \tag{17}$$

in which C_β is a covariant vector that satisfies the condition $C_0 = 1$. The equation $\overline{(17)}$ is identically satisfied for $\beta = 0$.

Instead of equation $\overline{(17)}$ we now have a number of equations (17) that one gets by substituting all possible systems of four functions for C_β. The totality of these equations is invariantly associated with the connection.

We could now write equations (17) in the form

$$\frac{\partial X^{\alpha}}{\partial x^{\sigma}} + \Lambda^{\alpha}_{\beta\sigma} X^{\beta} = 0$$

in which

$$\Lambda^{\alpha}_{\beta\sigma} = \Pi^{\alpha}_{\beta\sigma} - \Pi^{\alpha}_{\beta 0} C_{\sigma} + M \delta^{\alpha}_{\beta} C_{\sigma} \tag{18}$$

and M is any index. The $\Lambda^{\alpha}_{\beta\sigma}$ are the components of a projective connection that satisfies the conditions $\Lambda^{\alpha}_{\beta 0} = M\delta^{\alpha}_{\beta}$. Hence we obtain a number of projective displacements for each projective connection.

The Associated Projective Connections of the tensors $G_{\alpha\beta}$.

There are two symmetrical projective connections associated with $G_{\alpha\beta}$, characterized by the vanishing of the projective derivatives of $G_{\alpha\beta}$ and $\gamma_{\alpha\beta}$. Both of these connections that we shall call Π and Γ are therefore defined by the equations

$$G_{\alpha\beta|\gamma} = \frac{\partial G_{\alpha\beta}}{\partial x^{\gamma}} - G_{\alpha\sigma}\Pi^{\sigma}_{\beta\gamma} - G_{\sigma\beta}\Pi^{\sigma}_{\alpha\gamma} = 0 \tag{19}$$

and

$$\gamma_{\alpha\beta;\gamma} = \frac{\partial \gamma_{\alpha\beta}}{\partial x^{\gamma}} - \gamma_{\alpha\sigma}\Gamma^{\sigma}_{\beta\gamma} - \gamma_{\alpha\beta}\Gamma^{\sigma}_{\alpha\gamma} = 0 \tag{20}$$

We denote projective differentiation with respect to Π and Γ by $|$ and; respectively. On solving the equations (20) one obtains the Christoffel formulae

$$\Gamma^{\alpha}_{\beta\gamma} = \tfrac{1}{2}\gamma^{\alpha\sigma}\left(\frac{\partial \gamma_{\beta\sigma}}{\partial x^{\gamma}} + \frac{\partial \gamma_{\sigma\gamma}}{\partial x^{\beta}} - \frac{\partial \gamma_{\beta\gamma}}{\partial x^{\sigma}}\right). \tag{21}$$

One gets the equations for Π on replacing Γ by Π and γ by G in (21).

The Π must evidently be calculable from Γ by means of the scalar Φ. By simple application of the product rule of differentiatial calculus one gets

$$\Pi^{\alpha}_{\beta\gamma} = \Gamma^{\alpha}_{\beta\gamma} + \left(\delta^{\alpha}_{\beta}\frac{\partial \log \Phi}{\partial x^{\gamma}} + \delta^{\alpha}_{\gamma}\frac{\partial \log \Phi}{\partial x^{\beta}} - \gamma_{\beta\gamma}\gamma^{\alpha\sigma}\frac{\partial \log \Phi}{\partial x^{\sigma}}\right).$$

Further, by setting

$$\Phi_\alpha = \frac{\partial \log \Phi}{\partial x^\alpha},$$

there follows

$$\Pi^\alpha_{\beta\gamma} = \Gamma^\alpha_{\beta\gamma} + \left(\delta^\alpha_\beta \Phi_\gamma + \delta^\alpha_\gamma \Phi_\beta - \gamma_{\beta\gamma}\Phi^\alpha\right). \tag{22}$$

The Associated Displacements of $G_{\alpha\beta}$

It is seen from (21) that the projective connection Γ satisfies the conditions

$$\Gamma^\alpha_{\beta 0} = \gamma^{\alpha\sigma}\varphi_{\sigma\beta},$$

in which we have set

$$\varphi_{\sigma\beta} = \tfrac{1}{2}\left(\frac{\partial \varphi_\alpha}{\partial x^\beta} - \frac{\partial \varphi_\beta}{\partial x^\sigma}\right).$$

The important formula

$$\varphi_{\alpha;\beta} = \frac{\partial \varphi_\alpha}{\partial x^\beta} - \varphi_\alpha \Gamma^\sigma_{\alpha\beta} = \varphi_{\alpha\beta}$$

is valid for $\varphi_{\alpha\beta}$. Thus if we multiply (21) with $\varphi_\sigma = \gamma_{\sigma 0}$, we obtain

$$\varphi_\alpha \Gamma^\sigma_{\alpha\beta} = \tfrac{1}{2}\delta^\tau_0\left(\frac{\partial \gamma_{\beta\tau}}{\partial x^\alpha} + \frac{\partial \gamma_{\alpha\tau}}{\partial x^\beta} - \frac{\partial \gamma_{\alpha\beta}}{\partial x^\tau}\right) = \tfrac{1}{2}\left(\frac{\partial \varphi_\beta}{\partial x^\alpha} + \frac{\partial \varphi_\alpha}{\partial x^\beta}\right).$$

From this, the stated formula,

$$\boxed{\varphi_{\alpha;\beta} = \varphi_{\alpha\beta}} \tag{23}$$

follows directly.

Now in general $$\Gamma^\alpha_{\beta 0} = \varphi^\alpha_\beta \neq \delta^\alpha_\beta .$$

We therefore need the formula (17) to effect a displacement. No arbitrary vector C is obviously used in this case, but the vector φ which is invariably linked to $G_{\alpha\beta}$. The resulting displacement is therefore itself linked invariably to Γ, while that only applies to the totality of displacements in the general case.

So we define the displacements in our case by

$$X^\alpha_{;\beta} = X^\alpha_{\;;0}\varphi_\beta . \tag{24}$$

By assuming that X^α is of index $-M$ we obtain for Λ

$$\Lambda^\alpha_{\beta\sigma} = \Gamma^\alpha_{\beta\sigma} + M\delta^\alpha_\beta\varphi_\alpha - \Gamma^\alpha_{\beta 0}\varphi_\alpha . \tag{25}$$

Λ still depends on M. For $\sigma = 0$, $\Lambda^\alpha_{\beta 0} = M\delta^\alpha_\beta$, so that again

(26) $$\frac{\partial X^{\alpha}}{\partial x^{\beta}} + \Lambda^{\alpha}_{\sigma\beta} X^{\sigma} = 0$$

is identically valid for $\beta = 0$.

We develop the theory of Λ displacements further in chap. VI.

V. Non-Euclidean Geometry

In chap. III we applied our general theory to the special case of classical projective geometry. We shall now take the specialization one step further and consider a quadric in ordinary projective space.

Equation of a Quadric.

We assume that a system of functions Π has been given in our underlying space with coordinates $(x^1, \ldots, x^4)$, such that the differential equations

(1) $$\frac{\partial^2 Z}{\partial x^{\beta} \partial x^{\gamma}} = \Pi^{\sigma}_{\beta\gamma} \frac{\partial Z}{\partial x^{\sigma}}$$

or

(2) $$\frac{\partial Z^{\alpha}}{\partial q^{\beta}} + \Pi^{\alpha}_{\lambda\beta} Z^{\lambda} = 0$$

can be solved. In this q indicates the point in which the Z^{α} as functions of x have the initial values

$$\frac{\partial Z^{\alpha}}{\partial x^{\beta}} = \delta^{\alpha}_{\beta} .$$

By focussing on an arbitrary coordinate system Z with fixed q, one obtains a quadric through the equation

(3) $$G_{\alpha\beta} Z^{\alpha} Z^{\beta} = 0 ,$$

in which the G are only determined to within a common factor. Two arbitrary homogeneous coordinate systems are separated from each other under a linear homogeneous substitution. Due to (3) the G therefore transform like the coefficients of a quadratic form in a linear substitution.

The $G_{\alpha\beta}$ as functions of q.

We now consider the Z as functions of q. The relations between the G must then also depend on $q^0, \ldots, q^4$, while of course they are constant in a given projective coordinate system.

Considered as functions of q the Z^α are contravariant vectors of index -1. It therefore follows from the considerations above that the G transform according to the formulae

$$\overline{G}_{\alpha\beta}(\bar{q}) = \lambda G_{\sigma\tau}(q)\frac{\partial q^\sigma}{\partial \bar{q}^\alpha}\frac{\partial q^\tau}{\partial \bar{q}^\beta} \tag{4}$$

in which λ can be an arbitrary function of q.

To investigate the dependence of G on the q we differentiate (3) and substitute the values from (2) for the $\partial Z/\partial q^\beta$. It is shown by (2) how the Z^α behave as functions of the origin of q.

Thus results the system of equations

$$\left(\frac{\partial G_{\alpha\beta}}{\partial q^\gamma} - G_{\alpha\sigma}\Pi^\sigma_{\beta\gamma} - G_{\sigma\beta}\Pi^\sigma_{\alpha\gamma}\right) Z^\alpha Z^\beta = 0\,,$$

valid for all values of Z^α that satisfy equation (3). Hence the expression

$$\frac{\partial G_{\alpha\beta}}{\partial q^\gamma} - G_{\alpha\sigma}\Pi^\sigma_{\beta\gamma} - G_{\beta\sigma}\Pi^\sigma_{\alpha\gamma}$$

must be proportional to $G_{\alpha\beta}$, in which the proportionality factor must of course in general be different for different γ. This way we obtain the system of equations

$$\frac{\partial G_{\alpha\beta}}{\partial q^\gamma} - G_{\alpha\sigma}\Pi^\sigma_{\beta\gamma} - G_{\beta\sigma}\Pi^\sigma_{\alpha\gamma} = G_{\alpha\beta}A_\gamma\,. \tag{5}$$

Equations (5) describe the variation of G on a change of origin for q.

Normalization of $G_{\alpha\beta}$.

It is now easily demonstrated that no geometrical constraint results from setting $A_\gamma = 0$. Because of

$$G^{\alpha\beta}\frac{\partial G_{\alpha\beta}}{\partial q^\gamma} = \frac{\partial \log G}{\partial q^\gamma}$$

multiplication of (5) with $G^{\alpha\beta}$ gives the equations

$$\frac{\partial \log G}{\partial q^\gamma} - 2\Pi^\sigma_{\sigma\gamma} = 5A_\gamma\,, \tag{6}$$

where $G = |G_{\alpha\beta}|$. It now follows immediately from chap. III (13) that $\Pi^\sigma_{\sigma\gamma}$ satisfies the equation

$$\frac{\partial \Pi^\sigma_{\sigma\gamma}}{\partial q^\delta} - \frac{\partial \Pi^\sigma_{\sigma\delta}}{\partial q^\gamma} = 0\,.$$

We could therefore fix $\Pi^\sigma_{\sigma\gamma}$ in the form

$$\Pi^\sigma_{\sigma\gamma} = \frac{\partial \log f}{\partial q^\gamma}$$

such that we need no further specification of the behaviour of f under particular transformation. f is a function of only $q^1, \ldots, q^4$ since q^0 does not feature in Π at all.

On the basis of (6) we therefore obtain

(7) $$A_\gamma = \tfrac{1}{5}\frac{\partial \log A}{\partial q^\gamma}$$

where $$A = \frac{G}{f_2}$$ is implied.

If we now set $$G_{\alpha\beta} = G^*_{\alpha\beta} A^{\frac{1}{5}}$$ in (5), there follows

$$\frac{\partial G^*_{\alpha\beta}}{\partial q^\gamma} A^{\frac{1}{5}} + G^*_{\alpha\beta}\frac{\partial A^{\frac{1}{5}}}{\partial q^\gamma} - A^{\frac{1}{5}} G^*_{\alpha\sigma}\Pi^\sigma_{\beta\gamma} - A^{\frac{1}{5}} G^*_{\beta\sigma}\Pi^\sigma_{\alpha\gamma} = A^{\frac{1}{5}} G^*_{\alpha\beta} A_\gamma$$

or

(8) $$\frac{\partial G^*_{\alpha\beta}}{\partial q^\gamma} - G^*_{\alpha\sigma}\Pi^\sigma_{\beta\gamma} - G^*_{\beta\sigma}\Pi^\sigma_{\alpha\gamma} = 0\,,$$

We could evidently replace $G_{\alpha\beta}$ by $G^*_{\alpha\beta}$ in (3):

(3*) $$G^*_{\alpha\beta} Z^\alpha Z^\beta = 0\,.$$

$G^*_{\alpha\beta}$ obviously also obey the transformation law (4).

Because $G^*_{\alpha\beta}$ has been chosen such that (8) remains valid in all coordinate systems it follows that the quantity λ must be constant in (4). By setting $q = \bar{q}$ in (4) we find the value of unity for λ. The asterisk is now deleted in the following.

We therefore have as transformation law for $G_{\alpha\beta}$

(4′) $$\overline{G}_{\alpha\beta}(\bar{q}) = G_{\sigma\tau}\frac{\partial q^\sigma}{\partial \bar{q}^\alpha}\frac{\partial q^\tau}{\partial \bar{q}^\beta}\,.$$

Calculation of $G_{\alpha\beta}$ from Φ.

From equations (8) with the conditions $\Pi^\sigma_{\alpha 0} = \delta^\sigma_\alpha$ there now follows the invariant equations

(9) $$\frac{\partial G_{\alpha\beta}}{\partial q^0} = 2G_{\alpha\beta}\,,$$

(10) $$\frac{\partial G_{00}}{\partial q^\gamma} = 2G_{0\gamma}$$

and

(11) $$\frac{\partial G_{\alpha 0}}{\partial q^\gamma} - G_{\alpha 0}\Pi^\sigma_{\alpha\gamma} = G_{\alpha\gamma}\,.$$

From the first equation it follows that $G_{\alpha\beta}$ has the form

$$G_{\alpha\beta} = e^{2q^0} f_{\alpha\beta}(q) .$$

Taking (4) in addition it follows that $G_{\alpha\beta}$ is a projective second rank tensor of index 2. The theory of a quadric is therefore contained in the theory of a second-rank tensor that obeys equation (8). At the same time Π is an integrable projective connection. The quantities Π relate to the existence of a projective space in the same sense as before. The $G_{\alpha\beta}$ can then be estimated by integration of equations (8).

From the equations (10) and (11) we see that the whole theory of our tensors depends on the scalar

$$G_{00} = \Phi^2 .$$

According to equation (10) that is to say

$$\frac{G_{0\alpha}}{G_{00}} = \varphi_\alpha = \frac{\partial \log \Phi}{\partial q^\alpha} . \tag{12}$$

Equation (11) has the same meaning as

$$\frac{\partial\left(\Phi^2 \varphi_\alpha\right)}{\partial q^\beta} - \Phi^2 \varphi_\sigma \Pi^\sigma_{\alpha\beta} = \Phi^2 \gamma_{\alpha\beta}$$

or as
$$\frac{\partial \varphi_\alpha}{\partial q^\beta} - \varphi_\sigma \Pi^\sigma_{\alpha\beta} = \gamma_{\alpha\beta} - 2\varphi_\alpha \varphi_\beta .$$

Hence
$$g_{ij} = \frac{\partial \varphi_i}{\partial q^i} - \varphi_\sigma \Pi^\sigma_{ij} + \varphi_i \varphi_j$$

or from (12)

$$g_{ij} = \frac{\frac{\partial^2 \Phi}{\partial q^i \partial q^j} - \frac{\partial \Phi}{\partial q^\sigma} \Pi^\sigma_{ij}}{\Phi} . \tag{13}$$

One therefore obtains φ_α and g_{ij} from Φ with the help of (12) and (13) and by using
$$\gamma_{\alpha\beta} = g_{\alpha\beta} + \varphi_\alpha \varphi_\beta$$
and
$$G_{\alpha\beta} = \Phi^2 \gamma_{\alpha\beta}$$
also $\gamma_{\alpha\beta}$ and $G_{\alpha\beta}$. By specific choice of the origin point of the coordinate system and the gauge a homogeneous projective coordinate system is completely specified. In this coordinate system the surface has the equation

$$G_{\alpha\beta} Z^\alpha Z^\beta = 0 ,$$

where the coefficients $G_{\alpha\beta}$ are the values of the solutions to (9), (10) and (11) in the point q.

(37)	The distinguished Gauge.

The Equation of the Surface in Inhomogeneous Projective Coordinates.

We now take an inhomogeneous projective coordinate system and such a gauge that $\Pi^{\alpha}_{\beta\gamma}$ takes the value

$$\Pi^{\alpha}_{\beta\gamma} = \delta^{\alpha}_{\beta}\delta^{0}_{\gamma} + \delta^{\alpha}_{\gamma} - \delta^{\alpha}_{0}\delta^{0}_{\beta}\delta^{0}_{\gamma}.$$

It then follows from (8), (10), (11) that

$$\begin{aligned} \frac{\partial G_{ij}}{\partial q^k} &= 0, \\ \frac{\partial G_{0i}}{\partial q^j} &= G_{ij}, \\ \frac{\partial G_{00}}{\partial q^i} &= 2G_{0i}. \end{aligned}$$

By integration we then find

$$G_{ij} = e^{2q^0} a_{ij}, \tag{14a}$$

$$G_{0j} = e^{2q^0}\left(a_{ij}q^i + a_{j0}\right), \tag{14b}$$

$$G_{00} = e^{2q^0}\left(a_{ij}q^i g^j + 2a_{i0}q^i + a_{00}\right), \tag{14c}$$

where the $a_{\alpha\beta}$ are constants.

The equation of the quadric is

$$0 = G_{\alpha\beta}Z^{\alpha}Z^{\beta}$$

or by using (14) and chap. III (17)

$$0 = e^{2(x^0-q^0)}\left(G_{ij}(x^i-q^i)(x^j-q^j) + 2G_{i0}(x^i-q^i) + G_{00}\right)$$

and eventually $\qquad 0 = e^{2x^0}\left(a_{ij}x^i x^j + 2a_{i0}x^i + a_{00}\right).$

This way we have established that the surface is described by a second-degree equation with constant coefficients in each homogeneous coordinate system.

The Distinguished Gauge.

As we saw in chap. II the existence of a projective scalar distinguishes a special gauge. In our case we can establish by a similar gauge that the scalar Φ has the form

$$\Phi = e^{x^0}\varrho(x). \tag{15}$$

If Φ in an arbitrary gauge has the form $\Phi = e^{x^0}\varrho(x)$, we only need to perform the transformation

$$\bar{x}^0 = x^0 + \log\varrho = \log\Phi.$$

The effect of (12) on the vector φ is that $\varphi_i = 0$. Because no other gauge transformations, that leave (15) invariant, exist, we no longer have a projective geometry, but only an affine geometry. Because of the tensor g_{ij} this affine geometry is metrical.

With respect to our distinguished gauge is

$$G_{ij} = e^{2q^0} g_{ij}\,, \tag{16a}$$

$$G_{0\alpha} = e^{2q^0} \delta_0^\alpha\,, \tag{16b}$$

and hence the equation of our quadric is

$$e^{2q^0}\left(g_{ij}Z^iZ^j + Z^0Z^0\right) = 0\,. \tag{17}$$

Furthermore, from (8) and (16) follow the equations

$$\frac{\partial g_{ij}}{\partial x^k} - g_{is}\Pi^s_{jk} - g_{js}\Pi^s_{ik} = 0\,.$$

If we solve these equations in the usual way, we obtain

$$\Pi^i_{jk} = \left\{ \begin{matrix} i \\ jk \end{matrix} \right\} = \frac{1}{2}\left(\frac{\partial g_{js}}{\partial x^k} + \frac{\partial g_{ks}}{\partial x^j} - \frac{\partial g_{jk}}{\partial x^s}\right). \tag{18}$$

Using (15) it follows from (13) that

$$\Pi^0_{jk} = -g_{jk}\,. \tag{19}$$

It is clear from (18) and (19) that Π^i_{jk} is an affine connection and Π^0_{jk} an affine tensor. These are simply special applications of the formulae (10a) and (10b) of chap. III.

Caley's Geometry.

In an underlying space the tensor g_{ij} gives a Riemannian metric. Since this metric is invariantly connected with our quadric we suspect that the Riemannian metric is exactly the non-Euclidean metric that appears as absolute image of our quadric.

A non-Euclidean or Caleyan metric is readily defined by means of a "tangential" Euclidean metric. The general concept of a tangential metric according to E. Cartan (Bibl. 1928, 1, chap.IV) is as follows: Two types of measurement with the same g_{ij} in a given point are called tangential in the point concerned.

In each point q of our space there exists a Euclidean metric with respect to which the quadric is exactly a sphere with radius 1 and centre q. The infinitely remote hyperplane of this Euclidean space is the polar hyperplane of q with respect to the quadric.

(39)	Cayley's Geometry.

The non-Euclidean infinitesimal distance in q must by definition correspond to the infinitesimal distance of the Euclidean metric.

We now seek the analytical expression for our Euclidean metric.

In the inhomogeneous projective coordinate system defined by

$$\frac{Z^i}{Z^0} = z^i$$

the equation of our quadric is

$$g_{ij}(q)z^i z^j + 1 = 0\,. \tag{20}$$

With respect to the Euclidean metric with arc element

$$ds^2 = -g_{ij}dz^i dz^j \tag{21}$$

(20) is the equation of a space with radius 1 and centre q. Hence (21) is also the Cayleyan metric in point q, given by the quadric.

To relate this definition to that given by Cayley himself we calculate the Cayleyan distance of point Z from O. The straight line through O intersects the sphere in A and B. The cross ratio of the four points O, Z, A and B is

$$\alpha = \frac{OA}{OB} : \frac{ZA}{ZB} = \frac{1 + \sqrt{-g_{ij}z^i z^j}}{1 - \sqrt{-g_{ij}z^i z^j}}.$$

The Cayley distance from O to Z then is $m \log \alpha$, where m is a constant. By crossing the boundary it follows that

$$ds = 2m\sqrt{-g_{ij}dz^i dz^j}\,.$$

The Cayley distance therefore corresponds, within a constant, to

$$ds^2 = -g_{ij}dz^i dz^j\,.$$

Obviously all formulae fail for points in the absolute surface

$$\Phi = 0.$$

On these grounds Herr Whitehead (Bibl. 1931, 15) in pioneering study of these matters proposed the term "missing $(n-1)$ space" for this surface. This transition from projective to affine geometry can evidently be achieved by fixing a gauge if the missing plane is not of second order. Further development of these ideas are found in Whitehead's work. It is quite possible that other interesting geometries may occur in this direction.

VI. Generalized Theory of Conic Sections

Metrical Part of the Geometry.

The theory of a general projective second rank tensor of index 2 may be considered as a generalization of the non-Euclidean geometry considered in the previous chapter. A second-rank tensor produces a metrical geometry, *viz.* the Riemannian geometry given by the affine tensor g_{ij}. The g_{ij} is defined by the equation

$$G_{\alpha\beta} = \Phi^2 \left(g_{\alpha\beta} + \varphi_\alpha \varphi_\beta\right) . \tag{1}$$

However, our geometry is not only a metrical geometry. It contains also many elements which are not metrical.

In each tangent space we have a Cayley metric associated with the quadratic form

$$G_{\alpha\beta} X^\alpha X^\beta .$$

So all the known laws and formulae of Caylean geometry are valid in tangent space. As an example we have the formula for the distance between points of tangent space

$$\cos id = \frac{G_{\alpha\beta} X^\alpha Y^\beta}{\sqrt{G_{\alpha\beta} X^\alpha X^\beta G_{\sigma\tau} Y^\sigma Y^\tau}} . \tag{2}$$

The affine tensor g_{ij} is unambiguously determined by the projective tensor $G_{\alpha\beta}$ and g_{ij} fixes a Riemannian metric with element of arc

$$ds = \sqrt{-g_{ij} dx^i dx^j} . \tag{3}$$

Each Riemann metric is known to produce a Euclidean metric in each tangent space. With respect to the mode of measurement (3), the surface

$$G_{\alpha\beta} X^\alpha X^\beta = 0$$

now occurs exactly as a sphere of unit radius, centred at the contact point. As we have seen in the previous chapter the infinitesimal Cayley distance corresponds to that given by (3).

The infinitesimal environment of the origin of the Cayley space can obviously influence the metric of the underlying space. Hence the influence of the Cayley metric of the tangent space on the formula for the arc length

$$\int ds$$

perhaps is exactly the same as the influence of the tangential metric at different points along the curve. According to a proposal of Herr Howe we could describe the Euclidean metric given in a point by (3) as the metric of the tangent space of the Cayleyan tangent space, such that both tangent spaces have the same contact point with the underlying space.

By limiting ourselves to the metric point of view we find only one Riemann space with curvature element

$$ds^2 = -g_{ij}dx^i dx^j \,. \tag{3}$$

However, this curvature element does not contain the entire influence of the Cayley space on the underlying space, because φ and Φ do not feature in (3). The non-metrical properties first come to light when one examines the associated projective displacements Γ and Π as we started to do in chap. IV.

In particular we shall find systems of curves invariantly related to these displacements. However, to describe these curves conveniently we must develop our formalism somewhat further.

Invariants of g_{ij}.

A list of the invariants of $G_{\alpha\beta}$ must specifically contain the invariants of the affine tensor g_{ij} and of the projective vector φ_α. The tensor g_{ij} contains several known invariants or associated geometrical objects.

Next the determinant

$$g = |g_{ij}|$$

is a relative scalar[5] of weight 2. The transformation law is

$$\bar{g} = g \left| \frac{\partial x}{\partial \bar{x}} \right|^2 ,$$

where $|\partial x/\partial \bar{x}|$ is the functional determinant of the coordinate transformation $x \to \bar{x}$.

Further we have to mention the contravariant tensor g^{ij} which is determined by the relation

$$g_{ij}g^{ik} = \delta^k_j \,. \tag{4}$$

Furthermore we gain the components of an affine connection from the Christoffel formulae

$$\left\{ \begin{matrix} i \\ jk \end{matrix} \right\} = \frac{1}{2} g^{ia} \left(\frac{\partial g_{aj}}{\partial x^k} + \frac{\partial g_{ak}}{\partial x^j} - \frac{\partial g_{jk}}{\partial x^a} \right) . \tag{5}$$

[5] What I call "relative scalars" are often described as "scalar densities". I might however, reserve this name for the relative scalars of weight 1, as, for instance, $\sqrt{g}$, because physical densities are always of unit weight.

We can covariantly differentiate all affine tensors with respect to this affine connection, for example, the covariant affine derivative of a mixed tensor φ^i_j is

$$\varphi^i_{j,k} = \frac{\partial \varphi^i_j}{\partial x^k} + \varphi^s_j \left\{ \begin{matrix} i \\ sk \end{matrix} \right\} - \varphi^i_s \left\{ \begin{matrix} s \\ jk \end{matrix} \right\} .$$

We denote the covariant affine derivative by a comma (,). The condition

$$g_{ij,k} = 0$$

is known from(5) to apply for the tensor g_{ij}.

One obtains the Riemann tensor fundamental to the theory of the tensor g_{ij}:

$$(6) \quad R^i_{jkl} = \frac{\partial}{\partial x^l} \left\{ \begin{matrix} i \\ jk \end{matrix} \right\} - \frac{\partial}{\partial x^k} \left\{ \begin{matrix} i \\ jl \end{matrix} \right\} - \left\{ \begin{matrix} i \\ rk \end{matrix} \right\} \left\{ \begin{matrix} r \\ jl \end{matrix} \right\} + \left\{ \begin{matrix} i \\ rl \end{matrix} \right\} \left\{ \begin{matrix} r \\ jk \end{matrix} \right\} ,$$

in the normal way from the Christoffel symbols. Contraction of the curvature tensor leads to the Ricci tensor

$$(7) \quad R_{ij} = R^k_{ikj}$$

and from the Riemann tensor, raised with g_{ij}, the curvature scalar

$$R = g^{ij} R_{ij} .$$

An endless number of new invariants can be derived by the process of covariant differentiation (compare *e.g.* Bibl. 1927, 22).

Affine invariants of φ_α.

A further affine invariant is established by the projective vector φ_α.

$$(8) \quad \frac{1}{2} \left(\frac{\partial \varphi_i}{\partial x^j} - \frac{\partial \varphi_j}{\partial x^i} \right) = \varphi_{ij}$$

must be an affine tensor, because

$$\frac{\partial \varphi_\alpha}{\partial x^\beta} - \frac{\partial \varphi_\beta}{\partial x^\alpha}$$

vanishes whenever α or β assumes the value 0. The important vector φ_{ij} is central in electromagnetic theory.

We agree that Latin indices are raised and lowered by g^{ij} and g_{ij}, while the same process is performed by $\gamma^{\alpha\beta}$ and $\gamma_{\alpha\beta}$ for Greek indices, as mentioned before. So for instance is

$$(9) \quad \varphi^i_j = g^{ik} \varphi_{kj} .$$

To establish our arrangement unambiguously we stipulate that the first index will always be raised when there are more than one lower index. The sign of

φ_{ij} is determined by this stipulation.

Because φ_i and $\varphi_j g^{ij}$ are not affine tensors we cannot raise the Latin index φ_i by g^{ij}. Rather, by our agreement, is

(10) $$\varphi^\alpha = \gamma^{\alpha\beta}\varphi_\beta = \delta_0^\alpha \,.$$

Hence, for the proper φ^i follows

$$\varphi^i = 0 \,.$$

Invariants of $\gamma_{\alpha\beta}$.

We can derive a row of invariants of $\gamma_{\alpha\beta}$ in a way that agrees formally with that utilized for g_{ij}. First of all the determinant

(11) $$\gamma = |\gamma_{\alpha\beta}| = \begin{vmatrix} 1 & \varphi_1 & \varphi_2 & \varphi_3 & \varphi_4 \\ \varphi^1 & & & & \\ \varphi_2 & (g_{ij} & + & \varphi_i\varphi_j) & \\ \varphi_3 & & & & \\ \varphi_4 & & & & \end{vmatrix} = g \,.$$

The defining equation of $\gamma^{\alpha\beta}$

(12) $$\gamma^{\alpha\sigma}\gamma_{\sigma\beta} = \delta_\beta^\alpha$$

has been given before.

The tensor $\gamma^{\alpha\beta}$ is simply related to g^{ij} and φ_i. Firstly from (12) is

$$\delta_k^i = \gamma^{\alpha i}\gamma_{\alpha k} = \gamma^{\alpha i}(g_{\alpha k} + \varphi_\alpha\varphi_k)$$

or due to (10) and $g_{0k} = 0$: $\delta_k^i = \gamma^{ji} g_{jk}$

or finally because of (4)

(13) $$\gamma^{ij} = g^{ij}.$$

Likewise we could calculate γ^{0i} from g^{ij} and φ_j:

(14) $$\gamma^{0i} = -g^{ij}\varphi_j \,.$$

(14) follows directly from $\delta_0^i = 0 = \gamma^{\alpha i}\gamma_{\alpha 0} = g^{ij}\varphi_j + \gamma^{0i}$.

Eventually we calculate γ^{00} as well. It is

$$1 = \gamma^{\alpha 0}\gamma_{\alpha 0} = -g^{ij}\varphi_j\varphi_i + \gamma^{00} \quad \text{or}$$

(15) $$\gamma^{00} = 1 + g^{ij}\varphi_i\varphi_j \,.$$

The Connection Γ

We have introduced the connection Γ before with the formulae

(16) $$\gamma_{\alpha\beta;\gamma} = 0$$

and

(17) $$\Gamma^{\alpha}_{\beta\gamma} = \frac{1}{2}\gamma^{\alpha\sigma}\left(\frac{\partial\gamma_{\beta\sigma}}{\partial x^{\gamma}} + \frac{\partial\gamma_{\sigma\gamma}}{\partial x^{\beta}} - \frac{\partial\gamma_{\beta\gamma}}{\partial x^{\sigma}}\right).$$

We write (17) in somewhat more detail!

(18) $$\begin{cases} \Gamma^{\alpha}_{\beta\gamma} & = \frac{1}{2}\gamma^{\alpha i}\left(\frac{\partial g_{i\beta}}{\partial x^{\gamma}} + \frac{\partial\gamma_{i\gamma}}{\partial x^{\beta}} - \frac{\partial g_{\beta\gamma}}{\partial x^{i}}\right) + \frac{1}{2}\left(\frac{\partial\varphi_{\beta}}{\partial x^{\gamma}} + \frac{\partial\varphi_{\gamma}}{\partial x^{\beta}}\right) \\ & +\gamma^{\alpha i}\left(\varphi_{i\gamma}\varphi_{\beta} + \varphi_{i\beta}\varphi_{\gamma}\right). \end{cases}$$

That implies

(19) $$\Gamma^{\alpha}_{00} = 0\,,$$

(20) $$\Gamma^{\alpha}_{\beta 0} = \gamma^{i\alpha}\varphi_{i\beta}\,,$$

(21) $$\Gamma^{i}_{jk} = \Gamma\left\{ \begin{matrix} i \\ jk \end{matrix} \right\} + \varphi^{i}_{j}\varphi_{k} + \varphi^{i}_{k}\varphi_{j}\,,$$

(22) $$\Gamma^{0}_{jk} = -\varphi_{i}\Gamma^{i}_{jk} + \frac{1}{2}\left(\frac{\partial\varphi_{j}}{\partial x^{k}} + \frac{\partial\varphi_{k}}{\partial x^{j}}\right).$$

Because of (14) equation (20) splits into

$$\Gamma^{i}_{j0} = \varphi^{i}_{j} \qquad \text{and} \qquad \Gamma^{0}_{j0} = -\varphi_{i}\varphi^{i}_{j}\,.$$

In closing we mention again the formula derived in chap. IV

(23) For $\beta = 0$ this means $$\varphi_{\alpha;\beta} = \varphi_{\alpha\beta}\,.$$

(23a) $$\varphi_{\alpha}\Gamma^{\sigma}_{\alpha 0} = 0\,.$$

The Curvature Tensor from Γ

As in affine theory one can also define a curvature tensor $B^{\alpha}_{\beta\gamma\delta}$ for the connection Γ. By contraction one obtains an analog of the Ricci tensor:

(24) $$\begin{cases} B^{\sigma}_{\alpha\sigma\beta} & = \frac{\partial\Gamma^{\sigma}_{\alpha\gamma}}{\partial x^{\beta}} - \frac{\partial\Gamma^{\sigma}_{\alpha\beta}}{\partial x^{\sigma}} + \Gamma^{\varepsilon}_{\alpha\sigma}\Gamma^{\sigma}_{\varepsilon\beta} - \Gamma^{\varepsilon}_{\alpha\beta}\Gamma^{\alpha}_{\varepsilon\sigma} \\ & = R_{ij}\delta^{i}_{\alpha}\delta^{j}_{\beta} - \varphi^{s}_{i,s}\left(\delta^{i}_{\alpha}\varphi_{\beta} + \delta^{i}_{\beta}\varphi_{\alpha}\right) + 2\varphi^{s}_{\beta}\varphi_{s\alpha} \\ & \qquad + \varphi^{s}_{t}\varphi^{t}_{s}\varphi_{\alpha}\varphi_{\beta} \\ & = B_{\alpha\beta}. \end{cases}$$

(45)	Invariance of the Non-Euclidean Distance.

We can then also derive from $B_{\alpha\beta}$ a scalar that corresponds to the curvature scalar:

$$B = \gamma^{\alpha\beta} B_{\alpha\beta} = R - \varphi_t^s \varphi_s^t \,. \tag{25}$$

The Projective Displacement Λ.

A projective displacement is associated with the connection Γ for each index through the differential equations

$$X^\alpha_{\ ;\beta} = X^\alpha_{\ ;0} \varphi_\beta$$

as was shown in chap. IV, p.31, 32. The tangent space is displaced along the curve $x^i = x^i(t)$ by the amount $\quad \dfrac{dX^\alpha}{dt} + \Lambda^\alpha_{\sigma j} X^\sigma \dfrac{dx^j}{dt} = 0\,.$

At the same time

$$\Lambda^\alpha_{\sigma\tau} = \Gamma^\alpha_{\sigma\tau} - \Gamma^\alpha_{\sigma 0}\varphi_\tau + M\delta^\alpha_\sigma \varphi_\tau \,. \tag{26}$$

Invariance of the Non-Euclidean Distance.

The displacement Λ satisfies the equations

$$\frac{\partial \gamma_{\alpha\beta}}{\partial x^\gamma} - \gamma_{\alpha\sigma}\Lambda^\sigma_{\beta\gamma} - \gamma_{\alpha\beta}\Lambda^\sigma_{\alpha\gamma} = -2M\gamma_{\alpha\beta}\varphi_\gamma \tag{27}$$

and, easy to check:

$$\frac{\partial \varphi_\alpha}{\partial x^\gamma} - \varphi_\alpha \Lambda^\sigma_{\alpha\gamma} = \varphi_{\alpha\gamma} - M\varphi_\alpha\varphi_\gamma \,. \tag{28}$$

An application of (27) is the proposition that the non-Euclidean distance of two points X^α and X^β of a tangent space is conserved on displacement Λ. The non-Euclidean distance of both points is determined by

$$D = \frac{\left(G_{\alpha\beta}X^\alpha Y^\beta\right)^2}{\left(G_{\alpha\beta}X^\alpha X^\beta\right)\left(G_{\alpha\beta}Y^\alpha Y^\beta\right)} = \frac{\left(\gamma_{\alpha\beta}X^\alpha Y^\beta\right)^2}{\left(\gamma_{\alpha\beta}X^\alpha X^\beta\right)\left(\gamma_{\alpha\beta}Y^\alpha Y^\beta\right)}$$

We differentiate $\gamma_{\alpha\beta}X^\alpha Y^\beta$ with respect to x^γ:

$$\frac{\partial}{\partial x^\gamma}\gamma_{\alpha\beta}X^\alpha Y^\beta = \left(\frac{\partial \gamma_{\alpha\beta}}{\partial x^\gamma} - \gamma_{\alpha\sigma}\Lambda^\sigma_{\beta\gamma} - \gamma_{\alpha\beta}\Lambda^\sigma_{\alpha\gamma}\right) X^\alpha Y^\beta \,.$$

Using (27) we obtain from this

$$\frac{\partial}{\partial x^\gamma}\gamma_{\alpha\beta}X^\alpha Y^\beta = -2M\gamma_{\alpha\beta}X^\alpha Y^\beta \varphi_\gamma$$

or

$$\frac{\partial \log \gamma_{\alpha\beta}X^{\alpha Y^\beta}}{\partial x^\gamma} = -2M\varphi_\gamma .$$

As the right side of this equation is independent of X and Y we obtain in fact

$$\frac{\partial \log D}{\partial x^\gamma} = 0 .$$

On displacement Λ each figure of a tangent space therefore goes over to an equivalent figure of another tangent space.

Displacement in Inhomogeneous Coordinates.

In chap. IV, p. 29 we derived the equation

$$(29) \quad \left\{ \begin{array}{l} \frac{dV^i}{dt} + \Lambda^i_{kj} V^k \frac{dx^i}{dt} + \Lambda^i_{0k} \frac{dx^k}{dt} (1 - \varphi_j V^j) + \varphi_{k:j} V^k V^i \frac{dx^j}{dt} \\ + \varphi_{0:j} \left(1 - \varphi_k V^k\right) \frac{dx^j}{dt} V^i = 0 . \end{array} \right.$$

The Π of chap. IV (14) is here replaced by Λ and the projective derivative of φ_α with respect to Λ is here denoted by $\varphi_{\alpha:\beta}$. Using (26) there follows from this

$$(30) \quad \left\{ \begin{array}{l} \frac{dV^i}{dt} + \Gamma^i_{kj} V^k \frac{dx^j}{dt} - \Gamma^i_{k0} V^k \varphi_j) \frac{dx^j}{dt} + M V^i \varphi_j \frac{dx^j}{dt} + (1 - \varphi_j V^j) \Gamma^i_{0k} \frac{dx^k}{dt} \\ + \varphi_{k:j} V^k V^i \frac{dx^j}{dt} + \varphi_{0:j} (1 - \varphi_k V^k) \frac{dx^j}{dt} V^i = 0 . \end{array} \right.$$

By replacing $\varphi_{k:j}$ and $\varphi_{0:j}$ in (30) by their values from (28) the equations for the displacement along a curve assume the form:

$$(31) \qquad \frac{dV^i}{dt} + \left\{ \begin{array}{c} i \\ kj \end{array} \right\} V^k \frac{dx^j}{dt} + \varphi_{kj} V^k \frac{dx^j}{dt} V^i + \varphi^i_j \frac{dx^j}{dt} = 0$$

writing in inhomogeneous style.

The World Line of an Electrical Particle.

On the basis of (31) we can now define a system of curves which may be considered as a generalization of the geodesic lines of a Riemann space.

A curve given in the parameter representation

$$x^i = x^i(t)$$

identifies in each of its points a point of the associated tangent space through the equation

$$(32) \qquad k \frac{dx^i}{dt} = V^i ,$$

in which k is an arbitrary constant.

We now look for curves along which the points V from (32) are displaced into themselves. For that purpose we substitute (32) into (31). Because of the skew symmetry of φ_{ij} this takes (31) into

$$(33) \qquad \frac{d^2x^i}{dt^2} + \left\{ \begin{array}{c} i \\ jk \end{array} \right\} \frac{dx^j}{dt} + \frac{1}{k} \varphi^i_j \frac{dx^j}{dt} = 0 .$$

(47)	The Connection Π.

Equations (33) specify conditions for both the parameter representation of the curve and also for the curve itself. If

$$\frac{1}{k} = 0 \tag{34}$$

the curves (33) are just the geodesic lines of the metric, characterized by g_{ij}.

If (34) is not satisfied the differential equations are not homogeneous in t. By fixing k, each point with given velocity vector completely determines a curve. Only for the geodesic lines are the curves conditioned by the initial point.

On replacing $1/k$ in (33) by e/m we obtain

$$\frac{d^2x^i}{dt^2} + \left\{ \begin{matrix} i \\ jk \end{matrix} \right\} \frac{dx^j}{dt}\frac{dx^k}{dt} + \frac{e}{m}\varphi^i_j \frac{dx^j}{dt} = 0\,. \tag{33a}$$

In the framework of general relativity we have to interpret these equations as equations of motion of an electrical particle. e and m are the charge and mass of the particle while g_{ij} and the φ_i are the gravitational and electromagnetic potentials of the field in which the particle moves.

The Connnection Π.

We shall also consider the connection Π defined in chap. IV. We have seen that Π could be calculated from Γ and Φ. Like we referred the $\Gamma^\alpha_{\beta\gamma}$ back to $\left\{ \begin{matrix} i \\ jk \end{matrix} \right\}$, φ_k, φ_{ij}, *etc.* in equations (19), (20), (21) and (22), we shall now express $\Pi^\alpha_{\beta\gamma}$ by $\Gamma^\alpha_{\beta\gamma}$, φ and another quantity Θ, to be defined.

We set
$$\Phi_\alpha = \frac{\partial \log \Phi}{\partial x^\alpha} \qquad \text{and}$$

$$\Theta_\alpha = N\varphi_\alpha - \Phi_\alpha\,. \tag{35}$$

Since $\Phi_0 = N$ so is $\Theta_0 = 0$. Θ_i therefore is an affine tensor. We now express Φ^α by the quantity $\Theta^i = g^{ij}\Theta_j$.

Firstly
$$\Phi^\sigma = \gamma^{\sigma\alpha}\Phi_\alpha = N\gamma^{\sigma 0} + \gamma^{\sigma j}\Phi_j\,.$$

From this follows, using (13), (14) and (15)

$$\Phi^i = N\gamma^{i0} + \gamma^{ij}\Phi_j = -g^{ij}(N\varphi_j - \Phi_j) = -\Theta^i \tag{36}$$

and
$$\Phi^0 = N\gamma^{00} + \gamma^{0j}\Phi_j = N - g^{ij}\varphi_i\Phi_j + Ng^{ij}\varphi_i\varphi_j$$

or

$$\Phi^0 = N + \varphi_i\Theta^i\,. \tag{37}$$

In a very simple manner $\Phi_\alpha\Phi^\alpha$ may be calculated from Θ:

$$\Phi_\alpha\Phi^\alpha = \left(N + \varphi_i\Theta^i\right) N + \Theta^i \left(\Theta^i - N\varphi_i\right) = N^2 + \Theta_i\Theta^i.$$

We now obtain the following formulae for $\Pi^\alpha_{\beta\gamma}$ from (35), (36) and (37) and from (22) chap. IV (p. 32):

(38a) $$\Pi^0_{00} = \Gamma^0_{00} + N - \varphi_i\Theta^i,$$

(38b) $$\Pi^i_{00} = \Gamma^i_{00} + \Theta^i,$$

(38c) $$\Pi^0_{j0} = \Gamma^0_{j0} - \Theta_j - \varphi_i\varphi_j\Theta^i,$$

(38d) $$\Pi^i_{j0} = \Gamma^i_{j0} + N\delta^i_j + \Theta^i\varphi_j,$$

(38e) $$\Pi^0_{jk} = \Gamma^0_{jk} - (g_{jk} + \varphi_j\varphi_k)(N + \varphi_i\Theta^i),$$

(38f) $$\Pi^i_{jk} = \Gamma^i_{jk} + \delta^i_j\Phi_k + \delta^i_k\Phi_j + \Theta^i(g_{jk} + \varphi\varphi_k).$$

The Covariant Derivative of Φ with respect to Γ.

With the help of the newly defined quantity Θ we can also write a row of simple formulae for the covariant derivative of Φ with respect to Γ. However, we shall not show all the elementary steps in all detail.

For the covariant derivative of Φ we have

$$\Phi_{\alpha;\beta} = \frac{\partial \Phi_\alpha}{\partial x^\beta} - \Gamma^\sigma_{\alpha\beta}\Phi_\sigma\,.$$

Especially from (19) is $\Phi_{0;0} = 0\,.$

From (20), (21), (36) and (37) we obtain

(39) $$\Phi^\alpha{}_{;\alpha} = \frac{\partial \Phi^\alpha}{\partial x^\alpha} + \Gamma^\alpha_{\sigma\alpha}\Phi^\sigma = -\frac{\partial \Theta^i}{\partial x^i} - \left\{ {j \atop ji} \right\}\Theta^i$$

or $$\Phi^\alpha{}_{;\alpha} = -\Theta^i{}_{,i} = \frac{1}{\sqrt{g}}\frac{\partial(\Theta^i\sqrt{g})}{\partial x^i}\,.$$

In this we denoted the (affine) covariant derivative with respect to $\left\{ {i \atop jk} \right\}$ with a comma (,). Finally we find

(40) $$\Phi^i{}_{;0} = -\varphi^i_j\Theta^j$$

and

(41) $$\gamma^{\sigma j}\Phi^i{}_{;\sigma} = g^{ja}(N\varphi^i_a - \Theta^i{}_{,a}) = \tfrac{1}{2}g^{ia}g^{jb}(\Theta_{a,b} + \Theta_{b,a})\,.$$

In this we have referred the covariant derivative of Φ with respect to Γ, to a certain extent, to the covariant derivative with respect to $\left\{ {i \atop jk} \right\}$.

(49)	The Displacement connected with Π.

The Curvature Tensor of Π.

We now return to the connection Π and calculate the curvature tensor constructed from Π. To express $P^{\alpha}_{\beta\gamma\delta}$ conveniently in terms of Γ and Φ we define a new quantity $T^{\alpha}_{\beta\gamma}$ by the formula

$$\Pi^{\alpha}_{\beta\gamma}-\Gamma^{\alpha}_{\beta\gamma}=T^{\alpha}_{\beta\gamma}\,. \tag{42}$$

From chap. IV T has the value

$$T^{\alpha}_{\beta\gamma}=\left(\delta^{\alpha}_{\beta}\Phi_{\gamma}+\delta^{\alpha}_{\gamma}\Phi_{\beta}-\gamma_{\beta\gamma}\Phi^{\alpha}\right). \tag{43}$$

By using (42) we now obtain the equation for $P^{\alpha}_{\beta\gamma\delta}$

$$P^{\alpha}_{\beta\gamma\delta}=B^{\alpha}_{\beta\gamma\delta}+T^{\alpha}_{\beta\gamma;\delta}-T^{\alpha}_{\beta\delta;\gamma}+T^{\sigma}_{\beta\gamma}T^{\alpha}_{\sigma\delta}-T^{\sigma}_{\beta\delta}T^{\alpha}_{\sigma\gamma} \quad \text{or} \tag{44}$$

$$\left\{\begin{aligned} P^{\alpha}_{\beta\gamma\delta}=B^{\alpha}_{\beta\gamma\delta} &+\delta^{\alpha}_{\gamma}(\Phi_{\beta;\delta}-\Phi_{\beta}\Phi_{\delta}+\Phi^{\sigma}\Phi_{\sigma}\gamma_{\beta\delta})\\ &-\delta^{\alpha}_{\delta}(\Phi_{\beta;\gamma}-\Phi_{\beta}\Phi_{\gamma}+\Phi^{\sigma}\Phi_{\sigma}\gamma_{\beta\gamma})\\ &-\gamma_{\beta\gamma}(\Phi^{\alpha}{}_{;\delta}-\Phi^{\alpha}\Phi_{\delta})\\ &+\gamma_{\beta\delta}(\Phi^{\alpha}{}_{;\gamma}-\Phi^{\alpha}\Phi_{\gamma}). \end{aligned}\right. \tag{45}$$

By contraction we further obtain a tensor $P_{\beta\delta}$ and a scalar P, that correspond to the quantities $B_{\alpha\beta}$ and B in the formulae (24) and (25):

$$P_{\beta\delta}=B_{\beta\delta}+(n-1)(\Phi_{\beta;\delta}-\Phi_{\beta}\Phi_{\delta}+\Phi^{\sigma}\Phi_{\sigma}\gamma_{\beta\delta})+\gamma_{\beta\delta}\Phi^{\sigma}{}_{;\sigma} \quad \text{and} \tag{46}$$

$$P=B-2n\Theta^{i}{}_{;i}+n(n-1)(N^2+\Theta^{i}\Theta_{i}). \tag{47}$$

From (46) we finally obtain the affine invariants

$$\begin{aligned} P^{ij}&=\gamma^{i\alpha}\gamma^{j\beta}P_{\alpha\beta}=B^{ij}+(n-1)\left[g^{ij}(N^2+\Theta^{p}\Theta_{p})-\Theta^{i}\Theta^{j}\right],\\ P^{i}_{0}&=\gamma^{i\alpha}P_{\alpha 0}=B^{i}_{0}+(n-1)[N\Theta^{i}-\varphi^{i}_{j}\Theta^{j}],\\ P_{00}&=B_{00}+(n-1)\Theta_{i}\Theta^{i}-\Theta^{i}{}_{,i}\,. \end{aligned}$$

The Displacement connected with Π.

Totally analogous to the invariant displacement Λ connected with Γ a displacement also arises from Π through the equations

$$X_{|j}{}^{\alpha}=X_{|0}{}^{\alpha}\varphi_j \quad \text{or} \tag{48}$$

$$\frac{\partial X^{\alpha}}{\partial x^{j}}+\Pi^{\alpha}_{\sigma j}X^{\sigma}=(-MX^{\alpha}+\Pi^{\alpha}_{\sigma 0}X^{\sigma})\varphi_j\,. \tag{49}$$

The equations (48) correspond to equations (24) of chap. IV. If we now write

$$\begin{aligned} \Sigma^{\alpha}_{\beta\gamma} &= \Pi^{\alpha}_{\beta\gamma}-\Pi^{\alpha}_{\beta 0}\varphi_{\gamma}+M\delta^{\alpha}_{\beta}\varphi_{\gamma}\\ &= \Lambda^{\alpha}_{\beta\gamma}+\delta^{\alpha}_{\beta}\Phi_{\gamma}+\delta^{\alpha}_{\gamma}\Phi_{\beta}-g_{\beta\gamma}\Phi^{\alpha}-N\delta^{\alpha}_{\beta}\varphi_{\gamma}-\delta^{\alpha}_{0}\Phi_{\beta}\varphi_{\gamma}\,, \end{aligned}$$

we obtain by analogy with (27) and (28)

(50) $$\frac{\partial G_{\alpha\beta}}{\partial x^\gamma} - G_{\alpha\sigma}\Sigma^\sigma_{\beta\gamma} - G_{\sigma\beta}\Sigma^\sigma_{\alpha\gamma} = 2(N-M)G_{\alpha\beta}\varphi_\gamma\,,$$

(51) $$\left\{\begin{aligned} \tfrac{\partial\varphi_\alpha}{\partial x^\beta} - \varphi_\alpha\Sigma^\sigma_{\alpha\beta} &= \varphi_{\alpha\beta} - M\varphi_\alpha\varphi_\beta + \varphi_\alpha\Theta_\beta + Ng_{\alpha\beta} \\ &= \varphi_{\alpha:\beta} + \varphi_\alpha\Theta_\beta + Ng_{\alpha\beta} \end{aligned}\right.$$

and analogous to (31)

(52) $$\left\{\begin{aligned} &\tfrac{dV^i}{dt} + \left\{ {i \atop kj} \right\} V^k \tfrac{dx^j}{dt} + \varphi_{kj}V^k\tfrac{dx^j}{dt}V^i + \varphi^i_j\tfrac{dx^j}{dt} \\ &\qquad = \left(\delta^i_j\Theta_k - g_{kj}\Theta^i\right) V^k\tfrac{dx^j}{dt} - Ng_{kj}V^kV^i\tfrac{dx^j}{dt} - N\delta^i_j\tfrac{dx^j}{dt}\,. \end{aligned}\right.$$

We are not going to make any use of the formulae for Π in the following chapters. However, it is not unlikely that they may be useful in a later extension of the theory (compare chap VII).

VII. Field Equations

We have seen in the earlier chapters that a second rank projective tensor contains the formalism for a theory of gravity and electromagnetism. We interpreted the quantity g_{ij} as the gravitational potential and the φ_α as electromagnetic potentials. Further we know that the motion of an electrical particle is described by the projective displacement that depends on $\Gamma^\alpha_{\beta\gamma}$.

In all of these considerations the projective scalar Φ had no role to play. We therefore set $\Phi = 1$ in this chapter, so that we only have to deal with the theory of the tensor $\gamma_{\alpha\beta}$ of index 0.

The Field Equations in Projective Form.

We now try to discover a suitably restricted form of $\gamma_{\alpha\beta}$ and would actually favour the simplest auspicious statement. As in the usual relativistic theory we also seek differential equations less restrictive than

$$B^\alpha_{\beta\gamma\delta} = 0\,.$$

It appears obvious to consider

$$\Gamma_{\alpha\beta} = B_{\alpha\beta} - \tfrac{1}{2}\gamma_{\alpha\beta}B = 0\,,$$

which yields 15 rather than 14 equations. More promising is:

(1) $$\boxed{\Gamma_{\alpha\beta} - \varphi_\alpha\varphi_\beta\Gamma = 0\,,}$$

where Γ is defined by the equation:

$$\Gamma = \gamma^{\alpha\beta}\Gamma_{\alpha\beta} = -\tfrac{3}{2}B\,.$$

The affine expression for B is: $B = R - \varphi^s_t\varphi^t_s$. Incidentally the tensor $\Gamma_{\alpha\beta}$ satisfies a series of equations that correspond to the supporting laws of Einstein's theory: $\Gamma^{\alpha}_{\beta;\alpha} = 0$. These propositions result directly from the five-dimensional affine meaning of our theory.

Equations (1) are the differential equations of a four-dimensional variation principle. By demanding first of all that the integral

$$\int Bg^{\frac{1}{2}}dx^1dx^2dx^3dx^4 \tag{2}$$

be stationary on variation of $\gamma_{\alpha\beta}$, with the auxiliary condition $\gamma_{00} = 1$, one obtains as Euler–Lagrange equations precisely (1). We derive this at the end of this chapter.

It is not excluded that this property is the physical interpretation of our equations. It is also of interest for the reason that it renders the contradiction of our equations improbable.

In the following we shall therefore assume that equations (1) are the differential equations of empty space. Empty in this case implies the absence of both matter and electric density. The field equations and the trajectory equations of electrical particles are only valid in this case.

Separation of the Field Equations.

We shall now separate equations (1) into their affine parts. The left side of (1) represents a projective tensor $T_{\alpha\beta}$ so that we can also write the field equations in the abbreviated form

$$T_{\alpha\beta} = 0\,. \tag{4}$$

By raising the indices with $\gamma^{\alpha\beta}$ we obtain two further systems of equations equivalent to (4):

$$T^{\alpha}_{\beta} = 0 \quad \text{and} \tag{5}$$

$$T^{\alpha\beta} = 0\,. \tag{6}$$

First of all one can decide from the form of the particular transformation that T^{ij} is an affine tensor. Likewise we know that T_{00} and T^i_0 are affine invariants. From (4) therefore follows the equations

$$T^{ij} = 0\,, \qquad T^i_0 = 0\,, \qquad T_{00} = 0\,.$$

Or in more detail:

$$R^{ij} - \tfrac{1}{2} g^{ij} R + 2S^{ij} = 0\,, \tag{7}$$

$$\varphi^{is}_{\ \ ,s} = 0\,, \tag{8}$$

$$R = 0\,. \tag{9}$$

In addition

$$S^{ij} = g^{st} \varphi^{i}_{s} \varphi^{j}_{t} + \tfrac{1}{4} g^{ij} \varphi^{s}_{t} \varphi^{t}_{s} \tag{10}$$

is the Maxwell voltage tensor, while

$$\varphi^{is}_{\ \ ,s} = J^{i}$$

represents the electric current vector. The equations (7), (8), (9) agree with those derived from relativity theory.

Restriction of the Solution.

Thereby a purely formal solution of the unification problem is given. Many physicists had the expectation that a suitable solution of the unification problem required the recognition of new physical evidence. However, that is not the case here. The solution in question only contains the classical theory of relativity and also the equation of motion of an electrical particle, exactly as in relativity and Maxwell theory. The theory is free of foreign elements.

Geometrical Limitations of the Theory.

We ask next to what extent we have actually utilized our geometrical tool.

By separation into theories of gravity and electromagnetism we have used the position of the contact point of the tangent space with the underlying space relative to the quadric.

On the other hand we subjected the φ to restrictive assumptions. In the first place we demanded, by the introduction of homogeneous coordinates by means of the equation

$$dx^{i} = \frac{X_i}{\varphi_\alpha X^\alpha}$$

from chap. II, that the polar hyperplane of the origin with respect to the quadric corresponds exactly with the hyperplane at infinity. That implied a restriction on the position of the quadric in the tangent space. The assumption has no effect on the field equations. It only features physically in the differential equation of the electrical particle.

Also in the definition of displacements constructed from Γ we have used the vector φ instead of an arbitrary vector C_γ of index 0. That is also meaningful for the motion of an electrical particle. Perhaps it was possible to replace this assumption by another suitable assumption. It leaves the field equations unaffected.

Generalizations of the Theory.

Further we should stress that no use has been made of Φ. A further projective vector of index 0,

$$\frac{\partial \log \Phi}{\partial x^\gamma} = \Phi_\gamma \,,$$

which deserves geometrical analysis, is determined by Φ.

To establish a framework for extended physical theories it is necessary to introduce new geometrical ideas. One can anticipate that extension of our geometrical framework by the introduction of Φ could lead to physical applications beyond projective relativity theory.

$\Gamma^\alpha_{\beta\gamma}$ does not produce the generalization of the displacement Π as introduced in the usual projective geometry. We should rather examine instead the displacement Π, which is built from $G_{\alpha\beta}$ in the same way as Γ from $\gamma_{\alpha\beta}$. In this way one succeeds in setting up field equations in some form like

$$\Gamma^*_{\alpha\beta} = 0 \,. \tag{11}$$

By this procedure $\Gamma^*_{\alpha\beta}$ is built from $\Pi^\alpha_{\beta\gamma}$ in the same way as $\Gamma_{\alpha\beta}$ is built from $\Gamma^\alpha_{\beta\gamma}$ (Bibl. 1930, 9). One of the equations so obtained is identical to Schrödinger's equation of quantum theory. It therefore seems possible that a unification of quantum theory and projective field theory can be achieved in this direction. Of course it is unsatisfactory that the consistency of equation (11) has still not been proven. Furthermore it has up to now not been possible to derive the equations from a four-dimensional variation principle. Finally the physical meaning of the equations causes problems.

It is not too surprising that an equation of the Schrödinger form emerges from this. It is first of all a fact that one can modify a general second-order differential equation such that its invariance theory becomes exactly a projective theory (Cotton, Wiener and Struik, Bibl. 1900, 1; 1927, 19,20).

The generalized non-Euclidean geometry is only one of a series of mathematically interesting geometries. It should, for instance, be possible to create a generalized conformal geometry from constructs similar to our projective tensors. Judging by the experience with relativity theory such geometries are likely to find physical application. However, the choice of one or the other of such geometries would require new physical insights and not only the unification of two field theories. New physical facts can enter the theory only by the introduction of new parameters.

Derivation of the Field Equations from a Variation Principle.

We now show, as announced before, that the field equations (1) are the Euler–Lagrange equations of the variation principle

$$\int Bg^{\frac{1}{2}}dx^1dx^2dx^3dx^4 \quad \text{stationary.}$$

In this we have to vary the $\gamma_{\alpha\beta}$ under the auxiliary condition $\gamma_{00}=1$.
In the calculation of the variation we use the elegant method of Palatini (Bibl. 1919, 2). It is

$$B = B_{\alpha\beta}\gamma^{\alpha\beta}\,. \tag{12}$$

We show, first of all that

$$\int \gamma^{\alpha\beta}g^{\frac{1}{2}}\delta B_{\alpha\beta}dx^1dx^2dx^3dx^4 = 0 \tag{13}$$

is valid. The functions

$$\delta\Gamma^{\alpha}_{\beta\gamma}$$

are the components of a projective tensor, because it is the differences of the components of two projective connections. We now form the covariant derivative of this tensor:

$$\left(\delta\Gamma^{\alpha}_{\beta\gamma}\right)_{;\delta} = \frac{\partial\left(\delta\Gamma^{\alpha}_{\beta\gamma}\right)}{\partial x^{\delta}} - \left(\delta\Gamma^{\alpha}_{\beta\epsilon}\right)\Gamma^{\epsilon}_{\gamma\delta} - \left(\delta\Gamma^{\alpha}_{\epsilon\gamma}\right)\Gamma^{\epsilon}_{\beta\delta} + \left(\delta\Gamma^{\epsilon}_{\beta\gamma}\right)\Gamma^{\alpha}_{\epsilon\delta}.$$

From this follows

$$\left(\delta\Gamma^{\alpha}_{\beta\gamma}\right)_{;\delta} - \left(\delta\Gamma^{\alpha}_{\beta\delta}\right)_{;\gamma} = \delta B^{\alpha}_{\beta\gamma\delta}$$

or by contraction with respect to α and γ

$$\left(\delta\Gamma^{\alpha}_{\beta\alpha}\right)_{;\delta} - \left(\delta\Gamma^{\alpha}_{\beta\delta}\right)_{;\alpha} = \delta B_{\beta\delta}\,.$$

(55) Derivation of the Field Equations from a Variation Principle.

Multiplication with $\gamma^{\beta\delta}$ gives

$$\gamma^{\beta\delta}\delta B_{\beta\delta} = \left\{\gamma^{\beta\delta}\left(\delta\Gamma^{\alpha}_{\beta\alpha}\right) - \gamma^{\beta\alpha}\left(\delta\Gamma^{\delta}_{\beta\alpha}\right)\right\}_{;\delta}.$$

By using $\gamma = g$ one gets

$$\int \gamma^{\beta\delta} g^{\frac{1}{2}} \delta B_{\beta\delta} dx^1 dx^2 dx^3 dx^4$$
$$= \int \frac{\partial}{\partial x^\delta}\left[\gamma^{\beta\delta}\left(\delta\Gamma^{\alpha}_{\beta\alpha}\right)g^{\frac{1}{2}} - \gamma^{\beta\alpha}\left(\delta\Gamma^{\delta}_{\beta\alpha}\right)g^{\frac{1}{2}}\right] dx^1 dx^2 dx^3 dx^4.$$

We could delete the term with $\delta = 0$ since the expression in square brackets does not depend on x^0. From the generalized Green function it now follows that

$$\int \frac{\partial}{\partial x^i}\left[\gamma^{\beta i}\left(\delta\Gamma^{\alpha}_{\beta\alpha}\right)g^{\frac{1}{2}} - \gamma^{\beta\alpha}\left(\delta\Gamma^{i}_{\beta\alpha}\right)g^{\frac{1}{2}}\right] dx^1 dx^2 dx^3 dx^4 = 0.$$

The expression $\delta\Gamma^{\alpha}_{\beta\gamma}$ vanishes on the boundaries of the integration region. That proves the validity of (13).

The extremum condition hence produces the equation

$$\delta\int B g^{\frac{1}{2}} dx^1 dx^2 dx^3 dx^4 = \int B_{\alpha\beta}\delta\left(\gamma^{\alpha\beta}g^{\frac{1}{2}}\right) dx^1 dx^2 dx^3 dx^4 = 0.$$

Since $\delta\gamma = -\gamma\gamma_{\alpha\beta}\delta\gamma^{\alpha\beta}$ it follows from there that

$$\int \left(B_{\alpha\beta} g^{\frac{1}{2}}\delta\gamma^{\alpha\beta} - B_{\sigma\tau}\gamma^{\sigma\tau}\left(\tfrac{1}{2}g^{-\frac{1}{2}}\right)\gamma\gamma_{\alpha\beta}\delta\gamma^{\alpha\beta}\right) dx^1 dx^2 dx^3 dx^4 = 0$$

or

$$(14)\quad \begin{cases} \int \left(B_{\alpha\beta} - \frac{1}{2}B\gamma_{\alpha\beta}\right) g^{\frac{1}{2}}\delta\gamma^{\alpha\beta} dx^1 dx^2 dx^3 dx^4 \\ = \int \Gamma_{\alpha\beta} g^{\frac{1}{2}}\delta\gamma^{\alpha\beta} dx^1 dx^2 dx^3 dx^4 = \int \Gamma^{\alpha\beta} g^{\frac{1}{2}}\delta\gamma_{\alpha\beta} dx^1 dx^2 dx^3 dx^4 = 0 \end{cases}$$

Since $\gamma_{00} = 1$, $\delta\gamma_{00} = 0$. However, the remaining $\delta\gamma_{\alpha\beta}$ are arbitrary. From that we can derive that $\Gamma^{\alpha\beta} = \delta^\alpha_0\delta^\beta_0 K$ in which K denotes a function of x^1, x^2, x^3, x^4. We multiply by $\gamma_{\alpha\beta}$ to obtain $\Gamma = K$.
Consequently we have the necessary condition for an extremum

$$\Gamma^{\alpha\beta} - \delta^\alpha_0\delta^\beta_0\Gamma = 0.$$

By lowering of the indices we obtain in fact the required equation (1) from that.

We could also derive the equation (1) from demanding

$$\int B g^{\frac{1}{2}} dx^1 dx^2 dx^3 dx^4 \quad \text{stationary}$$

in which the functions $G^{\alpha\beta}$ are however to be varied.

Since $Bg^{\frac{1}{2}}$ only depends on $\gamma_{\alpha\beta}$ and not on Φ it is also true that

$$(15)\quad \delta\int Bg^{\frac{1}{2}}dx^1dx^2dx^3dx^4 = \int \Gamma_{\alpha\beta}g^{\frac{1}{2}}\delta\gamma^{\alpha\beta}dx^1dx^2dx^3dx^4 = 0\,.$$

However
$$\gamma^{\alpha\beta} = \Phi^2 G^{\alpha\beta} = G_{00}G^{\alpha\beta}$$

and
$$\begin{aligned}\delta G_{00} &= \delta\big(G_{0\alpha}G_{0\beta}G^{\alpha\beta}\big)\\ &= G_{0\alpha}G_{0\beta}\delta G^{\alpha\beta} + 2G_{0\alpha}G^{\alpha\beta}\delta G_{0\beta}\\ &= G_{0\alpha}G_{0\beta}\delta G^{\alpha\beta} - 2G_{0\alpha}G_{0\beta}\delta G^{\alpha\beta}\\ &= -G_{0\alpha}G_{0\beta}\delta G^{\alpha\beta}\,.\end{aligned}$$

We therefore obtain for the variation of $\gamma^{\alpha\beta}$

$$\delta\gamma^{\alpha\beta} = G^{\alpha\beta}\delta G_{00} + G_{00}\delta G^{\alpha\beta} = G_{00}\delta G^{\alpha\beta} - G_{0\alpha}G_{0\tau}\delta G^{\sigma\tau}G^{\alpha\beta}\,.$$

Hence it follows from (15)

$$\int\big(\Gamma_{\alpha\beta}G_{00} - G_{0\alpha}G_{0\beta}\Gamma\big)g^{\frac{1}{2}}\delta G^{\alpha\beta}dx^1dx^2dx^3dx^4 = 0$$

or, because $\Phi \neq O$
$$\Gamma_{\alpha\beta} - \varphi_\alpha\varphi_\beta\Gamma = 0\,.$$

VIII. Five-dimensional Associated Space

Homogeneous Coordinates in the Tangent Space.

We fixed the connection between the homogeneous projective coordinates and the inhomogeneous coordinates of the tangent space in chapter II by:

$$(1)\qquad dx^i = \frac{X^i}{\varphi_\alpha X^\alpha}.$$

We have seen further that any projective vector X^α corresponds through (1) to a point dx^i of the tangent space. We can also represent the relationship between the projective vectors and the tangent space in a somewhat different form.

Associated Spaces.

Take as given a point x, a specific representation and five arbitrary numbers $X^0, X^1, \ldots X^4$. *We now combine the totality of all contravariant projective vectors of fixed index whose components assume the values $X^0, X^1, \ldots X^4$ in the given presentation, into a single geometrical object*[6]. The vectors of a

[6]The components of a vector in point x are functions of x^0 of the form e^{Nx^0}. Two different vectors in point x belonging to the same geometrical object can therefore assume the same values $X^0, X^1, \ldots X^4$ if we select only two suitable x^0 values. This remark is simply valid only for $N \neq 0$.

specific geometrical object can have completely arbitrary shapes in points of different x if they can assume the values $X^0, X^1, \dots X^4$ only in x.

The totality of these objects for a given point and a specific index is called a space, and precisely the associated space of index N in x.

This space is furnished with coordinate systems and stated more precisely $X^0, X^1, \dots X^4$ are the coordinates of those points, defined in terms of these five numbers. These coordinate systems of the associated spaces are associated with a specific representation of the underlying space. In this way each point of an underlying space corresponds to an associated space of index N and each representation of the underlying space to a specific coordinate system in each associated space.

It follows immediately from the basic properties of a projective vector that the definition of associated spaces is independent of coordinate system. Hence we can in fact consider associated spaces as geometrical objects.

Each particular transformation of the underlying space generates a transformation of the coordinates X^α of associated spaces:

$$(2) \qquad \begin{cases} \bar{X}^0 = X^0 + v^0_i X^i, \\ \bar{X}^i = v^i_j X^j. \end{cases}$$

The large number of particular transformations are therefore connected with the smaller number of linear transformations of X^α.

For $N \neq 0$ the associated spaces are projective spaces because all vectors which can assume the values $X^0, X^1, \dots X^4$ in a point x, are multiplied with a factor k on changing from x^0. The point

$$X^\alpha = \delta^\alpha_0$$

has a special role in each associated space because its coordinates remain unchanged in all transformations (2).

Correspondence between Projective Associated Spaces and Tangent Spaces.

To begin with our associated spaces have nothing to do with tangent spaces. We can however, relate them to tangent spaces by setting:

$$(3) \qquad X \to dx\,.$$

This way the point (1,0,0,0,0) in projective space corresponds to the origin of the tangent space. Should we further assume that the Q_α components of the polar hyperplane of the origin with respect to the quadric considered before, this hyperplane is exactly the hyperplane at infinity.

Thereby we have reached the connection with our earlier developments. The position is essentially the same as in chapter II, where we considered the X^α as homogeneous coordinates in the tangent spaces.

The associated projective spaces serve as an aid to the introduction of homogeneous coordinates. If the existence of the mapping (1) or rather (3) is not assumed beforehand we only retrieve a theory of the associated spaces. The setting up of our field equations is also independent of (1). However, the connection between the associated spaces and special curves, planes, etc. of the underlying space is not established without the mapping (3) or another suitable assumption. So, for instance, is the correspondence between a curve

$$x^i = x^i(t)$$

and the tangent spaces belonging to its points established through the equation

$$dx^i = \frac{dx^i}{dt}dt\,.$$

But then one needs for instance the mapping (1) to establish a relationship between the points of the curve and the associated spaces. We have used (1) in the derivation of the differential equations of moving electrical particles.

Five-dimensional Associated Spaces.

Our definition of associated spaces is also valid in case $N = 0$. However, the associated spaces are then no longer four dimensional because the components of a vector of index 0 are independent of x^0 and because their absolute magnitude is also established thereby.

According to our definition the associated spaces of index 0 are therefore affine spaces. By each particular transformation

$$(4)\qquad \begin{cases} \bar{x}^0 = x^0 + \log\varrho \\ \bar{x}^i = x^i(x) \end{cases}$$

of the underlying four-dimensional space, a coordinate transformation of the form (2) of the five-dimensional associated affine spaces is generated. Hence

$$(5)\qquad \begin{cases} v_i^0 = \frac{\partial \log\varrho}{\partial x^i} \\ \\ v_j^i = \frac{\partial \bar{x}^i}{\partial x^j}\,. \end{cases}$$

(59) First Mapping of the Five-dimensional Spaces onto the Tangent Spaces.

From the form of transformation (2) it follows that the straight line has the same equation

(6) $$X^i = 0\,, \quad X^0 \text{ arbitrary}$$

in each of the associated space. We call this straight line the special straight line.

First Mapping of the Five-dimensional Spaces onto the Tangent Spaces.

We can also use equation (1) to map the five-dimensional associated spaces in space-time-point x to the tangent space belonging to x. This mapping

$$X \to dx$$

is evidently not unambiguously invertible. Should a point X of the five-dimensional space, for instance, go into point dx of the tangent space, so would all points $(kX^0, kX^1, \ldots, kX^4)$, with arbitrary k, map to the same point dx. Equation (1) therefore implies a mapping of the points of a tangent space to the straight line through the origin (0,0,0,0,0) of the associated space. In this way the origin of the tangent space corresponds to the special straight line of the five-dimensional space.

The distinction between these considerations and the previous is of course not too big. It is well known that the straight lines through a fixed point of a five-dimensional affine space build a four-dimensional projective space.

A projective contravariant vector $A^\alpha(x)$ of index 0 describes a specific point in each five-dimensional associated space,

$$X^\alpha = A^\alpha(x)\,.$$

A projective contravariant vector $B^\alpha(x)$ with index different from 0 describes, on the contrary, a straight line through the origin in each five-dimensional associated space. The equations of these straight lines are

(7) $$X^\alpha = kB^\alpha\,, \qquad \text{where } k \text{ is arbitrary.}$$

A straight line through the origin of each five-dimensional associated space is also determined by an affine contravariant vector $V^i(x)$. Their points satisfy the equations

(8) $$\frac{X^i}{\varphi_\alpha X^\alpha} = V^i.$$

The connection between the straight lines through a fixed point of a five-dimensional affine space and the points of a four-dimensional affine space can also be visualized by the intersection of a straight line with a hyperplane, not through the fixed point. The hyperplane

$$\varphi_\alpha X^\alpha = p$$

is especially suitable for this purpose.

If we specify that the points determined through (8), by the assumed V, lie in this hyperplane, equations (8) or rather (1) produce an unambiguous mapping between the points of the hyperplane and the points of the tangent spaces. Homogeneous coordinates in this hyperplane are defined through (7).

We obtain a whole swarm of hyperplanes for different p. However, our mapping fails for $p = 0$.

Differentials as Coordinates of the Five-dimensional Space.

The differentials of the coordinates $x^1, \ldots, x^4$ and the differential of the gauge variable x^0 transform exactly like the components of a contravariant vector:

$$(9) \qquad \begin{cases} d\bar{x}^0 = dx^0 + \frac{\partial \log \varrho}{\partial x^i} dx^i \\ d\bar{x}^i = \frac{\partial \bar{x}^i}{\partial x^j} dx^j \, . \end{cases}$$

One can therefore consider the differentials dx^α as the original coordinates of the five-dimensional associated space.

One could get the impression that the hyperplane $dx^0 = 0$ can simply be considered as the tangent plane with coordinates $dx^1, \ldots, dx^4$. But that is not possible as $dx^0 = 0$ is not an invariance condition. The tangent space cannot be extended in this way into a five-dimensional associated space. In fact though there exists a clear unambiguous mapping between the points of the tangent space and those straight lines of the associated space which are parallel to the specially selected straight lines. One sees that immediately from (9). This connection can also be seen as a degenerate mapping of the points of the five-dimensional associated space onto the points of the tangent space. We represent this mapping by the equations

$$dx^i = \delta^i_\alpha \partial x^\alpha \, ,$$

where we now describe the five-dimensional coordinates by ∂x^α and the four-dimensional by dx^i. In the following $dx^1, \ldots, dx^4$ still represent the coordinates of the tangent space and $\partial x^0, \ldots, \partial x^4$ the coordinates of the five-dimensional associated space.

(61) Displacements of the Five-dimensional Associated Spaces.

Euclidean Metric of a Five-dimensional Associated Space.

One can interpret $\gamma_{\alpha\beta}\partial x^\alpha \partial x^\beta$
as the square of the Euclidean distance. In each of these spaces

$$\gamma_{\alpha\beta}\partial x^\alpha \delta_0^\beta = \varphi_\alpha \partial x^\alpha$$

is the normal projection of the vector ∂x^α on the special straight lines, as $\delta_0^\beta = \varphi^\beta$ is the unit vector of the special straight lines.

The formula

$$\gamma_{\alpha\beta}\partial x^\alpha \partial x^\beta = g_{ij}\partial x^i \partial x^j + (\varphi_\alpha \partial x^\alpha)^2 \tag{11}$$

specifies the square of the vector's length ∂x^α as the sum of squares of the perpendicular and parallel components of the special straight lines.

The hyperplane

$$\varphi_\alpha \partial x^\alpha = p \tag{12}$$

stands perpendicular on the special straight lines. Hence, from (11) the Euclidean distance in each such hyperplane is

$$g_{ij}\partial x^i \partial x^j \,.$$

The mapping (8) of the five-dimensional space to the tangent space hence implies some sort of perpendicular projection of the Euclidean space with metric $\gamma_{\alpha\beta}$ on the Euclidean plane with metric g_{ij}.

Displacements of the Five-dimensional Associated Spaces.

The projective connection Γ was defined (p. 44) through the equations

$$\gamma_{\alpha\beta;\gamma} = 0 \tag{13}$$

and the displacement that goes with it, through the equations

$$X^\alpha{}_{;\beta} = X^\alpha{}_{;0}\varphi_\beta \,.$$

We have also written (p. 45) these equations in the form

$$\frac{\partial X^\alpha}{\partial x^\gamma} + X^\sigma \Lambda^\alpha_{\sigma y} = 0 \tag{14}$$

such that Λ, here of interest in the case $M = 0$, has the value

$$\Lambda^\alpha_{\beta\gamma} = \Gamma^\alpha_{\beta\gamma} - \Gamma^\alpha_{\beta 0}\varphi_\gamma \,. \tag{15}$$

On the basis of the formula $\Gamma^{\alpha}_{\beta 0} = \gamma^{\alpha\sigma}\varphi_{\alpha\beta}$ derived in chapter IV (p. 32), one notices that an equation of the form (13) is also valid for Λ. It is namely

$$\frac{\partial \gamma_{\alpha\beta}}{\partial x^{\gamma}} - \gamma_{\alpha\beta}\Lambda^{\sigma}_{\alpha\gamma} - \gamma_{\alpha\sigma}\Lambda^{\sigma}_{\beta\gamma} = 0 \,. \tag{16}$$

This means that the length $\gamma_{\alpha\beta}XX^{\beta}$ of a vector X as well as the angle of two vectors in five-dimensional space remain unchanged during displacement.

The special straight lines are not carried into themselves by this displacement. For the projective derivative of φ^{α} with respect to Λ we namely find

$$\frac{\partial \varphi^{\alpha}}{\partial x^{j}} + \varphi^{\alpha}\Lambda^{\alpha}_{\sigma j} = \Gamma^{\alpha}_{0j} = \gamma^{i\alpha}\varphi_{ij} \,. \tag{17}$$

Because the process of projective differentiation with respect to Λ commutes with raising and lowering of the indices, it follows from (17) that

$$\frac{\partial \varphi_{\alpha}}{\partial x^{j}} + \varphi_{\sigma}\Lambda^{\sigma}_{\alpha j} = \varphi_{\alpha j} \,. \tag{17a}$$

We now calculate the covariant derivative of the parameters $p = \varphi_{\alpha}X^{\alpha}$ of the swarm of hyperplanes considered before. It is

$$\frac{\partial \varphi_{\alpha}X^{\alpha}}{\partial x^{j}} = \left(\frac{\partial \varphi_{\alpha}}{\partial x^{j}} - \varphi_{\alpha}\Lambda^{\sigma}_{\alpha j}\right)X^{\alpha} + \varphi_{\alpha}\left(\frac{\partial X^{\alpha}}{\partial x^{j}} + X^{\sigma}\Lambda^{\alpha}_{\sigma j}\right).$$

On the basis of (17a) and (14) we obtain from that

$$\frac{d(\varphi_{\alpha}X^{\alpha})}{dt} = \varphi_{ij}X^{i}\frac{dx^{j}}{dt} \,. \tag{18}$$

Equation (18) is valid for the displacement of p with respect to Λ along a curve $x^{i} = x^{i}(t)$.

We now ask when does the left side of (18) vanish, *i.e.* when a point, which is displaced with respect to Λ, stays in the same hyperplane. Due to the skew symmetry of φ_{ij} this clearly happens when

$$X^{i} = k\frac{dx^{i}}{dt} \,. \tag{18a}$$

From (11) it follows besides, that in the case (18a) the distance of a point from the special straight line is also invariant. We now write the formula (14) in somewhat more detail. If we separate the case $\alpha = 0$ out, we obtain, due to chapter VI (20) and (21),

$$\frac{dX^{i}}{dt} + \left\{ {i \atop jk} \right\} X^{k}\frac{dx^{j}}{dt} + \varphi^{i}_{j}\frac{dx^{j}}{dt}\varphi_{\sigma}X^{\sigma} = 0 \,.$$

There is a similar differential equation for X^{0}.

(63) Introduction of General Coordinates in the Associated Spaces.

The points of the five-dimensional space

$$X^i = k\frac{dx^i}{dt} \quad , \quad \varphi_\alpha X^\alpha = \text{const.}$$

are linked to the velocity vector of the curve. If we demand that these points on displacement along the curve go over the same type of points, the following differential equations for the curve and its parameters are obtained:

$$\frac{d^2x^i}{dt^2} + \left\{ {i \atop j\,k} \right\} \frac{dx^j}{dt}\frac{dx^k}{dt} + \frac{e}{m}\varphi^i_j \frac{dx^j}{dt} = 0\,. \tag{19}$$

We have set the constant $\varphi_\alpha X^\alpha$ equal to e/m.

These equations for the special curves of the underlying space correspond exactly with the equations of chap. VI (33), that we obtained for the case $M \neq 0$, therefore a four-dimensional associated space. Thereby we have found a geometrical meaning of the world line of an electrical particle also in the five-dimensional associated spaces.

Introduction of General Coordinates in the Associated Spaces.

We can introduce completely general affine coordinates in the tangent spaces through the non-degenerate transformation

$$W^i = M^i_j dx^j\,. \tag{20}$$

If we assume, for instance, that $M^1_j, \ldots, M^4_j$ are four covariant vectors, then so are the scalars W^i. The coordinates obtained in this way we call scalar coordinates. We are however, free also to choose other transformation laws for the M^i_j. Even so we can also introduce general affine coordinates in the five-dimensional associated space. We do this through the transformation

$$\partial x^\beta = N^\beta_\alpha U^\alpha\,. \tag{20a}$$

All of our formulae assume generalized form in the generalized coordinates. Their geometrical and physical meaning however, of course remains unchanged. The degenerate mapping (10), for instance, acquires the form

$$W^i = t^i_\alpha U^\alpha \tag{21}$$

in general coordinates, where

$$t^i_\alpha = N^\beta_\alpha \delta^j_\beta M^i_j\,.$$

Einstein and Mayer (Bibl. 1931, 3) always use these general coordinate systems. In their work the mapping (21) plays a sinificant role that we hope to clarify in the following. Einstein and Mayer used the notation γ^i_α instead of our t^i_α. They also write $g_{\alpha\beta}$ for $\gamma_{\alpha\beta}$ and A_α for φ_α.

Second Mapping of the five-dimensional Space to the Tangent Space.

For our present purpose it is significant to work with the coordinates dx^i and ∂x^α. Nevertheless we shall write (10) in the form

$$dx^i = t^i_\alpha \partial x^\alpha \tag{21}$$

to emphasize the fact that this transformation is a geometrical object. Therefore in our coordinates

$$t^i_\alpha = \delta^i_\alpha . \tag{22}$$

We now try to express the relationship between the coordinates of the associated space and the tangent space, given by (21), also in another geometrical form. For that purpose we must invert (21). The inversion of (21) is not determined unambiguously because (21) is not an unambiguous transformation. In any case the formula

$$\partial x^i = dx^i$$

must apply since ∂x^0 is not subject to the choice of $(dx^1, \ldots, dx^4)$. We determine the inversion of t^i_α in possibly closer reliance on the measure relationships of both spaces. This way we define

$$t^{\alpha i} = \gamma^{\alpha\beta} t^i_\beta\,, \quad t^\alpha_i = g_{ij} t^{\alpha j}\,, \quad t_{\alpha i} = g_{ij} t^j_\alpha\,. \tag{23}$$

By using (22) we obtain

$$\begin{aligned} t^{\alpha i} &= \gamma^{\alpha i}\,, && (24)\\ t^\alpha_i &= (\gamma_{i\beta} - \varphi_i\varphi_\beta)\gamma^{\alpha\beta} = \delta^\alpha_i - \delta^\alpha_0 \varphi_i\,, && (25)\\ t_{\alpha i} &= \delta^j_\alpha g_{ij}\,. && (26)\end{aligned}$$

Hence the following identities therefore occur

$$\begin{aligned} t^\alpha_i t^j_\alpha &= \delta^j_i\,, \quad t^\alpha_i t^i_\beta = \delta^\alpha_\beta - \delta^\alpha_0 \varphi_\beta\,, && (27)\\ t^{\alpha i} t_{\alpha j} &= \delta^i_j\,, \quad t^{\alpha i} t_{\beta i} = \delta^\alpha_\beta - \delta^\alpha_0 \varphi_\beta\,, && (28)\\ t^\alpha_i t_{\alpha j} &= g_{ij}\,, \quad t^{\alpha i} t^\beta_i = \gamma^{\alpha\beta} - \delta^\alpha_0 \delta^\beta_0\,. && (29)\end{aligned}$$

In the light of our definition (23) we consider the transformation

$$\partial x^\alpha = t^\alpha_i dx^i \tag{30}$$

as inversion of (21). Under assumption (22) then follows

$$\partial x^\alpha = \delta^\alpha_i dx^i - \delta^\alpha_0 \varphi_i dx^i$$

or

$$\left\{ \begin{aligned} \partial x^0 &= -\varphi_i dx^i\,, \\ \partial x^i &= dx^i\,. \end{aligned} \right. \tag{31}$$

(65) Second Mapping of the five-dimensional Space to the Tangent Space.

From (31) one sees immediately that all points dx of the tangent space are mapped unambiguously on the points of the special hyperplane

$$\varphi_\alpha \partial x^\alpha = 0 \tag{32}$$

of the five-dimensional space. By limiting (21) to the points of this hyperplane (21) becomes the unambiguous reciprocal of (30). It is remarkable that (32) is the only hyperplane of the swarms

$$\varphi_\alpha \partial x^\alpha = p$$

in which the previously given transformation (8) fails.

With the help of the quantities t^i_α and t^α_i we can restore the desired correspondence between the images of the five-dimensional and four-dimensional spaces.

So, for instance, does the projective covariant vector A_α correspond to the affine vector

$$B_i = t^\alpha_i A_\alpha = A_i - A_0 \varphi_i \,. \tag{33}$$

To interpret this formula geometrically we consider the hyperplane $B_i dx^i = k$ of the tangent space. It is

$$B_i dx^i = B_i t^i_\alpha \partial x^\alpha = (A_i t^i_\alpha - A_0 \varphi_i t^i_\alpha) \partial x^\alpha$$

or

$$B_i dx^i = A_\alpha \partial x^\alpha - A_0 \varphi_\alpha \partial x^\alpha \,.$$

If we limit ourselves again to the points of the hyperplane (32), we obtain the unambiguous mapping of the hyperplane

$$B_i dx^i = k$$

of the four-dimensional space on the various intersections of the hyperplanes

$$A_\alpha \partial x^\alpha = k \qquad \text{and} \qquad \varphi_\alpha \partial x^\alpha = 0$$

of the five-dimensional space. If specifically $A_\alpha = \varphi_\alpha$ the transformation (33) reduces to $0 = t^\alpha_i \varphi_\alpha$, as also to be expected geometrically.

Alternatively, an affine covariant vector corresponds to the projective vector

$$t^i_\alpha B_i = t^i_\alpha (A_i - A_0 \varphi_i) = A_\alpha - A_0 \varphi_\alpha \,.$$

The various cuts of $(A_\alpha - A_0 \varphi_\alpha) \partial x^\alpha = k$ with $A_\alpha \partial x^\alpha = 0$ agree with the various cuts of $A_\alpha \partial x^\alpha = k$ and $\varphi_\alpha \partial x^\alpha = 0$.

The correspondence between the fundamental tensors is mediated by the formulae $\gamma_{\alpha\beta} t_i^\alpha t_j^\beta = g_{ij}$ and $g_{ij} t_\alpha^i t_\beta^j = \gamma_{\alpha\beta} - \varphi_\alpha \varphi_\beta$.

Relationships between the Displacements.

We occupy ourselves now with the displacement of the tangent space induced by the displacement of five-dimensional space. Alternatively we could also enquire into the displacement of the five-dimensional space generated by an usual Levi–Civita displacement of the tangent space.

We proceed from the unambiguous mapping

$$V^i = t_\alpha^i X^\alpha \qquad \text{and} \qquad X^\alpha = t_i^\alpha V^i \tag{34}$$

of the points of the tangent space on the points of the hyperplane

$$\varphi_\alpha X^\alpha = 0 \tag{32}$$

of the associated five-dimensional space. By differentiation of (34) and use of (14) we obtain

$$\frac{\partial V^i}{\partial x^j} = \frac{\partial t_\alpha^i}{\partial x^j} X^\alpha - t_\alpha^i \Lambda_{\sigma j}^\alpha X^\alpha . \tag{35}$$

Because of (15) it follows from this that

$$\frac{\partial V^i}{\partial x^j} = -\Gamma_{\sigma j}^i X^\alpha + \Gamma_{\sigma 0}^i X^\alpha \varphi_j$$

or due to Chap. VI equation (20) and (21) p. 44

$$\frac{\partial V^i}{\partial x^j} = -\left\{ {i \atop k\,j} \right\} V^k - \varphi_j^i \varphi_\alpha X^\alpha . \tag{36}$$

By virtue of the limitation of X to the hyperplane (32) we obtain

$$\frac{\partial V^i}{\partial x^j} + \left\{ {i \atop k\,j} \right\} V^k = 0 . \tag{37}$$

Taking note of the mapping (34) as well, the affine displacement (37) corresponds to the displacement (14) of the five-dimensional space.

Substitution of (36) into (35) gives

$$\frac{\partial t_\alpha^i}{\partial x^j} - t_\alpha^i A_{\alpha j}^\sigma + t_\alpha^k \left\{ {i \atop k\,j} \right\} = -\varphi_j^i \varphi_\alpha . \tag{38}$$

In this we had to consider that X^α can assume any arbitrary value in the five-dimensional space. The equation (38) is first of all only one of the identities constructed from equations (15), Chap. VIII, and (20) and (21), Chap. VI, which one can of course also confirm immediately from these equations.

(67)	Relationships between the Displacements.

We can also interpret (38) as the equation for the covariant derivative of the mixed quantity t^i_α whose indices are related to the tangent space or rather to the five-dimensional space.

We emphasized before that the mapping of a tangent space on a five-dimensional space is a geometrical object with the components $t^i_\alpha = \delta^i_\alpha$ in our special coordinates. Equation (38) represents the displacement of this geometrical object. Given the quantities t^i_α, (38) produces the desired relationship between the displacements of both spaces.

Equivalent to (38) are the equations

$$\frac{\partial t^\alpha_i}{\partial x^j} + \Lambda^\alpha_{\sigma j} t^\sigma_i - \left\{ {k \atop ij} \right\} t^\alpha_k = \varphi_\alpha \varphi_{ij} \,. \tag{39}$$

Equations (39) can immediately be derived from (38) by raising and lowering α and i with the help of $\gamma^{\alpha\beta}$ and g_{ij}. Without using (38) one can also confirm (39) immediately from (25).

Einstein and Mayer defined the left-hand side of (37) as the "absolute derivative" of t^α_i. Thereby the equations (39) are introduced by geometrical assumptions. The displacement properties of the associated five-dimensional space can then be derived from (37) and (39). The resulting geometrical structure corresponds of course with what we gave above. The field equations also have the same meaning as ours.

The equations (39) can also be replaced by others. Einstein and Mayer proposed, for instance, in later work, the equations

$$\frac{\partial t^\alpha_i}{\partial x^j} + \Lambda^\alpha_{\sigma j} t^\sigma_i - \left\{ {k \atop ij} \right\} t^\alpha_k = \varphi^\alpha F_{ij} + \gamma^{\alpha r} V_{rij} \tag{40}$$

in which

$$F_{ij} = -F_{ji}$$

and

$$V_{rij} = -V_{irj} = -V_{rji} \,.$$

The quantities Λ in these equations must clearly be different from the Λ used before. The introduction of new equations such as (40) therefore also implies a new choice of displacement and hence the possibility of formulating new field equations. The physical meaning of equations (40) has still not been established. We shall not go into this any further.

IX. Bibliography

1900

1. Cotton, E.: Sur les invariants differentiels de quelques équations lineaires aux dérivées partielles du second ordre. Ann. École norm. (3) 17, S. 211–244.

1916

1. Einstein, A.: Die Grundlage der allgemeinen Relativitätstheorie. Ann. Physik Bd. 49, S. 769–822.

1918

1. Weyl, H.: Gravitation und Elektrizität. S.-B. preuß. Akad. Wiss. S. 465–480.
2. — Raum, Zeit, Materie. 1.Aufl. 1918, 5.Aufl. 1923. Berlin: Julius Springer.
3. — Reine Infinitesimalgeometrie. Math. Z. Bd. 2 S. 384–411.

1919

1. König, R.: Beiträge zu einer allgemeinen linearen Mannigfaltigkeitslehre. Jber. Deutsch. Math.-Vereinig. Bd. 28 S. 213–228.
2. Palatini, A.: Deduzione invariantiva delle equazione gravitazionali dal principio di Hamilton. Rend. Circ. mat. Palermo Bd. 43 S. 203–212.
3. Weyl, H.: Eine neue Erweiterung der Relativitätstheorie. Ann. Physik Bd. 59 S. 101–133.

1921

1. Bach, R.: Zur Weylschen Relativitätstheorie. Math. Z. Bd. 9 S. 110–135.
2. Eddington, A.S.: A generalisation of Weyl's theory of the electromagnetic and gravitational fields. Proc. Roy. Soc. London Bd. 99 S. 104–122.
3. Kaluza, Th.: Zum Unitätsproblem der Physik. S.-B. preuß. Akad.Wiss. S. 966–972.
4. Weyl, H.: Zur Infinitesimalgeometrie: Einordnung der projektiven und der konformen Auffassung. Nachr. Ges. Wiss. Göttingen, Math.-phys. Kl. S. 99–112.

1922

1. Cartan, É.: Sur les espaces généralisés et la théorie de la relativité. C.R. Acad. Sci., Paris Bd. 174 S. 734–737.
2. Eisenhart, L.P.: Spaces with corresponding paths. Proc. Nat. Acad. Sci. U.S.A. Bd. 8 S. 233–238.
3. Veblen, O.: Projective and affine geometry of paths. Proc. Nat. Acad. Sci. U.S.A. Bd. 8 S.347 –350.

1923

1. Cartan, É.: Sur les variétés à connexion affine et la théorie de la relativité généralisée. Ann. École norm. Bd. 40 S. 325–412.
2. Einstein, A.: Zur algemeinen Relatvitätstheorie. S-B. preuß. Akad. Wiss. S. 32–38 u. 76–77.
3. Einstein, A.: Zur affinen Feldtheorie. S-B. preuß. Akad. Wiss. S. 137–140.
4. Schouten, J.A.: On a non-symmetrical affine field theory. Proc. Akad. Wetensch. Amsterd. Bd. 26 S. 850–857.
5. Veblen, O., u. T.Y. Thomas: The geometry of paths. Trans. Amer. Math. Soc. Bd. 25 S. 551–608.
6. Weitzenböck, R.: Invariantentheorie. Groningen: Noordhoff.

7. Weyl, H.: Mathematische Analyse der Raumproblemes. Berlin: Julius Springer.

1924

1. Cartan, É.: Sur les variétés à connexion projective. Bull. Soc. Math. France Bd. 52 S. 205–241.
2. – Sur la connexion projective des surfaces. C.R. Acad. Sci., Paris Bd. 178 S. 750–752.
3. – Les récentes généralisations de la notion d'espace. Bull. Sci. math. Bd. 48, S. 294–320.
4. – Les groupes d'holonomie des espaces généralisés. Acta math. Bd. 48, S. 1–42.
5. Hlavatý, V.: Sur le déplacement linéaire du point. Vestnuku Kral. Ces. Spolec. Nauk Bd. 2.
6. Schouten, J.A.: On the place of conformal and projective geometry in the theory of linear displacements. Proc. Akad. Wetensch. Amsterd. Bd. 27 S. 407–424.
7. – Sur les connexions conformes et projectives de M. Cartan et la connexion linéaire générale de M. König. C.R. Acad. Sci., Paris Bd. 178 S. 2044–2046.
8. – Der Ricci Kalkül. Berlin: Julius Springer.

1925

1. Eddington, A.S.: Relativitätstheorie in mathematischer Behandlung. Berlin: Julius Springer.
2. Einstein, A.: Einheitliche Feldtheorie von Gravitation und Elektrizität. S.-B. preuß. Akad. Wiss. S. 414–419.
3. Eyraud, H.: Sur le caractère riemannien projectif du champ gravifique électromagnétique. C.R. Acad. Sci., Paris Bd. 180 S. 127–129
4. – La théorie affine asymétrique du champ électromagnétique et le rayonnement atomique. C.R. Acad. Sci., Paris Bd. 180, S. 1245–1248.
5. Thomas, J.M.: Note on the projective geometry of paths. Proc. Nat. Acad. Sci. U.S.A. Bd. 11 S. 207–209.
6. Thomas, T.Y.: On the projective and equi-projective geometries of paths. Proc. Nat. Acad. Sci U.S.A. Bd. 11 S. 199–203.
7. – Note on the projective geometry of paths. Bull. Amer. Math. Soc. Bd. 31 S. 318–322.
8. – Announcement of a projective theory of affinely connected manifolds. Proc. Nat. Acad. Sci. U.S.A. Bd 11 S. 588–589.
9. – On the equi-projective geometry of paths. Proc. Nat. Acad. Sci. U.S.A. Bd. 11 S. 592–594.
10. Veblen, O., u. J.M. Thomas: Projective normal coordinates for the geometry of paths. Proc. Nat. Acad. Sci. U.S.A. Bd 11 S. 204–207.
11. Veblen, O.: Remarks on the foundations of Geometry. Bull. Amer. Math. Soc. Bd. 31 S. 121–141.

1926

1. Ehrenfest, P., u. G.E. Uhlenbeck: Graphische Veranschaulichung der De Broglieschen Phasenwellen in der fünfdimensionalen Welt von O. Klein. Z. Physik Bd. 39 S. 495–498.
2. Eisenhart, L.P.: Riemannian Geometry. Princeton University Press.
3. Fock, N.: Über die invariante Form der Wellen- und der Bewegungsgleichungen für einen geladenen Massenpunkt. Z. Physik Bd. 39 S. 226–232.

4. Gamow, G., u. D. Iwanenko: Zur Wellentheorie der Materie. Z. Physik Bd. 39 S.865–868.
5. Klein, O.: Quantentheorie und fünfdimensionale Relativitätstheorie. Z. Physik Bd. 37 S. 895–906.
6. Mandel, H.: Zur Herleitung der Feldgleichungen in der algemeinen Relativitätstheorie. Z. Physik Bd. 39 S. 136–145.
7. Schouten, J.A.: Erlanger Programm und allgemeine Übertragungslehre: Neue Gesichtspunkte zur Grundlegung der Geometrie. Rend. Circ. mat. Palermo Bd. 50 S. 142–169.
8. – Über die Projektivkrümmung und Konformkrümmung halbsymmetrischer Übertragungen. Kanzani ,,In mem. Lobatchewsku" Bd. 2 S. 90–98.
9. – Projective and conformal invariants of half symmetrical connections. Proc. Akad. Wetensch. Amsterd. Bd. 29 S.334–336.
10. Thomas, J.M.: Asymmetric displacement of a vector. Trans. Amer. Math. Soc. Bd. 28 S. 658–670.
11. – On normal coordinates in the geometry of paths. Proc. Nat. Acad. Sci. U.S.A. Bd. 12 S. 58–63.
12. – On various geometries giving a unified electric and gravitational theory. Proc. Nat. Acad. Sci. U.S.A. Bd. 12 S. 187–191.
13. Thomas, T.Y.: A projective theory of affinely connected manifolds. Math. Z. Bd. 25 S. 723–733.
14. Veblen, O., u. J.M. Thomas: Projective invariants of affine geometry of paths. Ann. of Math. Bd. 27 S. 279–296.

1927

1. Broglie, L. de: L'Univers à cinq dimensions et la mécanique ondulatoire. J. de Phys. Bd. 8 S. 65–73.
2. Donder, Th. de: Sur les extrémales décrites par les électrones et les particules électrisées. Bull. Acad. Roy. de Belgique Bd. 13 S. 27–30 u. 79–88.
3. – Signification et generalization de l'équation de Schrödinger. Bull. Acad. Roy. de Belgique Bd. 13 S. 103–113, 185–200.
4. Einstein, A.: Zu Kaluzas Theorie des Zusammenhangs von Gravitation und Elektrizität. S.-B. preuß. Akad. Wiss S. 23–30.
5. Eisenhart, L.P.: Non-Riemannian Geometry. New York: Amer. Math. Soc. Colloquium VIII.
6. Gonseth, F., u. G. Juvet: Sur les équations de l'électromagnétisme. C.R. Acad. Sci., Paris Bd. 185 S. 341–343.
7. – Sur la métrique de l'espace à cinq dimensions de l'électromagnétisme et de la gravitation. C.R. Acad. Sci., Paris Bd. 185 S. 412–413.
8. – Sur l'équation de M. Schrödinger. C.R. Acad. Sci., Paris Bd. 185 S. 448–450.
9. – Les équations de l'électromagnétisme et l'équation de M. Schrödinger dans l'univers à cinq dimensions. C.R. Acad. Sci., Paris Bd. 185 S. 535–538.
10. Hlavatý, V.: Sur les déplacements isohodoïques. L'Einsegn. Math. Bd. 26 S. 84–97.
11. Klein, O.: Zur fünfdimensionalen Darstellung der Relativitätstheorie. Z. Physik Bd. 46 S. 188–208.
12. Knebelman, M.S.: Groups of collineations in a space of paths. Proc. Nat. Acad. Sci. U.S.A. Bd. 13 S. 396–400.
13. London, F.: Quantenmechanische Deutung der Theory von Weyl. Z. Physik Bd. 42 S. 375–389.

14. Mandel, H.: Zur Herleitung der Feldgleichungen in der algemeinen Relativitätstheorie II. Z. Physik Bd. 45 S. 285–306.
15. Newman, M.H.A.: A gauge-invariant tensor calculus. Proc. Roy. Soc. London Bd. 116 S. 603–623.
16. Rosenfeld, L.: L'Univers à cinq dimensions et la mécanique ondulatoire. Bull. Acad. Roy. de Belgique Bd. 13 S. 304–325.
17. Schrödinger, E.: Der Energieimpulssatz der Materiewellen. Ann. Physik Bd. 82 S. 265–272.
18. Struik, D.J.: On the geometry of linear displacement. Bull. Amer. Math. Soc. Bd. 33 S. 523–564.
19. Struik, D.J., u. N. Wiener: Sur la théorie relativiste des quanta. C.R. Acad. Sci. Paris Bd. 185 S. 42–44.
20. – A relativistic theory of quanta. J. Math. Physics, Massachusetts Inst. Technol. Bd. 7 S. 1–23.
21. Thomas, T.Y.: The replacement theorem and related questions in the projective geometry of paths. Ann. of Math. Bd. 28 S. 549–561.
22. Veblen, O.: Invariants of Quadratic Differential Forms. London: Cambridge Tract.

1928

1. Cartan, É.: Leçons sur la géométrie des espaces de Riemann. Paris: Gauthier-Villars.
2. Douglas, J.: The general geometry of paths. Ann. of Math. Bd. 29 S. 143–168.
3. Einstein, A.: Riemann-Geometrie mit Aufrechterhaltung des Begriffes des Fernparallelismus. S.-B. preuß. Akad. Wiss. S. 217–221.
4. – Neue Möglichkeit für eine einheitliche Feldtheorie von Gravitation und Elektrizität. S.-B. preuß. Akad. Wiss. S. 224–227.
5. Gonseth, F., u. G. Juvet: Sur la relativité à cinq dimensions et sur une interprétation de l'équation de Schrödinger. Atti Congresso Internazionale Bologna Bd. 5 S. 75–78.
6. Hattori, K.: Über eine formale Erweiterung der Relativitätstheorie und ihren Zusammenhang mit der Theorie der Elektrizität. Physik. Z. Bd. 29 S. 538–549.
7. Hlavatý, V.: Bemerkung zur Arbeit von Herrn T.Y. Thomas: A projective theory of affinely connected manifolds. Math. Z. Bd. 28 S. 142–146.
8. Infeld, L.: Bemerkungen zu der Arbeit von Herrn K. Hattori. Physik. Z. Bd. 29 S. 810–811.
9. Robertson, H.P.: Note on projective coordinates. Proc. Nat. Acad. Sci. U.S.A. Bd 14 S. 153–154.
10. Veblen, O.: Projective tensors and connections. Proc. Nat. Acad. Sci. U.S.A. Bd. 14 S. 154–166.
11. – Differential Invariants and Geometry. Atti Congresso Internazionale Bologna Bd. 1 S. 181–189.

1929

1. Einstein, A.: Zur einheitlichen Feldtheorie. S.-B. preuß. Akad. Wiss. S. 2–7.
2. Mandel, H.: Zur Axiomatik der fünfdimensionalen Relativitätstheorie. Z. Physik Bd. 54 S. 564–566.
3. – Zur tensoriellen Form der wellenmechanischen Gleichungen des Elektrons. Z. Physik Bd. 54 S. 567–570.

4. Robertson, H.P., u. H. Weyl: On a problem in the theory of groups arising in the foundations of infinitesimal geometry. Bull. Amer. Math. Soc. Bd. 35 S. 686–690.
5. Schouten, J.A.: Über nichtholonome Übertragungen in einer L_n. Math. Z. Bd. 30 S. 149–172.
6. Schouten, J.A., u. V. Hlavatý: Zur Theorie der allgemeinen linearen Übertragung. Math. Z. Bd. 30 S. 414–432.
7. Veblen, O.: Generalized projective geometry. J. London Math. Soc. Bd. 4 S.140–160.
8. Weyl, H.: On the foundations of general infinitesimal geometry. Bull. Amer. Math. Soc. Bd. 35 S. 716–725.

1930

1. Bargmann, V.: Über eine Veralgemeinerung des Einsteinschen Raumtyps. Z. Physik Bd. 65 S. 830–847.
2. Bortolotti, E.: Sulla geometria delle varietà a connessione affine. Teoria invariantiva delle trasformazioni che conservano il parallelismo. Ann. Mat. pura appl. Bd. 8 S. 53–101.
3. – Connessioni proiettive. Boll. Un. Mat. Ital. Bd. 9 S. 288–294.
4. Eisenhart, L.P.: Projective normal coordinates. Proc. Nat. Acad. Sci. U.S.A. Bd. 16 S. 731–740.
5. Schouten, J.A., u. St. Golab: Über projektive Übertragungen und Ableitungen. I. Math. Z. Bd. 32 S. 192–214 — II. Ann. Mat. pura appl. Bd. 8 S. 141–157.
6. Schouten J.A., u. E.R. van Kampen: Zur Einbettungs- und Krümmungstheorie nichtholonomer Gebilde. Math. Ann. Bd. 103 S. 752–783.
7. Thomas, T.Y.: On the unified field theory. I, II. Proc. Nat. Acad. Sci. U.S.A. Bd. 16 S. 761–776, 830–835.
8. Veblen, O: A generalization of the quadratic differential form. Quart. J. Math., Oxford Ser. Bd. 1 S. 60–76.
9. Veblen, O, u. B. Hoffmann: Projective Relativity. Physic Rev. Bd. 36 S. 810–822.
10. Whitehead, J.H.C.: A method of obtaining normal representations for a projective connection. Proc. Nat. Acad. Sci. U.S.A. Bd. 16 S. 754–760.

1931

1. Bortolotti, E.: Differential invariants of direction and point displacements. Ann. of Math. Bd. 32 S. 361–377.
2. – Connessioni proiettive. Boll. Un. Mat. Ital. Bd. 10 S. 28–34, 83–90.
3. Einstein, A., u. W. Mayer: Einheitliche Theorie von Gravitation und Elektrizität. S.-B. preuß. Akad. Wiss. S. 541–557.
4. Hoffmann, B.: Projective relativity and the quantum-field. Physic. Rev. Bd. 37 S. 88–89.
5. Lanczos, C.: Elektromagnetismus als natürliche Eigenschaft der Riemannschen Geometrie. Z. Physik Bd. 73 S. 147–168.
6. – Die neue Feldtheorie Einsteins. Erg. exakten Naturwiss. Bd. 10 S. 97–132.
7. Schouten, J.A., u. D. van Dantzig: Über eine vierdimensionale Deutung der neuesten Feldtheorie. Proc. Akad. Wetensch. Amsterd. Bd. 34 S. 1398–1407.
8. Schouten, J.A.: Dirac equations in general relativity. J. Math. Physics, Massachusetts Inst. Technol. Bd. 10 S. 239–283.
9. Slebodzinski, W.: Sur les équations canoniques de Hamilton. Bull. Acad. Roy. de Belgique Bd. 17 S. 864–870.

10. Straneo, P.: Teorie unitarie della gravitazione e dell'electricita. Nuovo Cimento Bd. 8 S. 125–145.
11. – Gleichungen zu einer einheitliche Feldtheorie. S.-B. preuß. Akad. Wiss. S. 319–325.
12. Thomas, T.Y.: On the unified field theory. III–VI. Proc. Nat. Acad. Sci. U.S.A. Bd. 17 S. 48–58, 111–119, 119–210, 325–329.
13. Veblen, O., u. J.H.C. Whitehead: A set of axioms for differential geometry. Proc. Nat. Acad. Sci. U.S.A. Bd 17 S. 551–561.
14. Weyl, H.: Geometrie und Physik. Naturwiss. Bd. 19 S. 49–58.
15. Whitehead, J.H.C.: On a class of projectively flat affine connections. Proc. London Math. Soc. Bd. 32 S. 93–114.
16. – The representation of projective spaces. Ann. of Math. Bd. 32 S. 327–360.

1932

1. Bargmann, V.: Bemerkungen zur allgemein-relativistischen Fassung der Quanthentheorie. S.-B. preuß. Akad. Wiss. S. 346–354.
2. Bortolotti, E.: Sulle connessioni proiettive. Rend. Circ. mat. Palermo Bd. 56 S. 1–58.
3. Dantzig, D. van: Theorie des projektiven Zusammenhangs n-dimensionaler Räume. Math. Ann. Bd. 106 S. 400–454.
4. – Zur allgemeinen projektiven Differential-Geometrie: I. Einordnung in die Affin-Geometrie. II. X_{n+1} mit eingliedriger Gruppe. Proc. Akad. Wetensch. Amsterd. Bd. 35 S. 524–534, 535–542.
5. Einstein, A., u. W. Mayer: Einheitliche Theorie von Gravitation und Elektrizität. Zweite Abh. S.-B. preuß. Akad. Wiss. S. 130–137.
6. Kosambi, D.D.: Affin-geometrische Grundlagen der einheitliche Feldtheorie. S.-B. preuß. Akad. Wiss. S. 342–345.
7. Lanczos, C.: Zum Auftreten des Vektorpotentials in der Riemannschen Geometrie. Z. Physik Bd. 75 S. 63–77.
8. Schouten, J.A., u. D. van Dantzig: Zum Unifizierungsproblem der Physik, Skizze einer generellen Feldtheorie. Proc. Akad. Wetensch. Amsterd. Bd. 35 S. 642–656.
9. Generelle Feldtheorie. Z. Physik Bd. 78 S. 639– 667.
10. Veblen, O., u. J.H.C. Whitehead: The foundations of differential geometry. London: Cambridge Tract.
11. Weitzenböck, R.: Über projektive Differentialinvarianten. VII. Proc. Akad. Wetensch. Amsterd. Bd 34 S. 462–469.

1933

1. Schouten, J.A., u. D. van Dantzig: On projective connections and their application to the general field theory. Ann. of Math. Bd. 34.

Appendix B

The Gauge Principle

On the topic of Chemical Cosmology not many concepts are as relevant as gauge invariance – the most direct manifestation of space-time curvature. In an attempt to unify the electromagnetic and gravitational fields the idea of gauge transformation was first proposed by Herman Weyl (1920) as a space-time dependent change of scale, $S_\mu dx^\mu$, on displacement from point x^μ to $x^\mu + dx^\mu$. A displaced space-time function, f, may also change by an amount $(\partial f/\partial x^\mu)dx^\mu$ and with the change of scale the first-order increment is $(\partial_\mu + S_\mu)f dx^\mu$. Weyl could demonstrate that such a change of gauge generates the potentials of the electromagnetic field in four-dimensional space-time.

An obvious objection to Weyl's theory is that taking a clock through a closed loop in four-dimensional space-time, must change the speed of the clock and an atom carried around a closed path in an electromagnetic field must therefore radiate at a different wavelength on reaching the end of the loop. This is not confirmed experimentally.

It was demonstrated by Schrödinger (1922), in a paper never referred to again by him, that Weyl's prescription, with a complex gauge factor, predicted the correct quantized energy levels for the hydrogen electron, enabling London (1927) to reformulate the gauge transformation as a phase change

$$\Psi \rightarrow \Psi' = \Psi \exp(iq\alpha) \tag{B.1}$$

of the wave function of a charged particle. As in the original proposal the phase factor may have a globally fixed value, independent of position, in which case (B.1) is known as a *global* gauge transformation. This applies in the pseudo-Euclidean Minkowski space of special relativity and the transformation has the special property that the derivative of the field

$$\partial_\mu \Psi = e^{i\alpha} \partial_\mu \Psi$$

transforms like the field itself. If the complex field Ψ describes an electron, a charge q is associated with its wave field of charge density $\rho = |\Psi|^2$, which is clearly invariant under gauge transformation (B.1). This invariance implies, by Noether's theorem, the global conservation of the total charge, and shows that the overall phase change is not measurable.

The situation is entirely different within the theory of general relativity, which is based on a curved manifold rather than flat space with a globally fixed coordinate system. Each point now has its own coordinate system and hence its own (local) gauge factor. By doing away with the rigid coordinate system the gauge factor necessarily becomes an arbitrary function of the coordinates $x(\mu)$. Since the phase has no real physical significance, it may be defined locally by an arbitrary rotation, at every space-time point of the manifold, without changing the physical situation. This stipulation may seem to rule out local charge conservation, unless there is some compensating field that restores the invariance under local phase transformation.

In the case of an electron wave function the required gauge field must be the electromagnetic field. To find the correct form of this compensating field we look for the transformed wave function

$$\Psi' = \Psi e^{i\alpha(x,t)}$$

that satisfies Schrödinger's equation:

$$i\hbar\frac{\partial\Psi}{\partial t} = -\frac{\hbar^2}{2m}\left(\frac{\partial^2\Psi}{\partial x^2}\right) \tag{B.2}$$

The function $\Psi'(x,t)$ clearly does not satisfy the equation, yielding:

$$i\hbar\frac{\partial\Psi}{\partial t} = -\frac{\hbar^2}{2m}\left[\frac{\partial^2\Psi}{\partial x^2} + 2i\frac{\partial\alpha}{\partial x}\frac{\partial\Psi}{\partial x} - \Psi\frac{\partial^2\alpha}{\partial x^2} - \Psi\left(\frac{\partial\alpha}{\partial x}\right)^2\right] + \hbar\Psi\frac{\partial\alpha}{\partial t}$$

This expression is conveniently abbreviated in vector notation to read:

$$\begin{aligned} i\hbar\frac{\partial\Psi}{\partial t} &= -\frac{\hbar^2}{2m}\left[(\nabla + i\nabla\alpha)^2 - \nabla\cdot\nabla\alpha - \frac{2m}{\hbar}\frac{\partial\alpha}{\partial t}\right]\Psi \\ &\equiv -\frac{\hbar^2}{2m}\left[(\nabla + i\boldsymbol{A})^2 - \boldsymbol{\nabla}\times\boldsymbol{A} - V\right]\Psi \end{aligned} \tag{B.3}$$

The form of equation (B.2) is therefore recovered by defining the vector $\boldsymbol{A} = \partial\alpha/\partial x$ and the scalar

$$V = \frac{2m}{\hbar}\cdot\frac{\partial\alpha}{\partial t}\,.$$

Equation (B.3) is recognized as the well-known Pauli equation that describes the motion of an electron in an electromagnetic field. $\boldsymbol{A}$ and V are respectively known as the vector and scalar potentials of the electromagnetic field. The quantity $\boldsymbol{\nabla}\times\boldsymbol{A}$=curl$\boldsymbol{A}$=$\boldsymbol{B}$ represents the magnetic field strength. In order to leave the electromagnetic field invariant, the gauge field itself must transform like $A'_\mu \rightarrow A_\mu(x) + \partial_\mu\alpha(x)$.

The complex phase which is fundamental to gauge theory is commonly defined in terms of symmetry groups without consideration of its physical meaning, which emerges most clearly in its characterization of the quantum wave functions. Whereas phase relationships between point particles are hard to imagine, they appear naturally in wave structures. With respect to electrons and other chemical entities a wave model in terms of complex wave functions is therefore the most satisfactory physical model. The complex phase represents the fundamental attribute of non-classical systems and the major difference between classical particles and quantum waves. Simulation of chemical systems based on real basis sets is essentially classical. It is therefore wrong, although fashionable, to refer to such simulations as quantum chemistry.

In a sense, the terminology of global and local gauge invariance may be somewhat unfortunate. We distinguish between Riemannian underlying space-time and local Minkowski tangent space. The invariant gauge factor in tangent space is commonly said to define global gauge invariance whereas local gauge invariance is associated with the curved underlying space.

To summarize, the principle of local gauge invariance in a curved Riemannian manifold leads to the appearance of compensating fields. The electromagnetic field is the compensating field of local phase transformation and the gravitational field of local Lorentz transformations. Such compensating fields cannot be scalars, but are vector or tensor fields whose associated rest mass vanishes. The interaction between an electron and the electromagnetic field is said to be mediated by a massless photon, which leads to the local phase invariance. The four-vector $\boldsymbol{A}$ describes the electromagnetic field, or the wave functions of the photon.

The difference between global and local gauge invariance explains the apparent differences in behaviour of radiation and matter in tangent (field-free) space and underlying curved space. The conservation of electric charge in tangent space becomes subject to the generation of an electromagnetic field in curved space. This means that the stability of an electron depends on its interaction with the environment. This interaction is visualized as the superposition of advanced (in-coming) and retarded (out-going) waves to generate the wave packet recognized as the electron.

The behaviour of massive objects differs in an analogous way in tangent space and the underlying curved space. In tangent space the motion of a test object obeys Newton's law of inertia, but in curved space it becomes subject to gravitational effects.

By the same argument the transmission of radiation is perceived differently in tangent space and the underlying curved space. The photon wave that follows the curved geodesic is stretched, compared to that of a photon in Euclidean tangent space. This effect contributes to the observed redshift of galactic light. It has been analyzed, by the somewhat different, but equivalent, model based on time differences in the tangent and underlying spaces, by Segal (1976). We recognize this as the true *cosmological* component of observed redshifts. Unlike Doppler shifts it depends on the square of the distance from the source. It is augmented in all cases by a chemical redshift, which depends on the space-time curvature at the source, but differs from the smaller so-called gravitational redshift.

Gauge invariance, as formulated here, represents the elusive link between general relativity and quantum theory. It does not appear naturally in either theory and had to be introduced by special assumption. The amazing truth is that this link was in fact discovered many years ago and described in Veblen's (1933) monograph. It emerges naturally from relativity theory formulated in projective space-time.

The unique feature of projective geometry that generates the gauge principle is the absence of a metric. Any parameter may therefore be multiplied by an arbitrary factor without affecting its meaning. This gauge factor turns out to be the same as the complex phase of wave mechanics. We contend on these grounds that the best available model of cosmic space-time is provided by the topologically closed world of projective relativity.

Equally remarkable is the fact that the relativistic wave equation, derived in projective space by these arguments, is scaled to its affine counterpart by the golden ratio, which we now propose to define the relationship between our tangent space and the underlying projective space.

References

Schrödinger, E. (1922): *Über eine bemerkenswerte Eigenschaft der Quantenbahnen eines einzelnen Elektrons*, Z. Phys., 192 (12) 13–23.

Segal, I.E. (1976): *Mathematical Cosmology and Extragalactic Astronomy*, 1976, Academic Press, NY.

Veblen, O. (1933): *Projektive Relativitätstheorie*, Springer-Verlag, Berlin.

Veblen, O. & B. Hoffmann (1930): *Projective Relativity*, Phys. Rev., 1930 (36) 810–822.

Weyl, H. (1920): *Raum, Zeit und Materie*, 3rd ed., Springer-Verlag, Heidelberg.

London, F. (1927): *Quanten mechanische Deutung der Theorie von Weyl*, Z. Phys. 1927 (42) 375–389.

Appendix C

Abstracts

C.1 Introductory Synopsis

The introductory chapter summarizes the contentious issues which stimulate cosmological debate, including the chirality and self-similarity of space-time, the dark-matter postulate and the physical meaning of mathematical singularities. It cautions against the uncritical reliance on the rhetoric of authority and outlines the fundamental considerations which dictate the structure of the work that follows.

The mathematics of infinity is crucial and in order to avoid this unphysical situation projective geometry and a topologically closed cosmos are adopted at the outset. Projective topology is shown to satisfy the demands of both special and general relaticity. The periodic properties of both quantum and chemical systems arise naturally from closed topology and the gauge principle within projective relativity.

C.2 History

Cosmology in its present form developed as a by-product of Einstein's gravitational field equations, based on the astronomical data of the previous millenium, which established the heliocentric model of the solar system. The struggle against the authority of Ptolemy, Aristotle and the Inquisition, the rivalry between Kepler and Galileo, and the intrigue between Newton and his contemporaries, Descartes, Leibniz, Hooke and others, overshadow the important theoretical advances that produced the mechanical clockwork model of Laplace.

This model never matured into a coherent cosmology, stumbling upon the instability of an infinite static universe, without divine intervention.

Discovery of the Bode–Titius law was not enough to counter the teleological arguments of intelligent design which ended up in deadlock with the evolutionary models of Darwin and Kant. Without settling the argument, observational astronomy beyond the solar system added further confusion, which only started to clear up with the scientific discoveries of the late 19th century and the theory of an expanding universe.

C.3 World Geometry

Cosmology starts with our perception of the geometrical structure of the world. Like that of Aristotle and Ptolemy, the authority of Euclid stifled the development of a geometry, suitable for describing those features of cosmic structure, revealed by telescopic observation. Geometry is shown to be more general than the Euclidean, which implies a cosmology, more general than the expanding Euclidean universe. The most general, Projective Geometry, based on the concepts of projectivity, cross ratio and collineation, are described in outline to show its relationship with complex geometry, topology and Riemannian geometry, from which the theory of general relativity derives.

C.4 Physical Evidence

Geometry alone could not produce a theory of gravity, free of action at a distance, until physics managed to catch up with the ideas of Riemann. The development of special relativity, after discovery of the electromagnetic field, is described. It requires a holistic four-dimensional space-time, rather than three-dimensional Euclidean space and universal time. Accelerated motion, and therefore gravity, additionally requires this space-time to be non-Euclidean. The important conclusion is that relativity, more than a theory, is the only consistent description of physical reality at this time. Schemes for the unified description of the gravitational and electromagnetic fields are briefly discussed.

The other great theory of physics, in the formalism of wave mechanics, has equally important cosmological implications. It reveals the real meaning of gauge transformation as a change of complex phase, only associated with waves; thereby finding common ground for gravitational and electromagnetic fields in the curvature of space-time. It is significant to note that gauge invariance occurs as a natural phenemenon only in projective relativity theory.

The implied wave nature of elementary matter furthermore clarifies their mode of interaction through standing waves generated by the interference between advanced and retarded wave components. The negative-energy solutions of relativistic wave equations first indicated the existence of antimatter, as later confirmed experimentally. To avoid the annihilation of matter and antimatter on a cosmic scale an involuted structure of the vacuum, consistent with projective space-time, is inferred.

There is little evidence of quantum theory in standard cosmology, except for so-called quantum uncertainty, which is often invoked to rationalize illogical conclusions.

C.5 Chemical Evidence

Chemistry is the modern progeny of the ancient art of alchemy. In its heyday the theories of alchemy and cosmology were one. The structure and composition of heaven and earth, the geometry of circles and polygons, the processes and theories of alchemy, and the control of astrology were all connected inextricably. By demonstrating a proportional relationship between circle and polygon, Kepler effectively squared the circle and enabled the development of astronomy, unencumbered by astrology. At the same time, Boyle and Lavoisier recognized the true nature of chemical elements, effectively replacing alchemy by chemistry. The link between chemistry and cosmology, which was severed in the process, can be restored only now on recognizing the full implication of chemical elemental periodicity.

The periodic table of the elements is a subset of a more general periodic function that relates all natural nuclides in terms of integer numbers of protons and neutrons, the subject of elementary number theory. The entire structure is reproduced in terms of Farey sequences and Ford circles. The periodicity arises from closure of the function that relates nuclear stability to isotopic composition and nucleon number. It is closed in two dimensions with involution that relates matter to antimatter and explains nuclear stability and electronic configuration in terms of space-time curvature. The variability of electronic structure predicts a non-Doppler redshift in galactic and quasar light, not taken into account in standard cosmology.

The prominent role of the golden ratio that conditions the observed periodic table of the elements hints at a general self-similarity between atomic and celestial structures. By exploiting this similarity the Bode–Titius law is shown to be based on the same number theory as nuclide periodicity. All planets, moons and rings in the solar system obey the same rules of commensurability and move on quantized orbits like those assumed in planetary models of the atom. All of these orbits are correctly predicted at fixed divergence angles on a golden spiral.

A cosmologically important feature of chemical periodicity is the valid prediction of periodic variation in cosmic abundance, consistent with an equilibrium process of nucleogenesis by α-particle addition and totally at variance with the predictions of standard cosmology and the ΛCDM alternative.

C.6 Standard Cosmology

Standard cosmology is widely understood to be equivalent to relativistic cosmology in curved space that expands on comoving coordinates from a singular point at universal time zero. This definition is highly ambiguous. To clarify the issue the role of Einstein's field equations are scrutinized more closely. The first, Schwarzschild, solution of these equations has no cosmological significance except as a black-hole model. Einstein's original static-universe solution, with cosmological constant, was soon abandoned in favour of expanding-universe models. De Sitter's original solution, in contrast, assumed a proper relativistic space-time metric and was discarded on the basis of its vanishing stress tensor. From here on standard cosmology routinely assumed the Robertson–Walker metric with separate space and time coordinates and hence at variance with general relativity. Many cosmologists accept the Einstein–de Sitter model as standard. It describes an expanding universe with a universal time coordinate, orthogonal to three-dimensional Euclidean space.

The essence of standard cosmology is an expanding universe in which galactic redshifts are considered to be strictly of Doppler origin, quantified by Hubbles's linear law. The common view is that the expansion started at a singular point where time had its beginning. In reality there is no consensus amongst afficionades about the reason, time, nature and locality of the big bang, nor the nature of subsequent events. A major problem lies in the violation of the laws of physics.

The theory claims observational support from cosmological redshifts, microwave background radiation and nucleogenesis. Superficially it is supported by theoretical particle physics, but this forced consilience is rather uneasy, to say the least. The interpretation of crucial concepts such as background radiation, the cosmological constant and non-baryonic dark matter is often blatantly ignored. Auxiliary concepts such as inflation, antimatter, wormholes, large-number coincidences, many worlds and the anthropic principle are invoked *ad hoc* as required and ignored when bothersome. Of late the observation of modern stars, as old as the big-bang universe, is changing the complexion of standard cosmology into a simulation model, the so-called ΛCDM cosmology. This new scheme abandons or modifies the big-bang notions of inflation, nucleogenesis and galaxy formation by introducing the theoretical concepts of dark energy and population III stars.

C.7 Relativistic Cosmology

It is a common fallacy to assume that the use of Einstein's field equations

$$R_{\mu\nu} + \Lambda g_{\mu\nu} = -\kappa \left(T_{\mu\nu} - \tfrac{1}{2} T g_{\mu\nu}\right)$$

automatically defines a relativistic cosmology. As illustrated by some elementary examples, the resulting model is relativistic only if the metric tensor has the correct relativistic signature. Apart from de Sitter's empty universe the only other relativistic solution is that obtained by Gödel in 1949. It describes a system in rotation about a compass of inertia and never gained acceptance as it does not predict a Doppler frequency shift. As the solution remains valid in projective space it is of special importance in the present context. A null geodesic in the real projective plane defines the compass of inertia and the involution eliminates closed time loops in the double cover, which accommodates conjugate states of matter and time. In such a universe redshifts are the result of differently perceived null geodesics in the tangent space of the observer and the underlying cosmic space. These are the chronometric redshifts, shown by Segal to vary with the square of the separation, in preference to Hubble's law.

Detailed solution of the field equations of projective relativity is not known, but has been shown to give a unfied description of gravity, electromagnetism and wave mechanics, in which the golden ratio occurs as a descriptor of space-time curvature.

C.8 Reasoned Alternatives

Concepts in theoretical science, which may appear difficult to comprehend on first encounter, often become accepted through familiarity, rather than insight. Such concepts end up as dogmatic belief, which casts a shadow on the understanding of related theories. The notion of a quantum/classical limit is discussed as an example. Analysed as a measure of quantum potential it is shown to clear up the related concepts of Compton wavelength, fine-structure constant, wave structures and the nature of the vacuum. In a Riemannian manifold it automatically leads on to the concept of chirality, which defines the vacuum as an achiral interface in projective space, as before.

The role of black holes in dispersal of the conjugate forms of matter in a steady-state universe then suggests that globular clusters originate in anti-black (white) holes. Regular structures are easily recognized in Hubble-telescope images of such clusters as vortices or eddies, resulting from turbulent radial expansion. The same scenario provides interesting insight into the origin of cosmic rays, quasars, Seyfert galaxies, γ-ray bursts, and the different forms of X-ray and radio sources.

In the same spirit of enquiry, the dogmatically accepted Doppler origin of cosmological redshifts is shown to be seriously inadequate. A host of alternatives are examined, with the conclusion that no single effect can possibly account for all observations. The redshifts in quasar and galactic light are conclusively shown to be of two different kinds.

The flawed concept behind standard cosmology is the notion of an expanding universe. Most dissidents who seek to refute big-bang cosmology fail to recognize the fallacy of universal expansion and many so-called alternatives are simply variations on the theme. Serious alternatives, including plasma cosmology, the self-similar model and the modified big-bang model, known as population III cosmology, are briefly discussed. The latter is interpreted as the final abandonement of the big-bang lunacy by standard cosmology.

C.9 The Big Picture

Cosmolgy is culture-specific and unlike science and technology a single globally accepted cosmology cannot develop. Each cosmology, from the most primitive to the technologically most advanced, still consists of a mixture of magic, religion and science. Despite claims to the contrary, standard cosmology is no different. Its scientific content is fast approaching zero and the dream of a science-based cosmology can finally be dismissed.

Instead of cosmology, this treatise is concluded with an attempt to model the natural world, without futile dogmatic or teleological speculation on the origin, purpose, meaning or ultimate fate of the universe. The most reliable scientific guide, the theory of general relativity, points at a four-dimensional curved basis, called space-time, in which the entirety is embedded. The theory unequivocally identifies the curvature as positive. It refutes the notion of universal time and favours a closed manifold as the most likely topology.

Observed interaction at a distance is rationalized as contact, mediated by the gravitational and electromagnetic fields. Mathematical characterization of these fields in four-dimensional space-time requires solution of the field equations in projective space. The gauge principle, which emerges from this procedure, identifies elementary transplantation in space-time with a change of complex phase, as wave motion. Interaction is reduced to standing waves with advanced and retarted components.

The involution that occurs in projective geometry defines conjugate regions with time inversion and conjugate forms of matter. The function that describes the observed periodicity of atomic matter is of the same projective form and varies with local space-time curvature. This variation shows that spectroscopic analysis of light waves, stretched between sites of different curvature, must be frequency shifted, as observed.

The relationship between the locally experienced tangent Euclidean space and the underlying Riemann space is scaled by the golden ratio. Its well-known property of self-similarity, epitomized by the golden logarithmic spiral, is observed in all growth structures, from atoms, through molecules, biological organisms to the solar system, spiral galaxies and globular clusters.

The recognition of self-similarity simplifies the description of the physical world, all in response of the vacuum to the curvature of space-time, but complicates the perception of three-dimensional beings of their four-dimensional environment, which they are physically unable to visualize. The perceived infinity of space is the illusion created in simply-connected tangent space by the multiply-connected cosmic reality. The large-scale structure of the universe is destined to remain unknown for a long time.

Index